NEW TOOLS
IN
TURBULENCE MODELLING

Les Houches School, May 21-31, 1996

Editors

Olivier MÉTAIS

Joel FERZIGER

Les Éditions de Physique

France
Avenue du Hoggar
Zone Industrielle de Courtabœuf
B.P. 112
91944 Les Ulis cedex A

North America
PCG Inc.
875-81 Massachusetts Avenue
Cambridge MA 02139 USA

Springer-Verlag Berlin Heidelberg GmbH

Centre de Physique des Houches

Books already published in this series

1 Porous Silicon Science and Technology
Jean-Claude VIAL and Jacques DERRIEN, Eds. 1995

2 Nonlinear Excitations in Biomolecules
Michel PEYRARD, Ed. 1995

3 Beyond Quasicrystals
Françoise AXEL and Denis GRATIAS, Eds. 1995

4 Quantum Mechanical Simulation Methods for Studying Biological Systems
Dominique BICOUT and Martin FIELD, Eds. 1996

Book series coordinated by Michèle LEDUC

ISBN 978-3-540-63090-6   ISBN 978-3-662-08975-0 (eBook)
DOI 10.1007/978-3-662-08975-0

© Springer-Verlag Berlin Heidelberg 1997
Originally published by Springer-Verlag Berlin Heidelberg New York in 1997

LECTURERS

Aupoix B., ONERA/CERT, Aerothermodynamics Department, BP 4025, Toulouse cedex 4, France

Bailly C., Laboratoire de Mécanique des Fluides et d'Acoustique, CNRS, UMR 5509, École Centrale de Lyon, 69131 Écully, France

Besnard D., Commissariat à l'Énergie Atomique, Centre d'Études de Bruyères le Châtel, 91680 Bruyères le Châtel, France ; et Commissariat à l'Énergie Atomique, Centre d'Études de Limeil-Valenton, 94195 Villeneuve St Georges cedex, France

Blanc-Benon P., Laboratoire de Mécanique des Fluides et d'Acoustique, CNRS, UMR 5509, École Centrale de Lyon, 69131 Écully, France

Chevalier G., ONERA/CERT, Aerothermodynamics Department, BP 4025, 31055 Toulouse cedex 4, France

Comte P., Turbulence Simulation and Modelling Team, LEGI/IMG, BP 53, 38041 Grenoble cedex 9, France

Comte-Bellot G., Laboratoire de Mécanique des Fluides et d'Acoustique, CNRS, UMR 5509, École Centrale de Lyon, 69131 Écully, France

Cousteix J., ONERA/CERT, Aerothermodynamics Department, BP 4025, 31055 Toulouse cedex 4, France

Ducros F., Commissariat à l'Énergie Atomique, Centre d'Études de Limeil-Valenton, 94195 Villeneuve St Georges cedex, France

Ferziger J.H., Department of Mechanical Engineering, Stanford University, Stanford CA, USA

Galmiche D., Commissariat à l'Énergie Atomique, Centre d'Études de Limeil-Valenton, 94195 Villeneuve St Georges cedex, France

Gatski T.B., NASA Langley Research Center, Hampton, Virginia 23681, USA

Grand D., CEA-Grenoble DRN/DTP/STR, 17 Rue des Martyrs, 38054 Grenoble cedex 9, France

Lamballais E., LEA/CEAT, 43 Route de l'Aérodrome, 86036 Poitiers cedex, France

Lê T.H., 29 Avenue de la Division Leclerc, 92320 Châtillon, France

Lesieur M., LEGI/IMG, BP 53, 38041 Grenoble cedex 09, France

Loreaux Ph., Commissariat à l'Énergie Atomique, Centre d'Études de Limeil-Valenton, 94195 Villeneuve St Georges cedex, France

Malecki P., ONERA/CERT, Aerothermodynamics Department, BP 4025, 31055 Toulouse cedex 4, France

Meeder J.P., J.M. Burgers Centre, Delft University of Technology, Rotterdamse-weg 145, 2628 AL Delft

Métais O., LEGI/IMG, BP 53, 38041 Grenoble cedex 09, France

Meyer B., Commissariat à l'Énergie Atomique, Centre d'Études de Limeil-Valenton, 94195 Villeneuve St Georges cedex, France

Moscetti P., DGA/DRET, 26 Bd Victor 00460 Armées, France

Naudy M., Commissariat à l'Énergie Atomique, Centre d'Études de Limeil-Valenton, 94195 Villeneuve St Georges cedex, France

Nieuwstadt F.T.M., J.M. Burgers Centre, Delft University of Technology, Rotterdamseweg 145, 2628 AL Delft

Passot T., CNRS URA 1362, OCA, BP 4229, 06304 Nice cedex 4, France

Poinsot T., IMFT et CERFACS Av. Camille Soula, 31400 Toulouse cedex, France

Porter D.H., Department of Astronomy and Laboratory for Computational Science and Engineering, University of Minnesota, Minneapolis, USA

Pouquet A., CNRS URA 1362, OCA, BP 4229, 06304 Nice cedex 4, France

Rodi W., Institute for Hydromechanics, University of Karlsruhe, Kaiserstrasse 12, 76128 Karlsruhe, Germany

Sagaut P., 29 Avenue de la Division Leclerc, 92320 Châtillon, France

Urbin G., LEGI, Institut de Mécanique de Grenoble, BP 53, 38041 Grenoble cedex 9, France

Vazquez-Semadeni E., Instituto de Astronomía, UNAM, Apdo. Postal 70-264, México, D.F. 04510 México

Veynante D., Laboratoire E.M2.C., CNRS et École Centrale Paris, Grande Voie des Vignes, 92295 Chatenay-Malabry cedex, France

Viala S., ONERA/CERT, Aerothermodynamics Department, BP 4025, 31055 Toulouse cedex 4, France

Woodward P.R., Department of Astronomy and Laboratory for Computational Science and Engineering, University of Minnesota, Minneapolis, USA

PREFACE

Direct numerical simulations of turbulence, which calculate explicitly all the scales of motion (from the large energetic scales to the small dissipative scales) can treat only small values of the Reynolds number. Therefore, they cannot reproduce the turbulence encountered in the atmosphere, the ocean, or most industrial devices. Attention is focused on the large scales of motion since these are responsible for a significant part of heat and mass transfer. The large turbulent scales can be simulated numerically by Large-Eddy Simulation (LES). LES requires correct representation of energy exchanges with the small scales that are not explicitly simulated: "subgrid-scale" models have to be developed for this purpose. The first such model was proposed by Smagorinsky in 1963 for numerical studies of atmospheric turbulence.

Development of LES techniques is blooming at present. This is attributable to new ideas in subgrid-scale modelling and to tremendous progress in scientific computing. Once confined to very simple flow configurations such as isotropic turbulence or periodic flows, the field is evolving to include spatially growing shear flows, separated flows, pipe flows, riblet walls, and bluff bodies, among others. Classical turbulence modelling, based on one-point closures and a statistical approach allow computation of mean quantities. In many cases, it is necessary to have access to the fluctuating part of the turbulent fields such as the pollutant concentration or temperature: LES is then compulsory. Together with DNS, LES is indeed able to perform deterministic predictions (of flows containing coherent vortices, for instance) and to provide statistical information. The last is very important for assessing and improving one-point closure models, in particular for turbulent flows submitted to external forces (stratification, rotation, ...) or compressibility effects.

There is no doubt that the complexity of problems tackled by LES will continue to increase, and that this will almost certainly have a decisive impact on industrial modelling and flow control. The goal of the session held in May 1996 at the Center de Physique des Houches was to assess the state-of-the-art of LES applications for industrial and environmental flows and to highlight directions for future research.

The book starts with a presentation of the general frame work of Large-Eddy Simulations. M. Lesieur's lecture recalls the formalism of the LES both in physical and spectral space. He discusses important issues like the backscatter. Several applications of the subgrid-scale models to shear flows are presented with both statistical results and information on the topology of the coherent vortices. J. Ferziger's lectures presents other recent subgrid-scale models like the dynamic model. A particular focus is placed on the difficulties related to LES, such as numerical accuracy, initial and boundary conditions generation, correct modelling of the near-wall region. Complementary aspects of LES versus the Reynolds Averaged Navier-Stokes models (RANS) are discussed. This last aspect is the central topic of W. Rodi's lecture, which examines the relative performance and potential of statistical turbulence models and LES for calculating turbulent flows of practical interests. The criteria are the quality of the results and the computer ressources required. Practical examples of industrial interest are provided. T. Gatski's lecture gives a cohesive overview of the recent advances in compressible turbulence research that involve both direct and large-eddy simulations and closure models development. The part devoted to compressible turbulence modelling particularly focuses on the specific correlations due to compressibility. D. Veynante and T. Poinsot's lecture emphasizes the challenge attached to the modelling and simulations of turbulent combustion. The complexity of combusting flow field associated with strong heat release, large range of time and length scales, etc ... is discussed. The recent efforts in LES developments adapted to combustion applications are presented. G. Comte-Bellot, C. Bailly and P. Blanc-Benon's lecture provides a survey of the most efficient techniques which are presently available for theoretical analyses and engineering noise predictions. The various approaches allowing to obtain the far field noise generated by turbulent flows are presented. Examples of aeroacoustic predictions are given based upon DNS or LES results or upon estimations using modelled acoustic sources provided by one-point closure models. The remainder of the book is devoted to the most recent numerical, theoretical, and experimental achievements related to the modelling of the turbulence encountered in various industrial and environmental applications. Turbulence modelling and simulations of aeronautical flows are first presented. P. Comte's lecture presents the extension of the LES formalism to compressible flows in curvilinear co-ordinates. Applications to vortex dynamics in non-trivial geometries are then described. P. Sagaut, T.H. Le and P. Moschetti's lecture reviews the challenge to which LES are presently confronted to deal with applications of interest for the aeronautical and defense industries. B. Aupoix' lecture is devoted to modelling of wall flows such as on airfoil at high angles of attack, compressible boundary layers and three-dimensional boundary layers. Another domain in which a correct representation of turbulent phenomena is primordial, is the Inertial Confinement Fusion. Besnard's et al. lecture indeed shows that the design of optimized targets requires a good estimation of the complex instability mechanisms involved. D. Grand and G. Urbin's lecture demonstrates that thermohydraulics and flows in nuclear reactors are good candidates for the utilization of LES as an industrial tool. The chosen examples allow to show that physical complex flows are now accessible to the LES, and that this technique possesses a great potential for even more complex geometries. The following lectures are devoted to the LES of flows submitted to external forces of geophysical interest. O. Métais and E. Lamballais' lecture demonstrates the ability for the LES to correctly reproduce, in high Reynolds number flows, the detailed vortex topology in the presence of solid-body rotation and density gradients. It is in particular showed that standard LES techniques can be applied to the computation of atmospheric and oceanic mesoscale eddies resulting from different instability mechanisms such as baroclinic or convective instability. The recent efforts of LES of air pollution are reviewed in F. Nieuwstadt and J. Meeder's lecture. Several examples of simulations of turbulent atmospheric dispersion are presented, and the techniques that can be used in turbulent dispersion LES are discussed in detail. The Piecewise Parabolic Method (PPM) algorithm is used by D. Porter and P. Woodward to the numerical simulations of the compressible turbulence driven by convection in deep atmospheres, with applications to Stellar convection, in stars like the Sun. Other aspects of astrophysical turbulent motions have been studied by A. Pouquet, whose lecture deals with the dynamical evolution of the turbulence within the interstellar medium. This lecture clearly demonstrates the complexity of a flow involving magnetic-field, heating and cooling, self-gravity, star formation, rotation and shear.

Acknowledgements:

The Scientific Committee of the session was composed of M. LESIEUR (LEGI/IMG, France), P. MASON (Meteorological Office, Bracknell, UK), P. MOIN (Stanford University, USA), P. PERRIER (Avions Marcel Dassault, France), W. RODI (University of Karlsruhe, Germany).

The Physics School in Les Houches is affiliated to the University Joseph Fourier and the National Polytechnic Institute in Grenoble. It is subsidized by the Ministry of Superior Education and Research, CNRS and Atomic Energy Commission. This session has been granted by: the DRET, the CNRS, the programme TIM (CEA/CNRS), and "le Ministère des Affaires Étrangères".

Olivier MÉTAIS   Joel H. FERZIGER

CONTENTS

LECTURE 1

Recent approaches in large-eddy simulations of turbulence

by M. Lesieur

LECTURE 2

Large eddy simulation: an introduction and perspective

by J.H. Ferziger

LECTURE 3

Large-eddy simulation and statistical turbulence models: complementary approaches

by W. Rodi

LECTURE 4

Modeling compressibility effects on turbulence

by T.B. Gatski

LECTURE 5

Reynolds averaged and large eddy simulation modeling for turbulent combustion

by D. Veynante and T. Poinsot

LECTURE 6

Modeling tools for flow noise and sound propagation through turbulence

by G. Comte-Bellot, C. Bailly and P. Blanc-Benon

LECTURE 7

Vortices in compressible LES and non-trivial geometries

by P. Comte

LECTURE 8

LES-DNS: The aeronautical and defense point of view

by P. Sagaut, T.H. Lê and P. Moschetti

LECTURE 9

Turbulence modeling in aeronautical flows

by B. Aupoix, J. Cousteix, G. Chevalier, S. Viala and P. Malecki

LECTURE 10

Fluid instabilities in inertial confinement fusion

by D. Besnard, F. Ducros, D. Galmiche, Ph. Loreaux, B. Meyer and M. Naudy

LECTURE 11

Large eddy simulations in nuclear reactors thermal-hydraulics

by D. Grand and G. Urbin

LECTURE 12

Coherent vortices in rotating flows

by O. Métais and E. Lamballais

LECTURE 13

Large-eddy simulation of air pollution dispersion: a review

by F.T.M. Nieuwstadt and J.P. Meeder

LECTURE 14

Numerical simulations of compressible convection

by D.H. Porter and P.R. Woodward

LECTURE 15

Dynamical evolution of the turbulent interstellar medium at the kiloparsec scale

by A. Pouquet, T. Passot and E. Vazquez-Semademi

Recent Approaches in Large-Eddy Simulations of Turbulence

M. Lesieur

LEGI/IMG, BP53, 38041 Grenoble-Cedex 09, France.

Abstract

The general framework of large-eddy simulations (LES) is first presented with Smagorinsky's model [1]. Afterwards Kraichnan's spectral eddy-viscosity[2] is introduced, and how it can be handled for LES purposes in isotropic turbulence. The spectral eddy viscosity is generalized to a spectral eddy diffusivity. Using the nonlocal interaction theory, the backscatter issue is discussed, and a generalization of spectral eddy coefficients is presented allowing to account for non-developed turbulence in the subgridscales, the spectral-dynamic model. Utilization of these spectral models in physical space is envisaged in terms of respectively the structure-function and hyperviscosity models. Three applications of these models to shear flows are considered, namely the plane mixing layer, the channel flow and the backward-facing step, with statistical results and information on the topology of coherent vortices and structures. In the mixing layer case in particular, how longitudinal vorticity may be stretched into hairpins of spanwise wave length corresponding to Pierrehumbert and Widnall's translative instability[3] is explained.

1. THE NEED FOR LARGE-EDDY SIMULATIONS

A direct-numerical simulation (DNS) of a turbulent flow has to take into account explicitly all scales of motion, from the largest, imposed by the existence of boundaries or the periodicities, to the smallest. The latter may be for instance the Kolmogorov dissipative scale $(\epsilon/\nu^3)^{1/4}$ in three-dimensional isotropic turbulence, or the viscous thickness ν/v_* in a turbulent boundary layer.

It is well known (see e.g. [4]) that the total number of degrees of freedom necessary to represent a turbulent flow through this whole span of scales is of the order of $R_l^{9/4}$ in three dimensions for isotropic turbulence, where R_l is the turbulent Reynolds number based on the integral scale. Right now, the calculations done to the expense of reasonable computing times on the biggest machines take about 256^3 grid points. The improvement is very slow, and the Reynolds numbers thus attained with the aid of DNS are still several orders of magnitude lower than the huge Reynolds numbers encountered in natural situations.

The conclusion one can draw is that, for a weakly viscous fluid, it is not possible in the near future (and perhaps not in the distant future either) to simulate explicitly all the scales of motion from the smallest to the largest. Generally, scientists or engineers are more interested in the description of the large scales of the flow, which often contain the desired information about turbulent transfers of momentum or heat for example: it is these large scales which will be simulated on the computer. This is done using the LES techniques.

The paper is organized as follows. Section 2 presents the general framework of LES carried out in physical space, and the widely used model proposed by Smagorinsky[1]. Section 3 describes the concept of spectral eddy viscosity proposed by Kraichnan[2], and shows how it may be used for LES purposes. A generalization to a spectral eddy diffusivity for a scalar convected by the flow is discussed. This section presents results of the spectral eddy-viscosity model applied to decaying isotropic turbulence, with emphasis put upon infrared backscatter. It introduces also a new model, the spectral-dynamic model, where the spectral eddy coefficients take into account non-developed turbulence in the subgridscales. Section 4 shows how the spectral-space LES point of view may be recovered in physical space, in terms of respectively the structure-function model and hyperviscosity models. Section 5 will mainly present applications of the filtered structure-function model to a temporal or spatially-developing plane mixing layer. Section 6 will concern the plane turbulent channel studied with the spectral-dynamic model, and Section 7 the backward-facing step flow studied with the selective structure-function model. In the three cases, statistical results as well as informations regarding the topology of coherent vortices and structures will be given.

2. LES IN PHYSICAL SPACE: THE FORMALISM

2.1. Generalities

We present here the formalism for incompressible turbulence, with some indications of the way compressibility may be handled. We look at the philosophy of LES when the computation is carried out in physical space, using finite-difference or finite-volume methods. We assume first for sake of simplification that the spatial discretization is cubic, Δx being the grid mesh. To the fields defined in the continuous space $\vec{x}$, one will associate filtered fields obtained through the convolution with a filter $\bar{G}_{\Delta x}$, chosen so as to eliminate fluctuations in the motions of wavelength smaller than Δx. The filtered velocity and temperature are:

$$\bar{\vec{u}}(\vec{x},t) = \int \vec{u}(\vec{y},t)\bar{G}_{\Delta x}(\vec{x}-\vec{y})d\vec{y} \tag{1}$$

$$\bar{T}(\vec{x},t) = \int T(\vec{y},t)\bar{G}_{\Delta x}(\vec{x}-\vec{y})d\vec{y} \quad , \tag{2}$$

and the same for any quantity f (scalar or vectorial)

$$\bar{f}(\vec{x},t) = \int f(\vec{y},t)\bar{G}_{\Delta x}(\vec{x}-\vec{y})d\vec{y} = \int f(\vec{x}-\vec{y},t)\bar{G}_{\Delta x}(\vec{y})d\vec{y} \quad . \tag{3}$$

One can easily check that such a filter commutes with temporal and spatial derivatives, so that the continuity equation for the filtered field holds. This is however no more valid on irregular grids, where the filter has a variable width. Let $\vec{u}'$ and T' be the fluctuations of the actual fields with respect to the filtered fields

$$\vec{u} = \bar{\vec{u}} + \vec{u}' \; ; \; T = \bar{T} + T' \tag{4}$$

and more generally $f = \bar{f} + f'$. The fields "prime" concern fluctuations at scales smaller than Δx (the "grid scale"), and will then be referred to as subgridscale fields.

Let us write Navier-Stokes equations as

$$\frac{\partial u_i}{\partial t} + \frac{\partial}{\partial x_j}(u_i u_j) = -\frac{1}{\rho_0}\frac{\partial p}{\partial x_i} + \frac{\partial}{\partial x_j}\left\{ \nu\left(\frac{\partial u_i}{\partial x_j} + \frac{\partial u_j}{\partial x_i}\right)\right\} \quad . \tag{5}$$

After applying the filter, one gets

$$\frac{\partial \bar{u}_i}{\partial t} + \frac{\partial}{\partial x_j}(\bar{u}_i \bar{u}_j) = -\frac{1}{\rho_0}\frac{\partial \bar{p}}{\partial x_i} + \frac{\partial}{\partial x_j}\left\{ \nu\left(\frac{\partial \bar{u}_i}{\partial x_j} + \frac{\partial \bar{u}_j}{\partial x_i}\right) + T_{ij}\right\} \quad . \tag{6}$$

$$T_{ij} = \bar{u}_i \bar{u}_j - \overline{u_i u_j} \tag{7}$$

is the subgridscale tensor. The filtered fields do not need to be resolved at scales smaller than Δx, since they do not contain fluctuations under this scale. Therefore, they can be properly represented by the computer, provided proper numerical schemes are used. The main problem lies in the subgridscale tensor: when expressed in terms of fluctuations with respect to the filtered field, certain terms remain unknown. The equations of motion for the filtered field have analogies with Reynolds equations for the mean flow in non-homogeneous turbulence, but other terms than $-\overline{u_i' u_j'}$ arise in the LES.

In the LES, we have to solve Navier-Stokes equations for the filtered field (large scales) modified by supplementary subgridscale terms which we do not know. Reviews of the LES methods may be found in [5] [4].

The same analysis may be held for a scalar T (not necessarily passive) of molecular diffusivity κ convected by the flow. It satisfies

$$\frac{\partial T}{\partial t} + \frac{\partial}{\partial x_j}(T u_j) = \frac{\partial}{\partial x_j}\left\{\kappa \frac{\partial T}{\partial x_j}\right\} \quad . \tag{8}$$

If the low-pass filter $\bar{G}_{\Delta x}$ is applied to this equation, one finds

$$\frac{\partial \bar{T}}{\partial t} + \frac{\partial}{\partial x_j}(\bar{T}\bar{u}_j) = \frac{\partial}{\partial x_j}\left\{\kappa \frac{\partial \bar{T}}{\partial x_j} + \bar{T}\bar{u}_j - \overline{T u_j}\right\} \quad . \tag{9}$$

Here again, the question of modelling the subgrid scalar fluxes is posed.

The problem of the subgridscale modelling is a particular case of the passage from "micro" to "macro", where the laws governing a medium are known at a microscopic level, and one seeks evolution laws at a macroscopic level. Here in turbulence, the "microscopic" level corresponds to the individual fluid particle obeying Navier-Stokes equations; the "macroscopic" level represents the filtered field.

From a mathematical viewpoint, the LES problem is not very well posed. Indeed, let us consider the time evolution of the fluid as the motion of a point in a sort of phase space of very large dimension. At some initial instant, the flow computed with LES will differ from the actual flow, due to the uncertainty contained in the subgridscales. This initial difference between the actual and the computed flow will grow, due to nonlinear effects, as in a dynamical system having a chaotic behaviour. Therefore, the two points will separate in phase space, and, as time goes on, the LES will depart from reality. However, as will be seen below, LES permit to predict the statistical characteristics of turbulence, as well as the dynamics of coherent vortices and structures.

2.2. Eddy-Viscosity and Diffusivity

Now, an eddy-viscosity and eddy-diffusivity assumption is going to be made, in order to model the subgrid terms. More specifically, one writes

$$T_{ij} = 2\nu_t \, \bar{S}_{ij} + \frac{1}{3}T_{ll} \, \delta_{ij} \quad , \tag{10}$$

where

$$\bar{S}_{ij} = \frac{1}{2}\left(\frac{\partial \bar{u}_i}{\partial x_j} + \frac{\partial \bar{u}_j}{\partial x_i}\right) \tag{11}$$

is the deformation tensor of the filtered field. The LES momentum equation becomes

$$\frac{\partial \bar{u}_i}{\partial t} + \bar{u}_j \frac{\partial \bar{u}_i}{\partial x_j} = -\frac{1}{\rho_0}\frac{\partial \bar{P}}{\partial x_i} + 2\frac{\partial}{\partial x_j}\left\{(\nu + \nu_t)\bar{S}_{ij}\right\} \quad , \tag{12}$$

$\bar{P} = \bar{p} - (1/3)\rho_0 T_{ll}$ being a modified pressure, determined with the help of the filtered continuity equation (still valid on a regular mesh). For the scalar, one introduces an eddy diffusivity κ_t, such that

$$\bar{T}\bar{u}_j - \overline{Tu_j} = \kappa_t \frac{\partial \bar{T}}{\partial x_j} \tag{13}$$

to yield

$$\frac{\partial \bar{T}}{\partial t} + \bar{u}_j \frac{\partial \bar{T}}{\partial x_j} = \frac{\partial}{\partial x_j} \left\{ (\kappa + \kappa_t) \frac{\partial \bar{T}}{\partial x_j} \right\} \quad . \tag{14}$$

The eddy diffusivity is related to the eddy viscosity by the relation

$$P_r^{(t)} = \frac{\nu_t}{\kappa_t} \quad , \tag{15}$$

$P_r^{(t)}$) being a turbulent Prandtl number which will be specified below.

For compressible turbulence, one will assume that the subgrid scales are not far from incompressibility, and write compressible Navier-Stokes equations for the filtered fields, with the above incompressible eddy-coefficients supplementing the molecular (viscous and diffusive) ones. This is certainly justified away from the shocks. In the shock region, and although the eddy viscosity is no more physically justified, it may have a favourable effect by contributing to stabilize the shock.

The question is now to determine the eddy viscosity $\nu_t(\vec{x}, t)$. Notice that this eddy-viscosity assumption, in the framework of which we are going to work here, is rather questionable. One expects, however, that the informations about the physics of turbulence derived using this concept may help to improve it.

2.3. Smagorinsky's Model

The most widely used eddy-viscosity model was proposed by Smagorinsky[1]. He was looking for an eddy-viscosity simulating some sort of three-dimensional Kolmogorov energy cascade in the subgridscales. A local mixing-length assumption is made, in which the eddy viscosity is assumed to be proportional to the subgridscale characteristic length Δx, and to a characteristic turbulent velocity $v_{\Delta x} = \Delta x |\bar{S}|$. Here $|\bar{S}|$ is typical velocity gradient at Δx, determined with the aid of the second invariant of the filtered-field deformation tensor $\bar{S}_{ij}$ defined in (11). Smagorinsky's eddy viscosity writes

$$\nu_t = (C_S \Delta x)^2 |\bar{S}| \quad , \tag{16}$$

with $|\bar{S}| = \sqrt{2\bar{S}_{ij}\bar{S}_{ij}}$. If one assumes that $k_C = \pi/\Delta x$, the cutoff wavenumber in Fourier space, lies within a $k^{-5/3}$ Kolmogorov cascade, one can adjust the constant C_S so that the ensemble averaged subgrid kinetic-energy dissipation is identical to ϵ. It is found:

$$C_S \approx \frac{1}{\pi} \left(\frac{3C_K}{2} \right)^{-3/4} . \tag{17}$$

This yields $C_S \approx 0.18$ for a Kolmogorov constant $C_K = 1.4$. In fact, the dynamic-model ideas (see [6]) will consist in adjusting locally C_S, basically to reduce the eddy-viscosity in places where turbulence is not totally developed, or during transition.

Remark that a theorem of existence, uniqueness and regularity concerning the LES equations with Smagorinsky's model was derived by Lions[7].

We will talk of the performances of Smagorinsky's model concerning various applications which will be presented below. We are now going to present a different point of view of LES when working in Fourier space.

3.   LES IN SPECTRAL SPACE

3.1.   Spectral Eddy Viscosity and Diffusivity

The formalism of spectral eddy viscosity is due to Kraichnan ([2], see also [8] [9] [4]) in the case of a Kolmogorov subgridscale spectrum. We adapt it here to a wider range of spectra, and to a scalar convected by the flow.

We assume that Navier-Stokes is written in Fourier space (which requires periodicity in the three spatial directions), and consider the cutoff wave number $k_C = \pi \Delta x^{-1}$ already envisaged above. We define a sharp low-pass filter by setting equal to zero the velocity and scalar amplitudes at wave vectors whose modulus is larger than k_C. Let us first consider the kinetic-energy and scalar transfers given by the EDQNM theory[1] ([4]). One assumes first $k << k_C$ (both modes being larger than k_i, the kinetic-energy peak. Then one can write the spectral evolution equations for the supergrid-scale velocity, $\bar{E}(k,t)$, and scalar, $\bar{E}_T(k,t)$ spectra as

$$\left(\frac{\partial}{\partial t} + 2\nu k^2 \right) \bar{E}(k,t) = T_{<k_C}(k,t) + T_{sg}(k,t) \tag{18}$$

$$\left(\frac{\partial}{\partial t} + 2\kappa k^2 \right) \bar{E}_T(k,t) = T^T_{<k_C}(k,t) + T^T_{sg}(k,t) , \tag{19}$$

with

$$T_{sg}(k,t) = -2\nu_t^\infty k^2 \bar{E}(k,t) \tag{20}$$

$$\nu_t^\infty = \frac{1}{15} \int_{k_C}^\infty \theta_{0pp} \left[5E(p,t) + p\frac{\partial E(p,t)}{\partial p} \right] dp \tag{21}$$

---

[1] Kraichnan [2] considered the Test-Field Model (TFM) instead of the Eddy-Damped Quasi-Normal Markovian theory (EDQNM), but results are the same in a Kolmogorov inertial range.

$$T_{sg}^T(k,t) = -2\kappa_t^\infty \; k^2 \; \bar{E}_T(k,t) \tag{22}$$

$$\kappa_t^\infty = \frac{2}{3} \int_{k_C}^\infty \theta_{0pp}^T \; E(p,t) \; dp \;\;, \tag{23}$$

where θ_{kpq} and θ_{kpq}^T are nonlinear triple-correlation relaxation times of the EDQNM theory. The supergrid-scale transfers $T_{<k_C}(k,t)$ and $T_{<k_C}^T(k,t)$ correspond to triad interactions whose wave numbers lie in the supergrid range, and hence do not need any modelling, since they can be calculated exactly in the large-eddy simulation. Here, the nonlocal transfers from the supergrid to the subgrid scales have been evaluated with the aid of leading-order expansions with respect to the small parameter k/k_C.

Let us start by assuming a $k^{-5/3}$ inertial range at wave numbers greater than k_C. We obtain:

$$\nu_t^\infty = 0.441 \; C_K^{-3/2} \left[\frac{E(k_C)}{k_C} \right]^{1/2} \tag{24}$$

$$\kappa_t^\infty = \frac{\nu_t^\infty}{P_r^{(t)}} \tag{25}$$

with

$$P_r^{(t)} = 0.6 \;\;. \tag{26}$$

The latter value is in fact the highest value permitted by the EDQNM theory (see [4], for a discussion on this point). If one assumes for instance a Kolmogorov constant of 1.4 in the energy cascade, the constant in front of (24) will be 0.267. When k is close to k_C, the above concept of spectral eddy viscosity and eddy diffusivity can be generalized for a $k^{-5/3}$ inertial range extending over wave numbers larger than k_C. It is possible, with the aid of the EDQNM approximation, to calculate the subgridscale transfers, corresponding to triadic interactions where at least one of the wave numbers p and q is greater than k_C. This allows us to define two functions $\nu_t(k|k_C)$ and $\kappa_t(k|k_C)$, respectively the eddy viscosity in spectral space ([2] and the eddy diffusivity in spectral space ([9]), such that

$$T_{sg}(k,t) = -2\nu_t(k|k_C) \; k^2 \; \bar{E}(k,t) \tag{27}$$

$$T_{sg}^T(k,t) = -2\kappa_t(k|k_C) \; k^2 \; \bar{E}(k,t) \;\;. \tag{28}$$

The functions $\nu_t(k|k_C)$ and $\kappa_t(k|k_C)$ are such that

$$\nu_t(k|k_C) = K(k/k_C)\nu_t^\infty \tag{29}$$

$$\kappa_t(k|k_C) = C(k/k_C)\kappa_t^\infty \tag{30}$$

where ν_t^∞ and κ_t^∞ are the asymptotic values given by (24), (25), (26). As shown by Kraichnan[2], $K(x)$ is approximately constant and equal to 1, except in the vicinity of $k/k_C = 1$ where it displays a strong overshoot (cusp-behaviour), due to the predominance of semi-local transfers across k_C. It was shown in [9] that $C(x)$ behaves qualitatively as $K(x)$ (plateau at 1 and positive cusp), and that the spectral turbulent Prandtl number $\nu_t(k|k_C)/\kappa_t(k|k_C)$ is approximately constant (and thus equal to 0.6).

In fact, the function $K(x)$ can be put under the form

$$K(x) = 1 + \nu_n^* x^{2n} \quad , \tag{31}$$

with $2n \approx 3.7$ ([8][9]). We propose to determine ν_n^* by considering the energy balance between explicit and subgridscale transfers. This yields:

$$\int_0^{k_C} 2\nu_t k^2 E(k,t)dk = \epsilon \quad ,$$

which, in an infinite Kolmogorov inertial range, leads to

$$1 + \frac{1}{1 + (3n/2)}\nu_n^* = \frac{2}{3 \times 0.441} \quad . \tag{32}$$

We will come back to this expression below, when working in physical space in terms of generalized hyperviscosities.

Let us mention that the use of the subgridscale transfers (27), (28) allow one to solve numerically the EDQNM kinetic energy and passive scalar evolution equations at zero molecular viscosity and conductivity in the self-similar decaying regime (for $k \leq k_C$), as shown by [8][9].

3.2. LES of Isotropic Turbulence

Let us now come back to the evolution equations (in spectral space) of the instantaneous filtered fields (for $|\vec{k}| < k_C$)

$$(\frac{\partial}{\partial t} + \nu k^2)\bar{u}_i(\vec{k},t) = t_{<k_C}(\vec{k},t) + t_{sg}(\vec{k},t) \tag{33}$$

$$(\frac{\partial}{\partial t} + \kappa k^2)\bar{T}(\vec{k},t) = t^T_{<k_C}(\vec{k},t) + t^T_{sg}(\vec{k},t) \tag{34}$$

with the usual distinction between the explicit supergrid transfers, still calculated by a truncation for $k, p, q \leq k_C$ of the nonlinear terms involved in Navier-Stokes equations in Fourier space (see [4]), and the unknown subgridscale transfers. We propose to model the latter with the aid of $\nu_t(k|k_C)$ and $\kappa_t(k|k_C)$ introduced in (27) and (28) namely

$$t_{sg}(\vec{k},t) = -\nu_t(k|k_C)k^2\bar{u}_i(\vec{k},t) \tag{35}$$

$$t^T_{sg}(\vec{k},t) = -\kappa_t(k|k_C)k^2\bar{T}(\vec{k},t) \quad . \tag{36}$$

This subgridscale modelling is justified at the energetic transfer level, in the sense that, when one writes the exact evolution equations for the spectra of

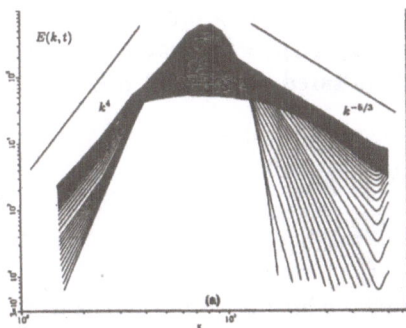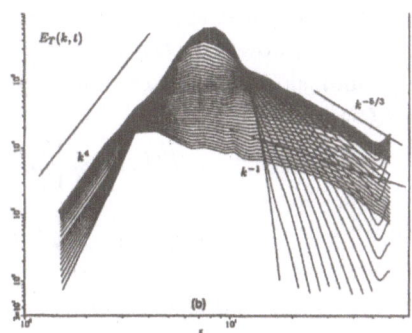

Fig. 1. — 3D isotropic decaying turbulence, resolution 128^3; decay of kinetic-energy (a) and passive-scalar (b) spectra, calculated from the LES of [10] using the spectral-cusp eddy viscosity

$\overline{\underline{u}}$ and $\overline{\overline{T}}$ as they arise from (33), (34), one obtains the EDQNM subgridscale transfers calculated in (27), (28). The results of the spectral-cusp eddy viscosity applied to LES of decaying three-dimensional isotropic turbulence are satisfactory. The first calculations of this type at a very low resolution were done by [8][9] respectively for the momentum equation and the passive scalar. Figures 1-a and 1-b show the decaying kinetic-energy and passive scalar spectra obtained in analogous LES at a resolution of 128^3 Fourier modes carried out by [10] (see also [11]). The initial velocity and scalar spectra are analogous, with a Gaussian ultraviolet behaviour and a k^8 infrared spectrum. It can be checked that Kolmogorov and Corrsin-Oboukhov $k^{-5/3}$ cascades establish. Afterwards, the kinetic-energy spectrum decays self-similarly, with a slope comprised between $-5/3$ and -2. The scalar spectrum seems to have a very short inertial-convective range close to the cutoff, and a very wide range shallower than k^{-1} in the large scales. This range was explained by [13] as due to the quasi two-dimensional character of the scalar diffusion in the large scales, leading to large-scale intermittency of the scalar.

3.3. The Spectral Backscatter

Let us look now at the infrared ($k \to 0$) spectra obtained in Figures 1-a and 1-b. The initial slope is k^8, as already stressed, and one sees k^4 spectra which form. These spectra had been predicted with the aid of two-point closures such as EDQNM(^{2})(see [14]) many years before they could be observed in the LES of [10] [11]. The derivation is the following: let k_i and k_i^T be respectively the peaks of the kinetic-energy and scalar spectra. The nonlocal interactions theory,

---

(^{2}) This is also valid for TFM, DIA (Direct-Interaction Approximation) and even Quasi-Normal theories.

where the nonlocal transfers are calculated to the leading order in terms of expansions in powers of the small parameters k/k_i (resp. k/k_i^T) permits to show that nonlocal transfers are dominant in this infrared range, and respectively equal to

$$T(k,t) \approx \frac{14}{15}k^4 \int_{k_i}^{\infty} \theta_{0pp} \frac{E(p)^2}{p^2} dp \qquad (37)$$

$$T^T(k,t) = \frac{4}{3}k^4 \int_{k_i^T}^{\infty} \theta_{0pp}^T \frac{E(p)}{p^2} E_T(p) dp \qquad (38)$$

for the velocity and the scalar. These transfers, which arise in the rhs of the evolution equations for the spectra

$$(\frac{\partial}{\partial t} + 2\nu k^2)E(k,t) = T(k,t) \qquad (39)$$

$$(\frac{\partial}{\partial t} + 2\kappa k^2)E_T(k,t) = T^T(k,t) \quad , \qquad (40)$$

inject a k^4 spectrum in low wave numbers, through some kind of nonlinear resonant interaction between two modes $\approx k_i$. This spectral backscatter is responsible for the sudden appearance of k^4 infrared spectra when the initial spectra are sharply peaked, or simply $\propto k^s$ with $s > 4$. This spectral backscatter phenomenon is important, since energy is thus injected in very low wave numbers, and, to our knowledge, two-point closure theories are the only ones which permit to predict it analytically[3] Violent backscatter phenomena occur also in the statistical unpredictability theory both in two and three dimensions, as shown by [12].

Notice finally that, in the case of decaying isotropic three-dimensional turbulence, and for an initial infrared spectrum $\propto k^2$, an eddy-viscous term $\approx -2\nu_t(k|k_i)k^2 E(k,t)$ is also present in $T(k,t)$, together with the k^4 backscatter. But both terms being of order $(k/k_i)^4$, they do not influence the evolution of the lhs of equation (39), which is of order $(k/k_i)^2$. This implies exact permanence of big eddies, with a k^2 infrared spectrum which is time invariant. The same holds for the scalar. If, on the other hand, turbulence is fed by some forcing concentrated around k_i, a stationary solution will imply an infrared balance between the backscatter and the eddy-viscous transfers, yielding a k^2 equipartition spectrum. A last remark is that the k^4 backscatter is negligible in LES where the cutoff k_C lies in the middle of a Kolmogorov range. Backscatter effects exist, which send back energy from the filtered to the subgridscales. But, from an energetic point of view, they are contained in the cusp-part of the plateau-cusp eddy viscosity considered above.

3.4. Spectral Dynamic Model

We will present below a local generalization of the plateau-cusp spectral eddy viscosity to the physical space (structure-function model), which gives better

---

[3] In two-dimensional turbulence, a k^3 backscatter also arises (see [4])

results for isotropic turbulence as far as the Kolmogorov cascade is concerned. But let us show now an adaptation of the spectral-cusp model to kinetic-energy spectra $\propto k^{-m}$ for $k > k_C$, when the exponent m is not necessarily equal to 5/3. The eddy viscosity given by (21) is now

$$\nu_t^\infty = \frac{1}{15a_1} \frac{5-m}{m+1} \sqrt{3-m} \left[\frac{E(k_C)}{k_C}\right]^{1/2} , \qquad (41)$$

for $m \leq 3$. The constant $1/(15a_1)$ in front of (41), coming from the EDQNM theory for the kinetic-energy spectrum, is equal to $0.31\, C_K^{-3/2}$. This expression, derived by [13], was used by Lamballais ([15] [16]) for LES of a plane channel (see below). The associated eddy diffusivity is

$$\kappa_t^\infty = \frac{4}{3a_3} \frac{\sqrt{3-m}}{m+1} \left[\frac{E(k_C)}{k_C}\right]^{1/2} , \qquad (42)$$

and the turbulent Prandtl number

$$P_r^{(t)} = \frac{5-m}{20} \frac{a_3}{a_1} , \qquad (43)$$

where the constant a_3 arises in the EDQNM scalar-spectrum equation. It is such that one recovers $P_r^{(t)} = 0.6$ for $m = 5/3$, so that one finds finally

$$P_r^{(t)} = 0.18\,(5-m) . \qquad (44)$$

For $m > 3$, this scaling is no more valid, and the eddy-viscosity and diffusivity coefficients will be set equal to zero. In the spectral dynamic model, the exponent m is determined through the LES with the aid of least-squares fits of the kinetic-energy spectrum close to the cutoff. The asymptotic eddy viscosity (41) is multiplied by the plateau-cusp function $K(k/k_C)$ defined above.

4. RETURN TO PHYSICAL SPACE

4.1. Structure-Function Model

Now, let us consider the EDQNM eddy viscosity (still scaling on $\sqrt{E(k_C)/k_C}$) with no cusp, and adjust the constant as proposed by [17], by balancing in the inertial range the subgridscale flux with the kinetic energy flux ϵ in the energy spectrum evolution equation([4]). This yields

$$\nu_t(k_C) = \frac{2}{3} C_K^{-3/2} \left[\frac{E(k_C)}{k_C}\right]^{1/2} . \qquad (45)$$

The problem with such an eddy-viscosity (if the energy spectrum may be computed) is that it is uniform in space when used in physical space. Obviously,

---

([4]) The same was done in oder to obtain (32).

the eddy viscosity should take into account the intermittency of turbulence: there is no need for any subgridscale modelling in regions of space where the flow is laminar or transitional. On the other hand, it is essential to dissipate in the subgridscales the local bursts of turbulence if they become too intense. Considering also that turbulence in the small scales may not be too far from isotropy, it was proposed by [13] to come back to the classical formulation (10) in the physical space, where the eddy viscosity is determined with the aid of (45). $E(k_C, \vec{x})$ is now a local kinetic energy spectrum, calculated in terms of the local second-order velocity structure function of the filtered field

$$F_2(\vec{x}, \Delta x) \doteq \langle \|\bar{\vec{u}}(\vec{x}, t) - \bar{\vec{u}}(\vec{x} + \vec{r}, t)\|^2 \rangle_{\|\vec{r}\| = \Delta x} \tag{46}$$

as if the turbulence is three-dimensionally isotropic, with Batchelor's formula

$$F_2(\vec{x}, \Delta x) = 4 \int_0^{k_C} E(k) \left(1 - \frac{\sin(k\Delta x)}{k\Delta x}\right) dk \quad . \tag{47}$$

This yields for a Kolmogorov spectrum

$$\nu_t^{SF}(\vec{x}, \Delta x) = 0.105 \, C_K^{-3/2} \, \Delta x \, [F_2(\vec{x}, \Delta x)]^{1/2} \quad . \tag{48}$$

F_2 is calculated with a local statistical average of square velocity differences between $\vec{x}$ and the six closest points surrounding $\vec{x}$ on the computational grid. In some cases, the average may be taken over four points parallel to a given plane; in a channel, for instance, the plane is parallel to the boundaries. Notice also that if the computational grid is not regular, interpolations of (48) have been proposed by [5], using the fact that the second-order velocity structure function scales like $(\epsilon r)^{2/3}$ in a Kolmogorov inertial range.

The structure-function model (SF) works well for decaying isotropic turbulence, where it yields a fairly good Kolmogorov spectrum ([13]), better than Smagorinsky's model (with $C_S = 0.2$) and Kraichnan's spectral-cusp model.

The SF model gives also good results for free-shear flows, where it is able to stretch secondary thin longitudinal hairpin vortices between primary vortices (see [5]). However, selective or filtered versions of it work better in this case (see below). The SF model permits also to go beyond transition in a temporal[5] compressible boundary layer upon an adiabatic wall at Mach 4.5 ([18]). But it does not work for transition in a boundary layer at low Mach (or incompressible) where, like Smagorinsky, it is too dissipative and prevents TS waves to degenerate into turbulence. This is still true within the four-point formulation in planes parallel to the wall, which eliminates the effect of the mean shear at the wall on the eddy viscosity. In fact, the spectrum $E_{\vec{x}}(k_C)$ determined by the isotropic formula (45) is too sensitive to the inhomogeneous low-frequency oscillations caused by the TS waves.

---

[5] periodic in the flow direction

4.2.   Selective and filtered SF models

To overcome the difficulty of dissipating too much the large quasi two-dimensional vortices or transitional waves, two improved versions of the SF model have been developed: the selective structure-function model (SSF), and the filtered structure-function model (FSF). The dynamic model in physical space (see [6]) is another way of adapting the eddy viscosity to the local conditions of the flow.

The SSF model was developed by [19]. The idea is to switch off the eddy viscosity when the flow is not three-dimensional enough. The three-dimensionalization criterion is the following: one measures the angle between the vorticity at a given grid point and the average vorticity at the six closest neighbouring points (or the four closest points in the four-point formulation). If this angle exceeds 20^0, the most probable value according to simulations of isotropic turbulence at a resolution of $32^3 \sim 64^3$, the eddy viscosity is turned on. Otherwise, only the molecular dissipation is active. The constant arising in (48) is changed, and determined with the aid of LES of freely-decaying isotropic turbulence: one requires that the eddy viscosity averaged over the computational domain should be the same in a selective structure-function model and a SF model simulation. It is found that the constant in (48) has to be multiplied by 1.56.

The SSF model works very well for isotropic turbulence and free-shear flows, as well as for a compression ramp at Mach 2.5 (see [19]). An example of application to a flow above a backward-facing step will be given at the end of the paper. The SSF model depends however upon the most probable angle of the next neighbours average vorticity, chosen above equal to 20^0. In fact, this angle is a function of the resolution of the simulation, since it should go to zero with Δx, and may be with the type of flow considered. Progresses in this model should be made by adjustment of this angle to the local grid.

The FSF model was developed by Ducros ([20]) and applied to a boundary layer at Mach 0.5 ([21]). Here, the filtered field $\bar{u}_i$ is submitted to a high-pass filter in order to get rid of low-frequency oscillations which affect $E_{\vec{x}}(k_C)$ in (45). The high-pass filter is a Laplacian discretized by second-order centered finite differences and iterated three times. It was shown by [20] that, for some 3D random or turbulent isotropic test fields, the eddy-viscosity (48) can be written as

$$\nu_t^{FSF}(\vec{x}, \Delta x) = 0.0014 \, C_K^{-3/2} \, \Delta x \, [\tilde{F}_2(\vec{x}, \Delta x)]^{1/2} \; , \qquad (49)$$

where $\tilde{F}_2$ is the second-order velocity structure function of the high-pass filtered field. This model gives good results for transition in a spatially-developing boundary layer. This simulation (see [21]) was done in a weakly-compressible case at $M_\infty = 0.5$, for an adiabatic plate. It is seen how a TS wave, to which a three-dimensional white noise perturbation is superposed upstream, propagates downstream. First, quasi two-dimensional billows of relatively low pressure and high vorticity form, then undergo some sort of staggered unstability, as predicted by [22], while weak longitudinal high- and low-speed streaks develop at the wall. Finally the staggered pattern breaks down into turbulence, within which the classical high- and low-speed streaks are observed, with ejection of hairpins above the latter. The FSF model is however not "perfect" for the prediction of average quantities. It overestimates in particular of about 15% the mean velocity in the logarithmic profile. The same happens when it is applied to the incompressible channel ([15]). In the latter case, the spectral dynamic model presented above works better, as will be seen below.

4.3. Generalized Hyperviscosities

One of the common drawbacks of the different versions of the SF model is the absence of cusp near k_C. We go back to (32), where we take $n = 2$, which yields $2n = 4$, not far from the EDQNM 3.7 value. Then (32) yields $\nu_2^* = 2.044$, so that an equivalent of the spectral-cusp eddy viscosity in physical space is for the subgridscale dissipative operator in the filtered Navier-Stokes equation:

$$2\frac{\partial}{\partial x_j}[0.661\,\nu_t^{SF}\bar{S}_{ij}] + 1.351\,\nu_t^{SF}(\nabla^2)^3\bar{u}_i \quad , \tag{50}$$

where ν_t^{SF} is given by (48). This model([6])was used by [23] in LES of a rotating stratified jet submitted to baroclinic instability. They could show developments of primary and secondary instabilities of the thermal fronts very similar to what is observed in the atmosphere (see Métais and Lamballais lectures notes in the present volume).

Notice that, as for Smagorinsky's, a theorem of existence and uniqueness has been demonstrated by Lions[7] for a subgridscale dissipation of the type (50), where ν_t is replaced by ν, provided the exponent arising in the Laplacian should be ≥ 2.

If one wants to have now in physical space a model equivalent to the spectral-dynamic model, the dissipative operator (50) has to be multiplied by the constant:

$$\tilde{A} = \frac{\sqrt{12}}{5}\,\frac{5-m}{m+1}\,\sqrt{3-m} \quad , \tag{51}$$

where m is the slope of the kinetic-energy spectrum at the cutoff, which has to be obtained dynamically in some way. This might be possible if one periodicity direction exists at least in the flow. For a scalar, the corresponding turbulent Prandtl number is given by (44).

Such a spectral-dynamic generalized hyperviscosity model should be tested on various shear flows, such as those presented below.

([6]) with an equivalent formulation for the density and a turbulent Prandtl number of 0.6

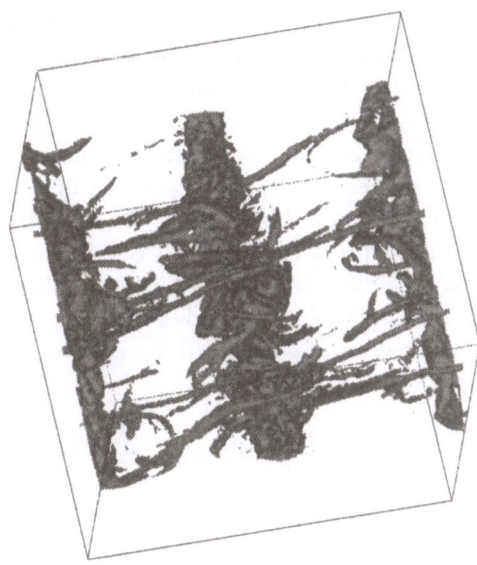

Fig. 2. — Vorticity field obtained in the LES of a temporal mixing layer forced quasi two-dimensionally

5. MIXING LAYER

We show now three-dimensional results of the filtered structure function model applied to a plane mixing layer, respectively in the temporal and spatial cases.

5.1. Temporal Mixing Layer

We consider in a fluid of constant density a mixing layer periodic in the streamwise and spanwise directions, initiated by a hyperbolic-tangent velocity profile, to which is superposed a small random perturbation. LES using the FSF model show the following results. If the perturbation is quasi two-dimensional, the mixing layer evolves into a set of big quasi two-dimensional vortices which both undergo pairing and stretch intense longitudinal hairpin vortices in the stagnation regions between them. Such a pattern is shown on Figure 2, taken from [24], and presenting a map of the vorticity modulus. This stretching of longitudinal vortices, observed experimentally for a long time (see e.g. [25] and [26]), may be explained as follows: let us consider the vorticity equation, written for a perfect fluid of uniform density as

$$\frac{D\vec{\omega}}{Dt} = \vec{\nabla}\vec{u} : \vec{\omega} = \overline{\overline{S}} : \vec{\omega} + \frac{1}{2}\vec{\omega} \times \vec{\omega} = \overline{\overline{S}} : \vec{\omega} \ , \qquad (52)$$

where D/Dt is the substancial derivative following the flow motion, and $\bar{\bar{S}}$ the deformation tensor, introduced in (11) for the filtered field. If one supposes that the vorticity in the stagnation region between the vortices is weak in front of the deformation, one can assume (at least initially) that the deformation tensor will not vary while the vorticity is stretched. Since the deformation tensor is real and symmetric, it admits eigenvectors (principal axes of deformation) which are orthogonal and can form a basis. Let s_1, s_2, s_3 be the three eigenvalues. Due to incompressibility, their sum is zero, so that one is positive (called here s_1) and another one at least is negative. Let s_2 be the smallest eigenvalue, always negative. Working in the orthonormal frame formed by the eigenvectors $\vec{l}, \vec{t}, \vec{s}$ respectively associated to s_1, s_2, s_3, the vorticity components $\omega_1, \omega_2, \omega_3$ satisfy the following equations

$$\frac{D\omega_1}{Dt} = s_1\omega_1 \quad , \quad \frac{D\omega_2}{Dt} = s_2\omega_2 \quad , \frac{D\omega_3}{Dt} = s_3\omega_3 \quad , \tag{53}$$

and the vorticity will be stretched in the direction of the first principal axis, and compressed in the direction of the second. Generally, $\bar{\bar{S}}$ is not far from a pure deformation in the stagnation region, so that, approximately, $\vec{l}$ will be inclined 45^0 with respect to the mean flow, $\vec{s}$ will be spanwise, $s_3 = 0$ and $s_2 = -s_1$. Such a derivation unifies explanations given by [27] and [28]. Intense longitudinal hairpins had also been found by [29] in temporal mixing-layer LES using the spectral-cusp eddy viscosity. An interesting feature of these simultions is to show that longitudinal vorticity stretched between the primary Kelvin-Helmholtz vortices is rolled up within the cores of the big vortices, thus producing intense longitudinal vorticity fluctuations in the cores themselves. In Figure 2, the maximum longitudinal vorticity stretched is of the order of $4\omega_i$, which might be larger than the effective values reached experimentally. Actually, the efficiency of the longitudinal stretching could depend upon the amplitude of the initial perturbation and the Reynolds number.

Experiments in a developed mixing layer show that the spanwise wavelength of the longitudinal vortices is of the order of two thirds of the longitudinal wavelength of the primary Kelvin-Helmholtz vortices between which they are stretched. This is precisely the most-amplified spanwise wavelength within a secondary-instability analysis of Stuart's vortices, as shown by Pierrehumbert and Widnall [3]. This instability, called the translative instability, corresponds in fact to a global in-phase oscillation of the big billows in the spanwise direction, and cannot explain the formation of the thin longitudinal hairpins. A plausible explanation could be that they do form according to the mechanisms corresponding to (53), but with a preferred spanwise wavelength imposed by the translative instability. Numerical simulations in the temporal or spatial cases need larger domains in the spanwise direction in order to validate the value of the preferred spanwise wavelength.

In fact Squire's theorem (at infinite Reynolds number) and the numerical resolution of the three-dimensional Orr-Sömmerfeld equation at large Reynolds show that the most-amplified mode in the 3D temporal mixing layer is indeed

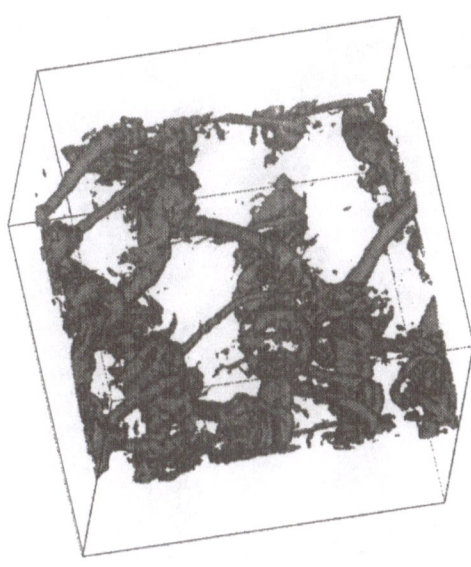

Fig. 3. — Vorticity field obtained in the LES of a temporal mixing layer undergoing helical pairing

two-dimensional. By a naive application of this result, one might have believed that two-dimensional Kelvin-Helmholtz vortices would emerge from a weak three-dimensional random isotropic perturbation superposed upon the basic shear. But this is not at all what happens. Instead, Comte et al. ([30][31]), using DNS with pseudo-spectral methods at a resolution of 128^3 Fourier wave vectors and a Reynolds number $U\delta_i/\nu = 100$, displayed the evidence for helical pairing, where vortex filaments oscillate out-of-phase in the spanwise direction, and reconnect, yielding a vortex-lattice structure. We have recovered the same dislocated pattern in LES (using the FSF model) with the same forcing. Figure 3 show the vorticity modulus obtained in such a simulation. Figure 4 show the low-pressure field from an analogous LES using the spectral-cusp eddy viscosity. It confirms that low pressure is a very good indicator of big or intense vortices. Notice that at the end of the FSF-based LES corresponding to Figure 3, the statistical data concerning velocity, rms velocity fluctuations and Reynolds stresses, are in very good agreement with the experiments of unforced mixing layers. The simulation with a quasi two-dimensional forcing is less good from this standpoint. Notice finally that a helical-pairing instability was found in the secondary-instability analysis of [3]. It corresponds here to a subharmonic instability whose amplification rate is, surprisingly, three times lower than its translative-instability counterpart. Looking at DNS or LES of helical-pairing, it turns out that this is not exactly a "secondary instability":

M. *Lesieur*

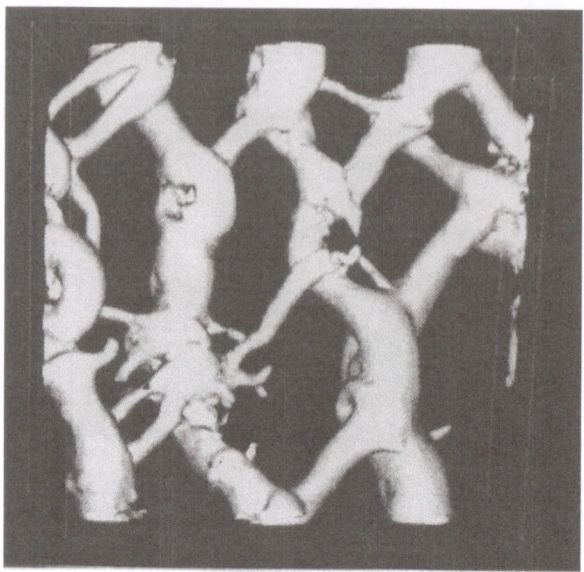

Fig. 4. — Low-pressure field obtained in a spectral-cusp LES in the helical-pairing case

one does not observe first the roll up of primary billows followed by a staggered deformation. Instead, oblique waves are seen to grow quickly, yielding directly the lattice structure of dislocated billows.

Finally, it may be interesting to look at what can be said from the point of view of a scalar gradient when a scalar is transported passively by the flow. This may be very important if the two currents in the mixing layer react chemically, such as in combustion for instance. If the scalar σ satisfies $D\sigma/Dt = 0$, its gradient follows the equation

$$\frac{D}{Dt}\vec{\nabla}\sigma = -\vec{\nabla}\vec{u}|^t : \vec{\nabla}\sigma \quad . \tag{54}$$

If we are in a stagnation region between two vortices[7], and assume that the vorticity is small in front of the deformation, (54) may be approximated by

$$\frac{D}{Dt}\vec{\nabla}\sigma = -\overline{\overline{S}} : \vec{\nabla}\sigma \quad , \tag{55}$$

which yields for the three components of $\vec{\nabla}\sigma$ in the principal axes of deformation

$$\frac{D}{Dt}\frac{\partial\sigma}{\partial l} = -s_1\frac{\partial\sigma}{\partial l} \quad , \quad \frac{D}{Dt}\frac{\partial\sigma}{\partial n} = -s_2\frac{\partial\sigma}{\partial n} \quad , \frac{D}{Dt}\frac{\partial\sigma}{\partial s} = -s_3\frac{\partial\sigma}{\partial s} \quad . \tag{56}$$

[7] situation where longitudinal hairpin vortices are likely to form

Fig. 5. — LES of an incompressible mixing layer forced upstream by a quasi two-dimensional random perturbation; the vorticity modulus is shown at a threshold $(2/3)\omega_i$

This shows that the scalar gradient is compressed along the first principal axis of deformation, and stretched in the transverse direction, so that scalar gradients across the mixing-layer interface will steepen. In case of chemical reaction between the two layers, this will enhance the molecular exchanges at the interface, and favour the reaction.

5.2. Spatial Mixing Layers

The temporal approximation is only a crude approximation of a mixing layer spatially developing, where one works in a frame traveling with the average velocity between the two layers. We present now LES using the FSF model of a spatial mixing layer, initiated upstream by a hyperbolic-tangent velocity profile superposed on the average flow, plus a weak random forcing regenerated at each time step. The numerical code combines pseudo-spectral methods in the spanwise and transverse directions, and compact finite-difference schemes of sixth order in the streamwise direction. This is a very precise code which has spectral accuracy (see [15]). With an upstream forcing consisting in a quasi two-dimensional random perturbation, intense longitudinal hairpins stretched between quasi 2D Kelvin-Helmholtz vortices are found again (Figure 5). When the forcing is a three-dimensional random white noise, helical pairing occurs upstream, as indicated by the low-pressure maps of Figure 6. But none of these simulations has reached self-similarity, since the kinetic-energy spectra in the downstream region are steeper than $k^{-5/3}$, and rms velocity fluctuations have a departure of about 20% with respect to the experiments. Thus calculations in longer domains are necessary, in order in particular to know in the helical-

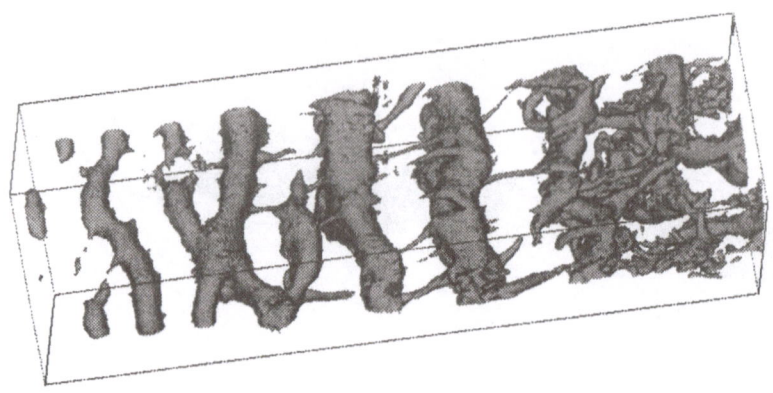

Fig. 6. — Same as Figure 5, but with a three-dimensional upstream white-noise forcing, low-pressure field

pairing case whether quasi two-dimensionality might not be restaured further
downstream.

6.   THE CHANNEL FLOW

The results presented here use the numerical code just described above, but
with the spectral-dynamic eddy viscosity presented in Section 3. Here, the
compact scheme is employed in the transverse direction, while pseudo-spectral
methods are used in the longitudinal and spanwise directions, which are periodic. Calculations start with a parabolic Poiseuille velocity profile, to which
a small white-noise perturbation is superposed, and are run up to complete
statistical stationarity, assuming a constant flow rate of average velocity U_m
across the section. The channel has a width $2h$, and the macroscopic Reynolds
number is defined by $R_e = 2hU_m/\nu$. It does not change with time during the
computation since U_m is conserved. We will define also a microscopic Reynolds
number $h^+ = v_* h/\nu$, based on the friction velocity in the turbulent regime. The
kinetic-energy spectrum allowing to determine the eddy-viscosity is calculated
in each plane parallel to the walls[8]
   We will present two LES at $R_e = 6666$ ($h^+ = 204$, case A) and $R_e =
14000$ ($h^+ = 389$, case B). They are respectively subcritical and supercritical

---

[8] In fact, the original formula for the spectral-eddy viscosity considered a three-
dimensional spectrum. It is possible, in the isotropic case and when spectra decrease
as a power law, to relate the two-dimensional to the three-dimensional spectrum. LES
of the channel seem to be insensitive to the particular spectrum chosen.

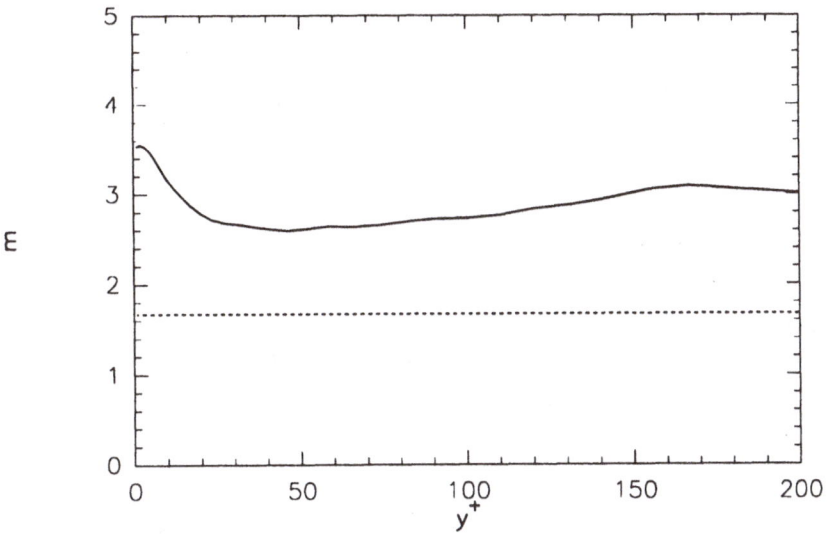

Fig. 7. — Spectral-dynamic LES of the channel flow (case A), exponent $m(y^+)$ of the kinetic-energy spectrum at the cutoff

with respect to the linear-stability analysis of the Poiseuille profile. In the two simulations there is a grid refinement close to the wall, in order to simulate accurately the viscous sublayer. Figure 7 shows for case A the exponent m arising in the energy spectrum at the cutoff, as a function of the distance to the wall y^+. Regions where $m > 3$ correspond to a zero eddy viscosity and hence a direct-numerical simulation. This is the case in particular close to the wall, up to $y^+ \approx 12$ where we know that longitudinal velocity fluctuations are very intense, due to the low- and high-speed streaks. Therefore, and since the first point is very close to the wall ($y^+ = 1$), our LES has the interesting property of becoming a DNS in the vicinity of the wall, which enables us to capture events which occur in this region. Figure 8 shows the mean velocity profile in case A, compared with the LES of Piomelli [32] using the dynamic model of Germano [6]. The latter is known to agree very well with experiments at these low Reynolds numbers. Our simulation coincides, with the right value for the Karman constant. On the other hand, a LES carried out with the classical spectral-cusp model with $m = 5/3$ gives an error of 20% for the Karman constant. Figure 9 shows for case A the rms velocity fluctuations, compared with Piomelli. The agreement is still very good, with a correct prediction of the longitudinal velocity fluctuations peak, corresponding to a maximum intensity

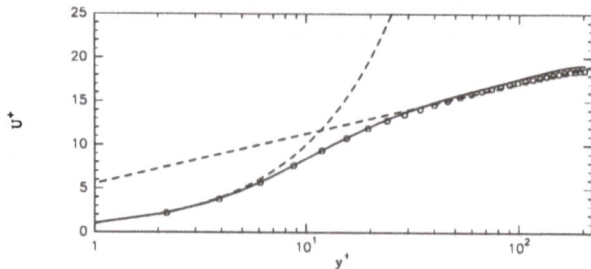

Fig. 8. — Same LES as Figure 7. Comparison of the mean velocity profile (straight line) versus Piomelli's [32] dynamic-model simulations (symbols)

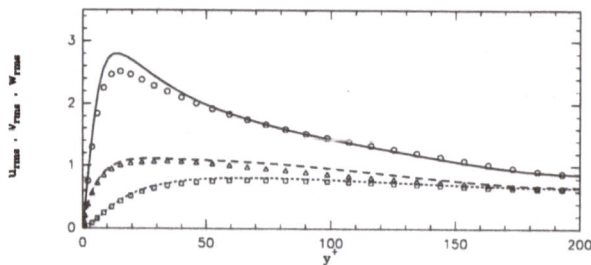

Fig. 9. — Same as Figure 8, but for the rms velocity fluctuations, from top to bottom longitudinal, spanwise and transverse

for the low an high-speed streaks. Concerning the supercritical case, the LES of case B are in very good agreement with a DNS at $h^+ = 395$ carried out by [33], both for the mean velocity and the rms velocity components. The latter are shown on Figure 10. Notice that the LES allows to reduce the computational cost by a factor of the order of 100, which is huge. We have checked that, in the LES of Figure 10 and very close to the wall, the spanwise vorticity fluctuations are much higher than their longitudinal and transverse counterparts. As shown by [21], this is due to the high shears existing under the high-speed streaks, and which are responsible for most of the drag at the wall.

It should be noticed also that the LES at these Reynolds numbers display much more vortical activity in the small scales than the DNS at lower Reynolds, as Figure 11 shows. The small-scale activity thus predicted is susceptible of enhancing mixing or chemical reactions in LES of turbulent transport or combustion for instance.

We stress finally that the spectral-dynamic model was applied with success to the rotating channel in [15][16] (see also Métais and Lamballais lectures notes in the present volume).

(a)

(b)

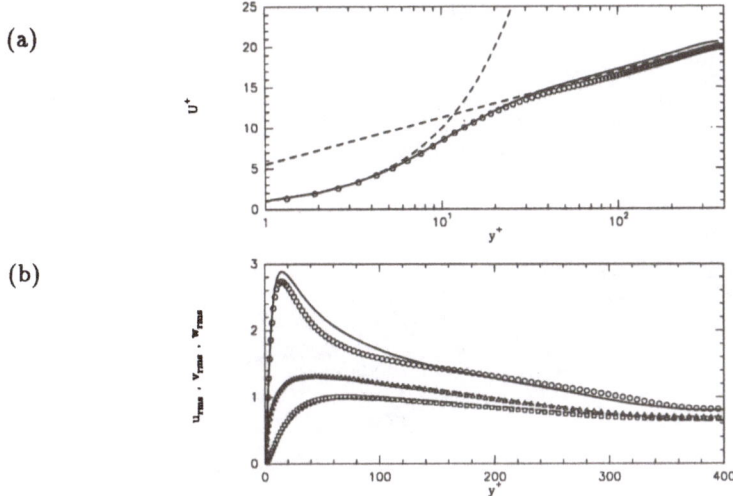

Fig. 10. — Turbulent channel flow, comparisons of the spectral-dynamic model (straight lines, $h^+ = 389$) with the DNS of Kim ([33], symbols, $h^+ = 395$); a) mean velocity, b) rms velocity components

7. BACKWARD-FACING STEP

Separated flows are very important in external and internal aerodynamics. One of the prototypes of such flows is the two-dimensional backward-facing step. Let us consider a step of finite height H, with an upstream velocity field corresponding to a boundary layer[9] developing upstream of the step. Let U_0 be its velocity on the upper bound of the domain, where a free-slip condition is assumed. The phenomenology of what happens is the following. In the average, a two-dimensional recirculation bubble sets up. Kelvin-Helmholtz type vortices are shed in the mixing layer behind the step. They grow by pairing and are three-dimensionalized in a way which will be discussed below. These vortices follow the upper part of the recirculation region, and impinge the lower wall at a position close to the reattachment length X_R. Afterwards, it is not clear to know in which proportions vortices are shed from the reattachment region and carried away downstream by the mean flow, or trapped in the recirculation and reinjected in the upstream mixing layer.

We have carried out LES of such a flow at a Reynolds number $U_0 H/\nu = 5100$, in conditions close to an experiment performed by Jovic and Driver (see [35] for details). The expansion ratio is 1.2. The subgrid model retained is the selective-structure function model in its four-point version in planes parallel to the upper and lower walls. We use a grid refinement in the high-shear region

---

(⁹) turbulent or possibly laminar

(a)

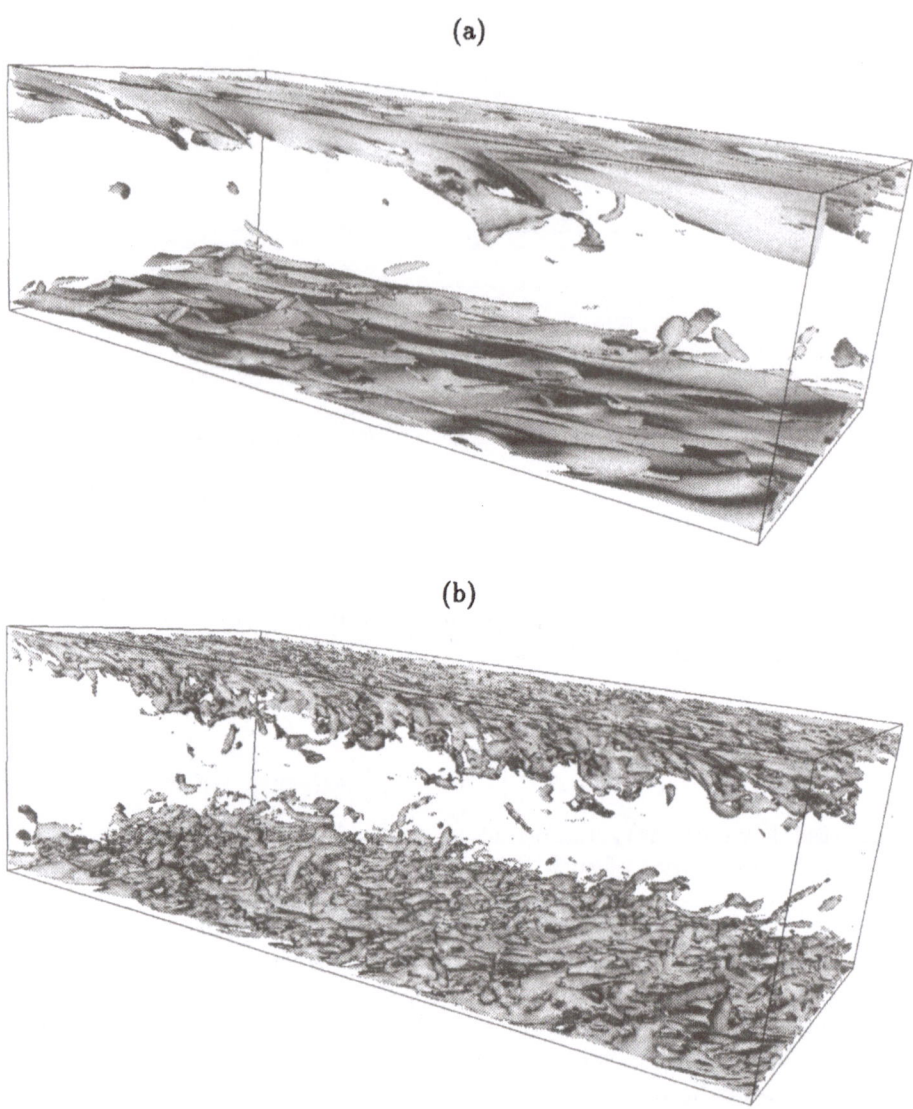

(b)

Fig. 11. — Channel flow, isosurface of the vorticity modulus (the flow goes from left to right); a) DNS at $h^+ = 165$, b) LES at $h^+ = 389$

downstream of the step, and on the lower wall (without any wall law) as in the DNS performed by [35] in the same case. The latter have shown the difficulty of obtaining a correct reattachment length: this requires to simulate deterministically the boundary layer upstream of the step. We have just taken an upstream velocity profile corresponding to the mean velocity measured experimentally, to which a small three-dimensional white noise is superposed. We

Fig. 12. — Vorticity modulus field obtained in a backward-facing step flow simulated in a LES using the filtered structure-function model with grid refinement (the step is figured on the left)

find $X_R \approx 7.5H$, which is far from the experimental measurement $\approx 6.1H$. However, the rms velocity fluctuations are in excellent agreement with the experiments, once the downstream coordinate is renormalized by X_R.

Here, we look rather at the dynamics of coherent vortices. Visualizations of the vorticity and pressure fields, as well as animations of the results, show the following features, confirmed by Figure 12, representing the vorticity field. First, quasi two-dimensional Kelvin-Helmholtz coherent vortices are shed behind the step. Then they undergo helical-pairing, yielding the formation of a a staggered pattern of big Λ-vortices which impinge the wall with a characteristic Strouhal number (based on H and U_0) $S_t \approx 0.2$. Since pressure spectra just upstream and downstream of the reattachment do have a peak at this frequency, it is clear that such a Kelvin-Helmholtz frequency corresponds to the passage of Λ-vortices which are carried away downstream. During this process, at least some of them transform into an arch, as Figure 12 indicates. Eaton and Johnston [34] have also found a peak at $S_t \approx 0.07$ in the kinetic-energy spectra of their experiments. This peak was recovered in the LES of [36] using the plain structure function model with a wall law, and the DNS of Moin [35]. Most authors agree now that this is due to a sort of flapping oscillations of the recirculation bubble itself, whose origin is not very well understood.

8. CONCLUSION

We have presented the general framework of large-eddy simulations (LES), where subgridscale motions are filtered out, their effect being represented by eddy-viscosity and eddy-diffusivity coefficients in the supergrid-scale motions. We have described Smagorinsky's model, which is the most-widely used for engineering applications. Afterwards, a different point of view has been considered, where the filtering is a sharp low-pass filter in Fourier space. Kraichnan's spectral eddy viscosity has been presented, and has been generalized to a spectral eddy diffusivity. It has been demonstrated, using the nonlocal interaction theory applied to a stochastic model of isotropic turbulence, how a k^4 backscatter arises in three dimensions, with a k^3 equivalent in two dimensions. The existence of such a spectral backscatter in large-eddy simulations has been confirmed. The spectral-dynamic model has also been presented , which is a generalization of the spectral eddy coefficients allowing to deal with laminar and transitional situations. Returning to physical space, these models have been reinterpreted in terms of velocity-structure functions and generalized hyperviscosity models.

The selective structure-function model has been applied to a temporal and spatially-growing mixing layer. Depending upon the quasi two-dimensional or three-dimensional character of the initial or upstream weak perturbation, it has been shown how the flow could bifurcate from a quasi two-dimensional state (where longitudinal hairpins are stretched between the Kelvin-Helmholtz vortices) to a helical-pairing configuration of dislocated vortices. In the former case, consideration of the vorticity equation permits to show how the vorticity is stretched in the direction of the first principal axis of deformation. A convected scalar gradient, on the contrary, is reduced in this direction, and intensified across the interface. Returning to the hairpin stretching, it has been explained how their spanwise wavelength may be imposed by Pierrehumbert and Widnall's [3] translative instability of the Kelvin-Helmholtz vortices.

The spectral-dynamic model has been applied to the turbulent channel flow at a subcritical ($h^+ = 204$) and supercritical ($h^+ = 389$) wall Reynolds number. In the two cases, the results are in excellent agreement with experiments and direct-numerical simulations. Compared with the latter at same Reynolds number, the LES reduces the computational cost by a factor of the order of hundred.

Finally, the selective-structure function model has allowed to elucidate the problem of coherent-vortex topology in turbulence generated behind a backward-facing step.

Acknowledgments

I am indebted to my collaborators, who contributed greatly to the numerical simulations of turbulence presented here: P. Begou, P. Comte, F. Delcayre, E.

Lamballais, G. Silvestrini. This work was sponsored by *Institut Universitaire de France*, CNRS, INPG and UJF. Part of the computations were carried out at the IDRIS (Institut du Développement et des Ressources en Informatique Scientifique, Paris).

References

[1] Smagorinsky J., *Mon. Weath. Rev.* **91 (3)** (1963) 99.
[2] Kraichnan R.H., *J. Atmos. Sci.* **33** (1976) 1521.
[3] Pierrehumbert R.T., Widnall S.E., *J. Fluid Mech.* **114** (1982) 59.
[4] Lesieur M., *Turbulence in fluids, third edition*, Kluwer Academic Publishers (1997).
[5] Lesieur M., Métais O., *Ann. Rev. Fluid Mech.* **28** (1996) 45.
[6] Germano M., *J. Fluid Mech.* **238** (1992) 325.
[7] Lions J.L., *Quelques méthodes de résolution des problèmes aux limites non linéaires.* Dunod (1969).
[8] Chollet J.P., Lesieur M., *J. Atmos. Sci.* **38** (1981) 2747.
[9] Chollet J.P., Lesieur M., *La Météorologie* **29-30** (1982) 183.
[10] Lesieur, M., Rogallo R., *Phys. Fluids A* **1** (1989) 718.
[11] Lesieur M., Métais O., Rogallo R., , *C.R. Acad. Sci. Paris* Ser II **308** (1989) 1395.
[12] Métais O., Lesieur M. *J. Atmos. Sci.* **43** (1986) 857.
[13] Métais O., Lesieur M., *J. Fluid Mech.* **239** (1992) 157.
[14] Lesieur M., Schertzer D., *Journal de Mécanique* **17** (1978) 609.
[15] Lamballais E., Simulations numériques de la turbulence dans un canal plan tournant. Thèse de l'Institut National Polytechnique de Grenoble (1995).
[16] Lamballais E., Lesieur M., Métais, O., *C. R. Acad. Sci. Paris* Ser IIb **323** (1996) 95.
[17] Leslie D.C., Quarini G.L., *J. Fluid Mech.* **91** (1979) 65.
[18] Ducros F., Comte P., Lesieur M., In: *Turbulent Shear Flows IX*. Springer-Verlag, (1995) 283.
[19] David E., Modélisation des écoulements compressibles et hypersoniques: une approche instationnaire. Thèse, Institut National Polytechnique de Grenoble (1993).
[20] Ducros F., Simulations numériques directes et des grandes échelles de couches limites compressibles. Thèse, Institut National Polytechnique de Grenoble (1995).
[21] Ducros F., Comte P., Lesieur, M., *J. Fluid Mech.* **326** (1996) 1.
[22] Herbert T., *Ann. Rev. Fluid Mech.* **20** (1988) 487.
[23] Garnier E., Métais O., Lesieur, M., *C.R. Acad. Sci. Paris* Ser II b **323** (1996) 161.
[24] Silvestrini, J., Simulation des grandes échelles des zones de mélange: application à la propulsion solide des lanceurs spatiaux. Thèse de l'Institut National Polytechnique de Grenoble (1996).

[25] Konrad J.H., An experimental investigation of mixing in two-dimensional turbulent shear flows with applications to diffusion-limited chemical reactions. Ph.D. Thesis, California Institute of Technology (1976).

[26] Bernal L.P., Roshko A., *J. Fluid Mech.* **170** (1986) 499.

[27] Corcos G.M., Lin S.J., *J. Fluid Mech.* **139** (1984) 67.

[28] Neu J.C., *J. Fluid Mech.* **143** (1984) 253.

[29] Comte P., Lesieur M., In: H.K. Moffatt (ed.) *Topological Fluid Dynamic.* Cambridge University Press (1989) 649.

[30] Comte P., Fouillet Y., Gonze M.A., Lesieur M. , Métais O., Normand X., Large-eddy simulations of free-shear layers. In: O. Métais and M. Lesieur (eds.) *Turbulence and coherent structures.* Kluwer Academic Publishers (1991) pp.45-73.

[31] Comte P., Lesieur M., Lamballais E., *Phys. Fluids A* **4** (1992) 2761.

[32] Piomelli U., *Phys. Fluids A* **5** (1993) 1484.

[33] Antonia R. A., Teitel M., Kim J., Browne L.W.B., *J. Fluid Mech.* **236** (1992) 579.

[34] Eaton J.K., Johnston J.P., Stanford University, Rep. MD-39 (1980).

[35] Le H., Moin P., Stanford University, Rep. TF-58 (1994).

[36] Silveira-Neto A., Grand D., Métais O., Lesieur M., *J. Fluid Mech.* **256** (1993) 1.

Large Eddy Simulation: An Introduction and Perspective

J. H. Ferziger

Department of Mechanical Engineering
Stanford University
Stanford CA

1. INTRODUCTION

A decade ago, it was possible to write comprehensive reviews of what had been accomplished using direct and large eddy simulation of turbulence [1, 2]. Since then, simulation techniques have been applied to many flows that a complete review is no longer possible. These methods are now well established as a complement to experimental methods.

In direct numerical simulation (DNS), all scales of motion of a turbulent flow are computed. Everything from the large energy-containing scale to the dissipative scale must be included; this requires a large number of grid points. Despite its cost, the detail that DNS produces has made it a valuable tool for investigating the physics of turbulence. For simple flows, it is the method of choice.

In large eddy simulation (LES), only the largest scales of motion are simulated explicitly and the small scales are modeled. LES is expensive but much less so than DNS. Whenever DNS is feasible, it is the method of choice. But, when DNS is too expensive, LES is now is a good alternative; indeed it is often inexpensive enough to consider for selected industrial applications.

In this article, we describe the methods employed in LES, concentrating on the differences with DNS, especially the models employed to represent the small scales and numerical methods that can be used in complex geometries.

We shall explore what is feasible with LES today. We begin with some remarks on turbulence and prediction methods.

2.   TURBULENCE AND ITS PREDICTION

It is difficult even for experts to agree on a definition of turbulence. Required elements include three dimensionality, unsteadiness, strong vorticity, unpredictability in detail, and a broad spectrum. Most people agree that turbulent flows are random and noisy. However, there are also coherent structures in most flows and they are important. They account for a small fraction of the turbulent energy, but are responsible for a large share of the transport of species, mass, momentum and energy. The coherent structures in a given flow are similar in form but not identical and do not appear regularly in either time or space, making them difficult to describe. Most of the turbulent energy is due to random motion which is probably responsible for much of the irregularity of the coherent structures. This helps to explain why turbulence is so difficult to study. Also, as coherent structures are different in each flow, the likelihood of finding a simple model applicable to all flows is exceedingly small.

Turbulence is not an equilibrium phenomenon. Without some means of producing it, turbulence decays via viscous dissipation. The traditional picture of turbulence is that the mean flow does work on the large scale motions thereby increasing their energy (*turbulence production*). This energy is transferred to the small scales where it is dissipated. Between the large and small scales, lies the *inertial subrange*. However, there are processes that transfer energy from small scales to large ones; they are obvious in two dimensional flows. Included in these processes are vortex roll-up and pairing [3]. In three dimensional turbulence energy is transferred in both directions, with the flow toward the small scales being about twice the reverse flow [4].

For homogeneous flows, the distribution of energy over the length scales is described by the wavenumber spectrum *i.e.*, in terms of its Fourier coefficients.

$$E(k) = \frac{1}{2}\hat{u}_i(k)\hat{u}_i^*(k) \tag{1}$$

where the asterisk (*) denotes a complex conjugate. A typical energy spectrum has a peak at low wave-numbers corresponding to integral scale, L. The peak is followed by the inertial subrange and, finally, a rapid decrease near the Kolmogoroff scale.

The most straight-forward and precise approach to turbulence simulation is to solve the Navier-Stokes equations without averaging or approximation. The result is equivalent to a short-duration laboratory experiment; this is direct numerical simulation (DNS).

In a DNS, the domain must be at least a few times the integral scale L. To capture all the dissipation, which occurs the grid size must be of the orde of the Kolmogoroff scale, η. For homogeneous turbulence, the number of grid points in each direction must therefore be at least L/η, which is proportional to $Re^{3/4}$. Since this number of points must be employed in each coordinate direction, and the time step is related to the grid size, the cost scales as Re^3.

¿From a practical point of view, the number of grid points that can be used is limited by the speed and memory of the machine so direct numerical simulation can be performed only at low Reynolds numbers. At present, DNS can reach the lowest Reynolds numbers of engineering interest.

Since DNS is essentially exact, it is a substitute for or complement to experiments. It has the advantage that all of the conditions imposed on the flow (initial conditions, boundary conditions, external forces, etc.) are treated exactly. This allows DNS to clarify problems and explain discrepancies in experimental results.

3. THE BASIS OF LES: FILTERING

In large eddy simulation the larger scales are simulated while the small ones are modeled *i.e.*, it is a compromise between DNS and RANS. The justification for LES is that the large eddies contain most of the energy, do most of the transporting of conserved properties, and vary most from flow to flow; the smaller eddies are more universal, less important and should be easier to model.

It is essential to define what is to be computed—the part of the velocity field that contains the large scales. This is best done by filtering [5]; the large or resolved scale field is a local average of the complete field. We use one dimensional notation for simplicity; the generalization to three dimensions is straight-forward. The filtered velocity is defined by:

$$\overline{u}_i(x) = \int G(x, x') u_i(x') dx' \tag{2}$$

where $G(x, x')$, the filter kernel, is a localized function.

In homogeneous flows, the filter depends only on the distance $|x - x'|$ so $G(|x - x'|/\Delta)$ where Δ is the length scale over which the averaging is done. Eddies larger than Δ are 'large' and those smaller than Δ are 'small'. There may be a connection between the filter width Δ and the grid size used in the numerical calculation but, in general, the two are not identical. Filter kernels which have been applied in LES of homogeneous flows include: the *Gaussian*, the *Box* and the *Fourier cutoff*.

Filtering is more complicated in inhomogeneous flows, especially near surfaces. In finite volume methods, it seems natural to define the filter as an average over a grid volume. This is often done but is not appropriate because, according to Nyquist's theorem, a grid of size Δx cannot resolve a structure smaller than $2\Delta x$.

When the Navier-Stokes equations are filtered, one obtains a set of equations similar in form to the RANS equations:

$$\frac{\partial \overline{u}_i}{\partial t} + \frac{\partial \overline{u_i u_j}}{\partial x_j} = -\frac{\partial \overline{p}/\rho}{\partial x_i} + \frac{\partial^2 \overline{u}_i}{\partial x_j \partial x_j} \tag{3}$$

The definitions of the averaged velocities in RANS and LES differ but the closure issues are similar. Since $\overline{u_i u_j} \neq \overline{u}_i \overline{u}_j$, a modeling approximation for the difference between the two sides of this inequality,

$$\tau_{ij} = \overline{u_i u_j} - \overline{u}_i \overline{u}_j \tag{4}$$

must be introduced. τ_{ij} is called the subgrid scale (SGS) Reynolds stress. It is similar to the RANS Reynolds stress but the physics it represents is different. The SGS energy is smaller than the RANS turbulent energy, so model accuracy may be less crucial in LES than in RANS. In comparing experimental data with LES results, it is important to take this into account. LES results should not agree with unfiltered data. Careful comparisons require either filtered experimental data or 'defiltered' LES results. Subgrid scale modeling is the distinctive feature of LES.

4.   SUBGRID SCALE MODELING

Models used to approximate the SGS Reynolds stress (4) are called *subgrid scale* (SGS) models, a nomenclature which suggests a connection betwen the filter and the grid used to solve the equations. As noted, there need not be such a connection. Note that filtering a second time does not give the original filtered field back again:

$$\overline{\overline{u}} \neq \overline{u} \tag{5}$$

except when the cutoff filter is used. This difference means that the modeling of the subgrid scale Reynolds stress (SGSRS) may be more complicated than for the RANS Reynolds stress.

The subgrid scale Reynolds stress is a local average of the small scale field so models for it should be based on the local velocity field or, perhaps, on the past history of the local fluid. The latter leads to models that require solution of partial differential equations for the parameters. Models that do not have these properties have been proposed; we expect their utility to be limited.

4.1.   Smagorinsky Model

By far the most commonly used subgrid scale model is the one proposed by Smagorinsky [6], an eddy viscosity model that is an adaptation of the Boussinesq concept to the subgrid scale. It is:

$$\tau_{ij} - \frac{1}{3}\tau_{kk}\delta_{ij} = -\nu_T\left(\frac{\partial \overline{u}_i}{\partial x_j} + \frac{\partial \overline{u}_j}{\partial x_i}\right) = -2\nu_T \overline{S}_{ij} \tag{6}$$

and can be derived in a number of ways.

The form of the subgrid scale eddy viscosity can be derived by dimensional arguments. We simply present it; it can be derived from theories [7]. These lead to:

$$\nu_T = C_S^2 \Delta^{4/3} L^{2/3} |\overline{S}| \tag{7}$$

where L is the integral scale of the turbulence. Theories provide estimates of the constant as well as the form of the model.

The presence of the integral scale L in (7) makes the model difficult to use. For this reason, the substitution $\Delta^{4/3} L^{2/3} \to \Delta^2$ is often used, leading to the usual form of the Smagorinsky model:

$$\nu_T = (C_S \Delta)^2 |\overline{S}| \tag{8}$$

The approximation introduces error into the model and one should not be surprised if the model parameter is not constant.

Various methods of evaluating the parameter agree that $C_S \approx 0.2$ for isotropic turbulence. LES shows that this value is nearly optimum. The range of Reynolds numbers for which simulations have been made is relatively narrow. The above substitution may make the parameter C_S a function of Δ/L which is, in turn, a function of Reynolds number. We should not be surprised to find that the parameter C_S is a function of Reynolds number or different in different flows as is observed to be the case.

The Smagorinsky model is satisfactory in many ways but is not without problems. To simulate shear flows, several modifications to it are required. The value of the parameter C_S in the bulk of the flow has to be reduced from 0.2 to approximately 0.065, reducing the eddy viscosity by nearly an order of magnitude. Close to a surface, the value has to be reduced even further; usually van Driest damping (which is difficult to justify in LES) is used.

Near a wall, the flow is very anisotropic. Long, thin regions of low speed fluid (streaks) are created. Resolving them requires a highly anisotropic grid and the length scale to be used in the SGS model is not obvious.

In stably stratified flow, the Smagorinsky model parameter becomes a function of the Richardson number, a non-dimensional parameter that represents the relative importance of stratification and shear. Similar effects occur in flows with rotation and/or curvature.

Thus, there are many difficulties with the Smagorinsky model. If we wish to simulate more complex and/or higher Reynolds number flows, a more accurate model is needed.

The traditional test of a turbulence model is to apply it to a series of problems and compare the results with experimental data. This is an obvious and practical way of testing models, The availability of DNS (or detailed experimental data) makes another kind of test possible, the *a posteriori* approach. In particular, it is possible to evaluate the terms which must be modeled in the LES and, at the same time, the model estimates of them; this kind of *a priori* test is a very severe test of a model.

Let us that the flow is homogeneous. Then, given the exact velocity field at an instant, u_i, we filter it to obtain its large eddy component, $\overline{u}_i$. Next, we compute the subgrid scale Reynolds stress tensor, $\overline{u_i u_j} - \overline{u}_i \overline{u}_j$. Finally, from the filtered velocity field, we can compute the model estimate of the Reynolds stress. We thus have the exact Reynolds stress and its model representation at every point in the flow. To test the accuracy of the model, we compare the two. Two methods of doing so are by computing a correlation coefficient and by producing a scatter plot. This yields a very severe test of a model. By comparing the magnitudes of the model and exact values, one can obtain a value of the model parameter. The values obtained are in good agreement with those found by other means, *cf.* [8].

In the precise sense of *a priori* testing, the Smagorinsky model is not very accurate. Most of the lack of correlation arises because the SGS Reynolds stress tensor and the strain rate of the resolved field, which the Smagorinsky model assumes to be proportional, actually have little in common. In particular, the principal axes of the two tensors are not well correlated. Despite the poor rating the Smagorinsky model receives in *a priori* tests, it performs reasonably well in some simulations. Meneveau *et al* [9] used experimental data to do *a priori* tests at high Reynolds number and found similar results.

There are two directions in which one may seek improved models. The first follows the lines established by RANS modelers. One introduces partial differential equations for quantities from which the modeled SGS Reynolds stresses are computed. Unfortunately, this approach has not been very successful. The other approach is unique to LES. The smallest resolved scales are similar to the unresolved scales so the information contained in them can be used to construct models for the unresolved scales. This is a more fruitful approach.

4.2. Scale Similarity and Mixed Models

The concept that the small scales of a simulation can be used to study modeling leads to a number of interesting modeling methods. The first of these is an alternative model for the subgrid scales, the scale similarity model [10]. The idea behind this model is that the important interactions between the resolved and unresolved scales involve the smallest eddies of the former and the largest eddies of the latter. This leads to the following subgrid scale model (we give the Galilean invariant form, not the original one):

$$\tau_{ij} = \overline{\overline{u}_i \overline{u}_j} - \overline{\overline{u}}_i \overline{\overline{u}}_j \tag{9}$$

This model correlates very well with the actual SGS Reynolds stress in *a priori* tests. However, this model hardly dissipates any energy and cannot serve as a 'stand alone' SGS model. It does transfer energy from the smallest resolved scales to the larger resolved scales, which is useful.

To correct for the lack of dissipation, it is necessary to combine the scale similarity models with the Smagorinsky model. The resulting 'mixed' model has been used with considerable success. Some of the difficulties of the Smagorinsky model disappear and the variability of the parameter is reduced. Even more importantly, the idea that the small scales of the simulation can be used for modeling purposes has led to further improvements given below.

Piomelli et al [11] showed that, when spectral methods are used, the Smagorinsky model is preferred but the mixed model performs better when finite difference or volume methods are used.

4.3. Dynamic Procedure

A very popular approach today is the dynamic procedure [12]. We shall give a simple version of it that illustrates the concept. Suppose we do a large eddy simulation but regard it as a DNS and obtain the the parameter by *a priori* estimation. The result is a kind of self-consistent subgrid scale model. The actual procedure of Germano *et al.* [12] is a bit more formal than the one we have just described but, as it is found in many places, we shall not present it in detail. We simply state that, if it is assumed that the same model (usually the Smagorinsky or mixed models are used) holds at both the original LES filtering level and the 'test filter' level, an identity can be derived from which the parameteer can be computed by the least squares method [13].

Although the dynamic model is very appealing, significant problems are encountered when it is put into practice. One difficulty is that the derivation assumes that the model parameter is constant but the computed parameter is a function of the spatial coordinates and time. In fact, C varies very rapidly, producing large eddy viscosities of both signs. Although a negative eddy viscosity may be considered as backscatter, if the eddy viscosity remains negative over too large a spatial region or for too long a time, the resulting numerical instability destroys the simulation.

One cure for the problem is set any eddy viscosity less than the molecular viscosity *i.e.* any $\nu_T < -\nu$, equal to $-\nu$; this is called clipping. It works but is not very satisfying. Another alternative is to employ averaging. For a homogeneous flow, we may average over the entire domain in estimating the parameter. This technique has been used with excellent results. Another approach is to use a combination of local spatial and temporal averaging, which is possible in any flow *cf.* [14].

Part of the problem is that it isassumed that the model is constant; this is contrary to what is actually found. To overcome this problem, one must remove this assumption. The result is an integral equation for the parameter. This helps considerably but does not completely cure the problem; it also increases the computational cost.

One can subject the integral equation referred to in the last paragraph to a constraint that the total viscosity be non-negative, leading to a constrained optimization problem. This can produce good results at a cost of some increase in computer time and is called the dynamic localization model [15].

The dynamic procedure is not restricted to using the Smagorinsky model as

the base model. One can use any other model. The mixed model was used by
Zang *et al.* [16] with considerable success; in particular, it reduces the range of
the eddy viscosity. Shah and Ferziger [17] found that by removing the smallest
scales ('prefiltering') prior to applying the dynamic procedure, simulations that
are both accurate and stable were obtained.

Another approach was suggested by Meneveau [18] who showed that using a
Lagrangian average in the dynamic procedure was capable of producing results
of quality equal to those of obtained with other kinds of averaging. This method
has the advantage of being applicable to any flow and has been used successfully
by a number of other authors.

4.4. Spectral Models

Turbulence theories provide guidance for model construction. In most theories,
the principal variables are the Fourier transforms of the velocity components
or *i.e.*, the energy spectrum. Theories of this kind are applicable only to homogeneous turbulence. Despite this, they provide insight into issues connected
with subgrid scale modeling.

¿From the Navier-Stokes equations, one can derive a dynamic equation for
the energy spectrum. In it, the effect of the non-linear (advective) term is to
transfer energy from one wavenumber to another; it does not produce or destroy
turbulent energy. The pressure gradient terms are absent entirely; they transfer
energy from one component of the turbulence to another and play no role in
the equation for the total kinetic energy. The viscous term is responsible for
the dissipation, an energy drain. For details of these theories, the roles of the
various terms, and results obtained from them, see [19].

Theories allow one to define a spectral eddy viscosity. Suppose we want to
do large eddy simulations with the cutoff filter, the natural one for application
to theories. We can ask how much energy is transferred from a given resolved
wavenumber k to wavenumbers beyond the cutoff. If we call this energy transfer
rate $T_>(k)$, we can define an effective spectral eddy viscosity:

$$\nu_T = T_>(k)/2k^2 E(k) \tag{10}$$

Such an eddy viscosity was given by Lesieur [19]. There is a decrease in ν_T at
low wave-numbers that is of little consequence. The rise at the high wavenumbers is due to the nature of the interactions.

The rise in the eddy viscosity at high wavenumber does not follow a simple
law it is useful to fit a model function to it. The most commonly used spectral
eddy viscosity was proposed by Chollet and Lesieur [20]:

$$\nu_T = .267 \left(\frac{E(k_c)}{k_c} \right)^{1/2} \tag{11}$$

Métais and Lesieur [21] showed that the coefficient depends on the slope of
the spectrum near the cutoff. For further details, see the review by the same

authors [22]. Alternatively, the curve can be approximated by a constant plus a term proportional to a power or exponential of k.

Spectral models can produce simulations with spectra of the form $k^{-5/3}$ up to the smallest resolved scales. These velocity fields contain more resolved energy than simulations based on Gaussian or box filters. This is often cited as a major advantage. In a real flow, the smallest resolved scales would interact with scales beyond the cutoff; in an LES, these interactions are represented by a model. Thus these scales are never accurately simulated. This, and the fact that most discretizations are not accurate at the smallest scales, cast doubt on the value of trying to produce a Kolmogoroff spectrum. Furthermore, the claim that that these simulations represent flow at infinite Reynolds number is incorrect. At infinite Reynolds number, there is not only a $k^{-5/3}$ spectrum but also a large number of large scale modes with which a given wavenumber mode may interact. In any LES, the number of such modes is restricted; this may have a strong effect on the results of a simulation.

These results suggest that a combination of the Smagorinsky model (to approximate the constant eddy viscosity) and a hyperviscosity (a term containing an even-order velocity derivative of order higher than second) might be a good subgrid scale model. The simplest such model uses a fourth order viscosity. Adding such a term to the model increases the order of the partial differential equation so additional boundary conditions might be needed. This can be avoided if the fourth order viscosity vanishes rapidly enough at the wall. A dynamic version of this model was used by Morris and Ferziger [23]; simulations of isotropic turbulence and channel flow showed this model to be superior to the dynamic Smagorinsky model.

We shall not discuss other models that cannot be expressed in physical space. An interesting model based on spectral ideas was given by Métais and Lesieur [21] who called it a *structure function model*; it uses the two point velocity correlation. In practice, this model is similar to the Smagorinsky model but the authors claim that it improves simulations. They have created a number of extensions to the model. Further details on it are provided elsewhere in this volume [24].

It is possible to use spectral eddy viscosity models in a dynamic context. This has not been done, but it is an interesting possibility.

Simulations claimed to be direct numerical simulations of complex flows have been made; an early example is Kawamura and Kuwahara [25]. The grids used are too coarse to resolve all of the turbulence so these simulations do not meet the definition of DNS. They use third order upwind approximations for the convective terms, thus producing errors that act as fourth order viscosities so these are best regarded as large eddy simulations. However, the eddy viscosity is determined by the grid so the solution may depend on the grid, an undesirable feature.

4.5.   Models Based on RANS Models

We now discuss subgrid scale models based on RANS models. Any RANS model can be adopted as an SGS model. These models use partial differential equations to find the subgrid scale quantities.

The first RANS model beyond the mixing length model (which corresponds to the Smagorinsky model) is the two-equation class of models that use equations for quantities that determine the turbulence velocity and length scales; the most popular such model is the $k - \epsilon$ model which solves equations for the turbulent kinetic energy (k) and its rate of dissipation (ϵ); the length scale is $\ell = k^{3/2}/\epsilon$. In LES, the natural length scale is the filter width (Δ) so the second equation is not needed; a partial differential equation for the subgrid scale kinetic energy suffices. There is only a little experience with such models but they seem to provide little or no improvement over the Smagorinsky model. This is hardly a surprise because the major deficiency of the Smagorinsky model is lack of alignment of the principal axes of the SGS Reynolds stress and rate of strain tensors. Some benefit is obtained in transitional flow but the dynamic model performs just as well. Ghosal *et al.* [26] constructed a dynamic model that includes a differential equation for the turbulent kinetic energy.

The most complex RANS models in common use today are Reynolds stress models which solve a set of equations for the Reynolds stress tensor. One can derive dynamic equations for the SGS Reynolds stress. LES with an SGS Reynolds stress model has been tried only once, in an early (and very daring) simulation of the atmospheric boundary layer by Deardorff [27]. It gave a huge increase in the cost but no improvement in the results.

In the turbulent kinetic energy and Reynolds stress subgrid scale models that have been used, the constants were taken from RANS models. This is not appropriate as the physics of subgrid scale turbulence is different from that of large scale turbulence.

Finally, we mention that non-linear RANS models have received considerable attention in recent years. In these, the Reynolds stress in allowed to be a non-linear funciton of the mean strain rate and rotation tensors. There is only a little experience with these models so it is too early to know whether these models are a significant improvement over the linear ones; they do appear to have promise. They have also been suggested as subgrid scale models [28] and may prove useful in that context.

5.   EFFECTS OF EXTRA STRAINS

The models described above work well for flows without 'extra strains' (stratification, rotation, compressibility and curvature). However, the dynamic model can handle a number of these issues without difficulty.

For LES, extra strains can be divided into two classes. Some, including rotation, curvature, and stratification, affect the large scales more than the small ones but their effect on the small scales cannot be neglected. For flows containing these strains, the Smagorinsky model may need to be modified. Thus, for LES of a stable planetary boundary layer Mason and Derbyshire [29] made the parameter dependent on the Richardson number. The results agree very well with DNS [30] and field data. However, equally good results can be obtained with the dynamic model without explicit modification [31].

For 'strains' whose action is on the small scales, the situation is more difficult. Effects of this kind occur in compressible and combusting flows. These 'strains' allow thin regions of near-discontinuity—shock waves and flames. For these, simple fixes will not work. Although DNS has been applied to these flows, work on LES is just beginning.

6. WALL MODELS

Another important issue is modeling of the near wall region. BWe start with a review some results of experiments and simulations with no-slip wall boundary conditions.

Shear flows near solid boundaries contain thin 'streaks' of high- and low-speed fluid. If these are not adequately resolved, the turbulence production near the wall (which is a large fraction of the total) is under-predicted [32], reducing the Reynolds stress and the skin friction.

Simulations suggest that the wall-region and the region far from the wall are loosely coupled. Chapman and Kuhn's [33] simulation with an artificial boundary condition imposed at the top of the buffer layer ($y^+ = 100$) displayed most of the characteristics of the wall layer. On the other hand, Schumann [34] and Piomelli et al. [11] showed that relatively crude boundary conditions can represent the effect of the wall region in a simulation of the central part of a channel flow. Thus, either region can be well-simulated with a reasonable condition to represent the interaction with the other zone.

Schumann's [34] model assumes that the velocity at the grid point nearest a wall is proportional to the shear stress at the wall directly below it. Mason and Callen [35] assumed that the logarithmic profile holds locally and instantaneously, an approximation that is reasonably at the high Reynolds numbers of meteorology but not accurate enough in engineering flows.

Piomelli et al. [11] used direct simulation results to test wall layer models and proposed two new models. One is based on the idea that Reynolds stress-producing events move at a small angle with respect to the wall; this leads to a shifted model. The second model notes that significant Reynolds stress-containing events involve vertical motion and is called the ejection model. Both models give better results that Schumann's model for channel flow. A model designed for both fully developed turbulent regions and regions of unsteady laminar flow, was presented by Werner and Wengle [36]. For channel flow at Reynolds number 15,000, use of these conditions reduces the time of a simulation by a factor of ten [11] so their value is unquestionable.

Bagwell *et al.* [37] used linear estimation to find the best estimate of the skin friction given the velocity distribution. The skin friction estimate becomes a weighted average of the velocity at various points on the plane closest to the wall. This method requires the two point correlation, which is difficult to use in complex flows.

These models have been applied only to flows on flat walls with mild pressure gradients. They are not applicable to flows with separation and/or reattachment but they may work in three-dimensional boundary layers.

A workshop on LES in which various participants presented simulations of the same well-defined flows was held recently [38]. Two flows were considered. The flow over a square cylinder had a laminar inflow, making it a transitional flow. The results exhibited considerable variation, probably because transitional flows are very difficult to treat with LES. The other case was flow over a cube mounted on the wall of a channel, for which the results demonstrated greater consistency and agree well with experiment. This may be due to the fact that the flow is fully turbulent.

7. NUMERICAL METHODS

A wide variety of numerical methods have been employed in LES. Because these methods are adequately described elsewhere (e.g., [39]), we shall not describe them in detail here. Instead, a few items peculiar to LES will be discussed.

The most important requirements arise from the need for accurate treatment of flows that contain a wide range of length scales. Time accuracy requires a small time step. The time-advance method must be stable for the chosen time step, a condition that can be met by explicit methods so most simulations use them. Near walls, fine grids must be used and the viscous terms may require very small time steps. In such a case, these tersm may receive implicit treatment. The methods for time advancement used in LES are usually second to fourth order accurate, including the Runge-Kutta, Adams-Bashforth and leapfrog methods. It is important to use versions of these methods that require a minimum of storage.

The need to handle a wide range of length scales decreases the value of the most common descriptor of accuracy of a discretization method, its order. To see why this so, it is useful to consider the Fourier decomposition of the velocity field. The highest wavenumber that can be resolved on a uniform grid of size Δx is $k_{max} = \pi/\Delta x$, so $0 < k < \pi/\Delta x\}$. Consider the effect of discretization on e^{ikx}, whose exact derivative is ike^{ikx}. Discretizations replace this by $ik_{eff}e^{ikx}$; k_{eff} is called the effective wavenumber. For example, the standard central difference approximation, when applied to e^{ikx}, gives:

$$\frac{\delta e^{ikx}}{\delta x} = i\frac{\sin k\Delta x}{\Delta x}e^{ikx} \qquad (12)$$

so $k_{eff} = \sin k\Delta x/\Delta x$ for this method. For small k, the Taylor series approximation to this function shows the second order nature of the approximation. This approximation is accurate only for $k < \pi/2\Delta x$, the first half of the wavenumber range. Since a spectral method is exact *i.e.*, $k_{eff} = k$, an N point spectral method is more accurate than a second order finite difference method with $2N$ points, which explains why spectral methods are preferred when they are applicable. Unfortunately, they are difficult to use.

Each discretization has a different effective wavenumber. Upwind approximations have complex effective wavenumbers, reflecting their dissipative nature. The reason why effective wavenumber is important in DNS and LES is that the spectra (the distribution of the energy in wavenumber) is large over the entire wavenumber range so the order of the method is no longer sufficient to define its accuracy. For example, Cain *et al.* [40] found that, for a spectrum typical of isotropic turbulence, a fourth order method has half the error of a second order method.

Because second order methods do not compute derivatives of modes with wavenumbers higher than $\pi/2\Delta x$ accurately, that part of the spectrum should not contain much energy and the filter should have a width of at least $2\Delta x$. In the zeal to include as much energy as possible, many authors ignore this point. Also, the subgrid scale model and numerical method need to be related. Piomelli *et al.* [11] showed that, with finite difference methods, the mixed model is best, while spectral LES is best carried out with the Smagorinsky model. The methods and step sizes in time and space should be chosen so that their errors are as nearly equal as possible.

Accuracy is difficult to measure in DNS and LES. The reason is inherent in the nature of turbulent flows. A small change in the initial state of a turbulent flow is amplified exponentially in time so, after a short time, the perturbed flow hardly resembles the original one. This makes determining the error by direct comparison of solutions impossible. Instead, one can do simulations with different grids and compare the statistics of the solutions. A simpler method is to look at the spectrum of the turbulence. If the energy at the smallest scales is at least two orders of magnitude smaller than the energy at the peak of the spectrum, it is safe to assume that the flow is well resolved.

In addition to discretization error, one must consider *aliasing*, an error that arises from the non-linearity of the equations. It is most easily discussed in terms of the Fourier series. When two functions are multiplied, the result contains terms of the form $exp(i(k_1 + k_2)x)$. If k_1 and k_2 are larger than $\pi/2\Delta x$, the resulting wavenumber is greater than $\pi/\Delta x$, the highest wavenumber that can be represented on the grid. The result is misinterpreted as belonging to wavenumber $(k_1 + k_2 - \pi)/\Delta x$ (aliased) and energy that should be transferred to wavenumbers outside the considered range is found at low wavenumbers.

8. INITIAL AND BOUNDARY CONDITIONS

Another difficulty in LES is generating initial and boundary conditions. Since the effects of these conditions may be 'remembered' for a considerable time, they can have a significant effect on the flow and the computation time. Initial conditions should represent a three dimensional divergence free velocity field. Because it is impossible to reproduce experimental conditions exactly, simulation results are not realistic for some time.

For homogeneous isotropic turbulence, it is easiest to construct the initial conditions in Fourier space. One chooses the initial spectrum which fixes the energy in each mode. Continuity is another constraint, leaving just one parameter to be selected randomly. The field in physical space is created by Fourier transformation. It is necessary to run the simulation for about two eddy turnover times before the simulated flow becomes realistic. The initial conditions for strained homogeneous turbulence are taken from a developed field of isotropic turbulence.

For channel flow, in the absence of results of another simulation, the initial conditions may be a combination of a mean velocity field, instability modes (with finite amplitude), and random noise. Again, a considerable time is required to produce a realistic flow.

For flows that are homogeneous (in the statistical sense) in a given direction, one can use periodic boundary conditions in that direction; these are easy to use, fit well with spectral methods, and are as realistic as possible.

On solid walls, no-slip boundary conditions may be used but very fine grids are required, especially in the direction normal to both the wall and the principal flow direction.

Inflow conditions should contain the velocity field on a plane of a turbulent flow. The best inflow conditions are obtained from other simulations. For example, for the flow at the inlet to a curved channel, a plane channel flow is ideal. A less satisfactory choice is a combination of a mean flow, finite amplitude instability modes, and random noise.

At outflow boundaries, one possibility is to require that all derivatives in the direction normal to the boundary be zero. This condition is often used in steady flows. For unsteady flows, including turbulence simulations, it is better to replace this condition by the unsteady convective condition:

$$\frac{\partial \phi}{\partial t} - U \frac{\partial \phi}{\partial \xi} = 0 \,. \tag{13}$$

where, U is a constant velocity chosen so that overall conservation is maintained. This condition eliminates most of the problems caused by pressure waves reflecting from the outflow boundary.

Symmetry boundary conditions, which are often used in RANS computations are not applicable in DNS or LES because, although the mean flow may be symmetric about some particular plane, the instantaneous flow is not. Symmetry conditions have, however, can be used to represent flat free surfaces.

9. ACCOMPLISHMENTS

Large eddy simulation has been applied to a range of flows too large to be covered in a single work. The purpose of most simulations is to study the physics of turbulence and the models used to approximate it, treating the results as experimental data. For that reason, one should consider the results together with experimental data for the same flow. We shall therefore give only a short overview of the flows that have been treated with LES and a discussion of what may be possible in the next few years.

Simple flows that have been treated include all homogeneous flows, plane channel flow, and free shear flows that are inhomogeneous in one direction (2D mixing layer, wake, and jet). These flows can treated with DNS, the preferred technique for this kind of investigation.

LES is applied to flows that are beyond the reach of DNS. An important engineering issue is flow separation and reattachment, common phenomena in technological flows that pose difficulties for RANS methods. The simplest separating flow is the backward facing step in which a plane channel flow encounters a one-sided sudden expansion. DNS of this flow was performed by Le [41]; good results were obtained but at a huge cost in computer time. An LES of this flow by Akselvoll and Moin [42] required only much less than one tenth the time. While the cost is still high, it is now at a point at which it may be sensible to do an occasional simulation to check the validity of models used in RANS calculations. Other flows of this kind which have been simulated recently include the two dimensional obstacle [43] and flow over a cube [44, 17]. The former introduces a flow-determined separation not found in the backward facing step flow. The latter introduces three dimensional separation. In all of these flows excellent agreement with experiment was obtained and new knowledge of the physics of the flow was produced.

For the near future, a sensible role for LES is to check RANS turbulence models. LES can and will be used as a complement and partial substitute for experiments. LES will continue to be used to test the accuracy of RANS models, a role it has played with distinction throughout its history. By using LES to tune RANS models, it is possible to obtain most of the benefits of LES at a small fraction of the cost.

Flows that are good candidates for LES are those which are difficult for RANS models. These include flows with extra strains that are different in each part of the flow and flows with strong unsteady effects. Particular cases include turbine blade passages and the internal combustion engine cylinders. RANS modeling of these flows is exceedingly difficult.

A word of caution is necessary. LES and its subgrid scale models have been validated only for simple flows at fairly low Reynolds numbers. Most of the energy is resolved in these simulations so, even if the subgrid scale model is not accurate, its effect may not be too important. If one uses the success of these simulations to justify applying LES to much more complex flows, reasonable-looking results may be obtained but placing trust in them may be risky. (Of course, qualitative results that help to explain observations can be useful if used judiciously.) The results of a recent conference on LES of complex flows [38] shows that success with LES is not automatic.

It would be valuable to have models that eliminate the need to specify no-slip conditions at a wall. Boundary conditions of this kind exist for attached flows and were mentioned above. It is not known is whether such conditions can be constructed for separated flows. Success in this area for RANS models has been limited and there is no reason not to expect the task to be less difficult for LES.

In flows in complex geometries, it is impossible to construct optimum grids prior to the calculation, even for steady flows. To compensate for the absence of this information, methods which modify the grid as the solution procedure converges have been developed. It is important to point out that adaptive gird methods have not yet been applied to LES or DNS, that doing so will be difficult, and that this remains an issue for future research.

10. COHERENT STRUCTURE CAPTURING

Until now, LES has been done by starting with simple flows, using them to learn about SGS modeling, and then proceed to more complex flows. These are mainly flows in which a large fraction of the turbulence energy is resolved so the importance of the model is limited. This does not assure equal success for LES of complex flows; success could be assured if sufficient computer resources are available, but, for flows of technological interest, this is not the case. At present, LES and DNS are not everyday engineering tools.

A better choice for the near term is to simulate 'building block' flows, ones that are structurally similar those of actual interest. The results can be used to test RANS models that can be applied to more complex flows. RANS computations can then be used as the everyday tool. LES need be performed only when there are significant changes.

As noted earlier, there have been attempts at simulation of complex flows. In most of these, the subgrid scale model was uncontrolled and the results were of uncertain value.

Since answers to questions involving technologically significant flows are required, the following questions arise. Is there a method that will enable more complex flows to be simulated on available machines? Are there good candidates for simulation in the relatively near future?

The answers to these questions appear to be qualified yeses. Good candidates are flows in which there are a small number of important, energetic, and easily identified coherent structures; usually, these are vortices. Let us consider two such cases.

Flows over bluff bodies produce strong vortices that lead to strong fluctuating forces on the body in both the stream- and span-wise directions; their prediction is important in wind and ocean engineering among other fields. If the vortices are sufficiently larger than the remainder of the 'turbulence', it should be possible to construct a filter that retains the vortices while removing all of the smaller scale motions. However, one must be careful not to convert an aperiodic flow into a periodic one.

We have called a method that accomplishes this 'coherent structure capturing' or CHC [45]. It was earlier called very large eddy simulation (VLES). Simulations of this type have not been attempted other than some which gave unsteady results when a steady flow was expected.

Internal combustion engine cylinders provide another example. The flow is inherently unsteady. Some interesting issues are the following. What does RANS mean in such a flow? How should RANS results be compared with experiments? Since the flow is unsteady, an LES represents a single realization; can such a simulation answer important questions? A partial answer to the first question is that the RANS mean velocity should probably be defined as an average over many cycles and the turbulence as the deviation from the multi-cycle average. LES should simulate a single cycle. Thus, after the intake stroke, the flow contains a strong vortex whose location, size, and strength varies from cycle to cycle. A RANS calculation should produce an average vortex, one that is relatively large and of average circulation. An LES vortex should be smaller, of similar circulation, but its location should vary from realization to realization; it should be possible to construct a filter that separates the vortex from the other turbulence.

CHC models should be different from those used in both RANS and LES. Indeed one needs to be very careful and experience will be required before this kind of simulation can be trusted.

According to the Smagorinsky model, the length scale is the filter width, Δ. But, in CHC, this width may be quite large, possibly even larger than the RANS length scale, which is approximately the integral scale, L. If this is so, the LES viscosity could exceed the RANS viscosity, violating the concept that the RANS viscosity should be large enough to remove all unsteadiness from the flow. To prevent this situation from occurring, it may be necessary to introduce an equation for the length scale into the subgrid scale model. LES would then inherit many of the difficulties that RANS models have with length scales.

11. CONCLUSIONS AND RECOMMENDATIONS

After years of being regarded as a method of second choice relative to DNS, LES is receiving increased attention. The principal reasons are dissatisfaction with RANS models and the inherent limitations of DNS.

Improved models for both the small-scale turbulence and the wall layer are needed. The dynamic model removes many of the difficulties that have plagued LES and give it an important advantage over RANS. Improved models for the wall region, especially for separating and reattaching flows, are needed just as badly and are an important subject for future research.

However, if LES is to prove useful in truly complex high Reynolds number flows, a great deal more work is needed. For the near future, it is probably best to use LES to understand the physics and to tune RANS turbulence models in a way that will enable them to produce more accurate predictions.

Some 'extra strains,' namely those that mainly affect the large scales, appear to be relatively easy to incorporate into large eddy simulations; little, if any, modification of the SGS models is required. Others, which act on scales smaller than the Kolmogoroff scale, for example, compressibility and combustion, are likely to require significant changes in the SGS model.

Acknowledgments

The author has been active in this field for over twenty years and the list of people who have helped him is too long to give here. I therefore mention only a few people who have influenced me in the past few years. These include Profs. P. Bradshaw, J.R. Koseff, P. Moin, S.G. Monismith and W.C. Reynolds and my students D. Briggs and K. Shah. The support received from a number of agencies over the years has also been very important; these include NASA, the Office of Naval Research and the Air Force Office of Scientific Research.

References

[1] Ferziger J.H., In *Computational Methods for Turbulent, Transonic, and Viscous Flows* J.-A. ffi Essers, ed. (Hemisphere, 1983).

[2] Rogallo R.S., Moin P., *Ann. Rev. Fluid Mech.* **16** (1984) 99.

[3] Lesieur M., *La Turbulence*, Grenoble Univ. Press, Grenoble (1994).

[4] Leslie D.C., Quarini G.L., *J. Fluid Mech.* **91** (1979) 65.

[5] Leonard A., *Adv. in Geophys.* **18A** (1974) 1974.

[6] Smagorinsky J., *Mon. Wea. Rev.* **91** (1963) 99.

[7] Lilly D.K., In *Proc. of the IBM Scientific Computing Symposium on Environmental Sciences*, H.H. Goldstine, ed., IBM Form No. 320-1951 (1967), 195.

[8] Clark R.A., Ferziger J.H., Reynolds W.C., *J. Fluid Mech.* **91** (1979) 92.

[9] Meneveau C., *Phys. Fluids* **6** (1994) 815.

[10] Bardina J., Ferziger J.H., Reynolds W.C., *AIAA paper 80-1357* (1980).

[11] Piomelli U., Ferziger J.H., Moin P., Kim J., *Phys. Fluids A* **1** (1989) 1061.

[12] Germano M., Piomelli U., Moin P., Cabot W.H., In *Proc. Summer Workshop, Center for Turbulence Research* (Stanford CA, 1990).

[13] Lilly D.K., *Phys. Fluids A* **4(3)** (1992) 633.

[14] Piomelli U., *Bull. Amer. Phys. Soc.* **35** No. 10 (1991).

[15] Moin P., Carati D., Lund T., Ghosal S., Akselvoll K., *74th AGARD Fluid Dynamics Panel*, Chania, Greece (April 1994).

[16] Zang Y., Street R.L., Koseff J.R., *Phys. Fluids* **5** (1993) 3186.

[17] Shah K., Ferziger J.H., Simulation of Flow over a Wall-mounted Cube,

in preparation (1996).

[18] Meneveau C., *Proc. Summer Workshop*, Center for Turbulence Research, (Stanford Univ., 1994).

[19] Lesieur M., *Turbulence in fluids, third edition* Kluwer Academic Publishers (1997).

[20] Chollet J.P., Lesieur M., *J. Atmos. Sci.* **38** (1981) 2747.

[21] Métais O., Lesieur M., *J. Fluid Mech.* **239** (1992) 157.

[22] Lesieur M., Métais O., *Ann. Rev. Fluid Mech.* **28** (1996) 45.

[23] Morris P., Ferziger J.H., *Bull. Am. Phys. Soc.* **39** (1994) 1896.

[24] Lesieur M., *in this volume* (1997).

[25] Kawamura T., Kuwahara K., *AIAA paper 85-0376* (1985).

[26] Ghosal S., Lund T.S., Moin P., Akselvoll K., *J. Fluid Mech.* **286** (1994) 229.

[27] Deardorff J.W., *Boundary Layer Meteorology* **1** (1974) 191.

[28] Horiuti K., *Phys. Fluids* **A2** (1990) 1708.

[29] Mason P.J., Derbyshire S.H., *Bound. Layer. Meteor.* **53** (1990) 117.

[30] Coleman G.N., Ferziger J.H., Spalart P.R., *Rept. TF-48, Thermosciences Div., Dept. Mech. Engr.* (Stanford Univ., 1990).

[31] Bohnert M.J., Ferziger J.H., In *Engineering Turbulence Modeling and Experiments 2* W. Rodi and F. Martelli eds. (Elsevier, 1993).

[32]

[33] Chapman D.R., Kuhn G.D., *J. Fluid Mech.* **70** (1986) 265.

[34] Schumann, U., Ein Untersuchung ueber der Berechnung der Turbulent Stroemungen im Platten- und Ringspalt-Kanalen, Dissertation (University Karlsruhe, 1973).

[35] Mason P.J., Callen N.S., *J. Fluid Mech.* **162** (1986) 439.

[36] Werner H., Wengle H., *Proc. 7th Conf. Turb. Shear Flows* (Stanford U., 1989).

[37] Bagwell T.G., Adrian R.J., Moser R.D., Kim J., In *Near Wall Turbulent Flows, R.M.C. So* C.G. Speziale and B.E. Launder eds. (Elsevier, 1993).

[38] Rodi W., Ferziger J.H., Breuer M., Porquié M., Large Eddy Simulation of Complex Flows: Report of a Workshop, it J. Fluid Engr. in press.

[39] Ferziger J.H., Perić M., In *Computational Fluid Dynamics*, M. Lesieur, P. Comte, & J. Zinn-Justin eds (Elsevier, 1996).

[40] Cain A.B., Reynolds W.C., Ferziger J.H., *Report TF-14, Dept. Mech. Engr.* (Stanford U., 1981).

[41] Le H., *Direct Numerical Solution of Turbulent Flow over a Backward Facing Step* Dissertation, Dept. Mech. Engr. (Stanford Univ., 1993).

[42] Akselvoll K., Moin P., *ASME Fluids Engr. Conf.* (Washington, DC, June 1993)

[43] Yang K.S., Ferziger J.H., *Fifth Asian Cong. Fluid Mech.*, DaeJon, Korea (1992).

[44] Mauch H., Berechnung der 3-D Umstroemung eines quadrefoermigen Koerpers in Kanal, Dissertation, (Univ. Karlsruhe, 1991).

[45] Ferziger J.H., In *Computational Wind Engineering 1* S. Murakami ed. (Elsevier, 1993).

Large-Eddy Simulation and Statistical Turbulence Models: Complementary Approaches

W. Rodi

Institute for Hydromechanics, University of Karlsruhe
Kaiserstrasse 12, 76128 Karlsruhe, Germany

Abstract

The relative performance and potential of statistical turbulence models and large-eddy simulations for calculating turbulent flows of practical interest is examined by considering the quality of the results and the computer resources required. For a variety of flows, including square-duct flow, 2D flows with separation, flows past bluff bodies and in a stirred tank reactor, the results of calculations obtained with the two methods are compared with each other and with available experimental data. The computing times are also compared. The suitability of the two methods for engineering applications is discussed for various types of flow phenomena

1. INTRODUCTION

Turbulence plays an important role in virtually all engineering problems involving fluid flow and in the environment. Turbulent fluctuations contribute significantly to the transport of momentum, heat and mass and therefore have a determining influence on the flow development, the forces exerted by the flow, the mixing, dilution and heat and mass transfer. Hence, in calculation procedures the effects of turbulence must be accounted for realistically and since there is a growing demand for powerful procedures for predicting flow problems, there is also an increased need for methods to simulate turbulence. There are three basic methods for simulating turbulence in a multidimensional numerical calculation: direct numerical simulations (DNS), large-eddy simulations (LES) and statistical turbulence models for solving the Reynolds-averaged Navier-Stokes equations (RANS). DNS is not suitable for solving practical flow problems in the foreseeable future because the number of grid points required to resolve numerically the motion of the small-scale dissipative motion increases as Re^3, and usually in practical flow problems the Reynolds number is fairly high. Hence, such calculations can be carried out only for fairly low Reynolds numbers and even then the computational effort is very large.

Until recently, virtually all practical flow calculations were carried out by solving the Reynolds-averaged Navier-Stokes equations (RANS) together with a statistical turbulence model. However, progress in the development of large-eddy simulation codes for complex geometries, more universal subgrid-scale models and above all the greatly increased computing power have brought the LES method within reach for solving practical flow problems. The motivation to consider and develop such methods for practical calculations is the fact that statistical turbulence models experience problems when large-scale structures dominate the turbulent transport, unsteady processes prevail like vortex shedding and bistable behaviour and when measures to manipulate the turbulence are to be investigated. For a number of practically relevant flows, calculations have been performed with both LES and RANS methods, and the paper compares the performance of the two methods for these flows in terms of predictive ability and computer effort required and draws conclusions on which problems are more suitable to be solved with LES and for which statistical models are to be preferred. To set the theme, a brief introduction of both methods is given first.

2. LARGE-EDDY SIMULATION METHODS

In large-eddy simulations, the three-dimensional time-dependent Navier-Stokes equations are solved numerically. Since at higher Reynolds numbers the small-scale turbulent motion cannot be resolved in such calculations, it is filtered out and only motions larger than the filter width, which is generally effectively the mesh size, are resolved. The effect of the unresolved small-scale fluctuations on the resolved larger-scale motions needs still be accounted for by a model. This effect is mainly dissipative, i.e. energy is withdrawn from the part of the spectrum that can be resolved. The effect can be achieved in two ways: the usual one is through a subgrid-scale model for determining the turbulent stresses introduced by the subgrid-scale fluctuations. The other possibility is to leave the energy withdrawal to damping by the numerical scheme that introduces a certain amount of numerical dissipation. The latter approach is advocated by Kuwahara and his group in Japan (see e.g. [1]): basically the Navier-Stokes and continuity equations also used in DNS calculations are solved with a finite-volume method on a grid as fine as can be afforded, and the necessary damping is introduced by a third-order upwind-differencing scheme for the convection terms.

Subgrid-scale models. In most LES calculations, the effect of the unresolved small-scale motion is simulated with a subgrid-scale model. The filtering-out of the small-scale motion introduces correlations between the unresolved fluctuating velocities in the equations to be solved which act as stresses on the resolved motions and need to be modelled. The two presently most popular models with which also most of the calculation examples presented below have been calculated will briefly be introduced. One is the Smagorinsky [2] eddy-viscosity model which introduces a subgrid-scale eddy viscosity and relates this to the strain rate of the resolved motion as velocity scale and to the filter width or effectively the mesh size as length scale. Near walls, this length scale has to be modified by a van Driest damping function. The model introduces one empirical constant C_S which was, however, found not to be universal but to depend on the flow considered.

Because the optimal value of C_S varies from flow to flow and even from point to point within one flow a special near-wall treatment is needed and the model is also not suitable for laminar-turbulent transition, recently the dynamic model proposed by Germano et al. [3] became popular. In this, the information available from the smallest resolved scales is used to determine the local (and time-dependent) value of parameters in the basic model to be applied. When the basic model is the Smagorinsky model, then the variation of the "constant" C_S with time and space results. The dynamic procedure applied with the Smagorinsky model produces often negative eddy viscosities which can make the simulation numerically unstable. Hence, some measures need to be taken to avoid this like averaging of the quantities over one or more homogeneous coordinate directions or clipping.

Near-wall treatment. Ideally, the no-slip condition should be used at walls in LES calculations, but this would strictly demand very high resolution near the wall at high Reynolds numbers as otherwise the scales of motion contributing most to the turbulent momentum transfer in the viscous sublayer cannot be resolved. Sufficient resolution is often not possible, but the no-slip conditions are nevertheless frequently used. More appropriate at high Reynolds numbers would be the employment of a near-wall model which corresponds to applying an artificial boundary condition at some distance from the wall. Various such models have been proposed; they all assume a phase coincidence between the velocity parallel to the wall at the first grid point and the wall shear stress at the adjacent wall. Schumann [4] then relates the average shear stress to the average velocity by the usual logarithmic law while Werner and Wengle [5] assume that the instantaneous tangential velocity inside the first cell has either a linear ($y^+ < 11.81$) or 1/7 power law ($y^+ > 11.81$) distribution. It should

be mentioned that both wall models were basically developed for attached flow
and their application in separated flow regions is somewhat questionable.

3. STATISTICAL TURBULENCE MODELS

The calculation methods used mostly in practice today are statistical methods. In these, all stochastic turbulent fluctuations (i.e. the whole spectrum)
are averaged out and the resulting Reynolds-averaged Navier-Stokes (RANS)
equations are solved. The averaging has to be carried out over a time which is
long compared with the time scale of the turbulent fluctuations, but in statistically unsteady flows the averaging time should be small compared with the
time scale of the mean flow variation; in situations with periodic mean flow
variation, phase-averaging is carried out. Instantaneous quanitites f are split
into an average value $\bar{f}$ and a fluctuating value f'. Introducing the separation $f = \bar{f} + f'$ for the various quantities into the original time-dependent
Navier-Stokes equations leads to the RANS equations. The averaging intro-
duces correlations between the fluctuating velocities $\overline{u'_i u'_j}$ acting like stresses
which are called turbulent or Reynolds stresses. A statistical turblence model
is necessary to determine these stresses. In contrast to subgrid-scale models in
LES calculations, statistical turbulence models simulate the entire spectrum of
turbulence.

Three main approaches have emerged in turbulence modelling: two of them
employ the eddy-viscosity concept in which the turbulent stresses are assumed
proportional to the gradients of the mean velocity with the eddy viscosity ν_t as
proportionality factor. ν_t is not a fluid property but depends on the structure
of the turbulence and its distribution over the flow field must be determined
by the turbulence model. An important concept in turbulence modelling is to
characterise the local state of turbulence by a few parameters, the minimum
being one velocity scale V and one length scale L. For dimensional reasons,
$\nu_t \propto V$. L and hence the eddy viscosity can be determined when the velocity
and length scale of the turbulence can be estimated. In the calculation examples
to follow, mainly various versions of the $k - \epsilon$ model have been used, and in
a few cases also a Reynolds-stress model or an algebraic stress model derived
from it. Hence, these models are now introduced briefly.

$\underline{k - \epsilon \text{ model}}$. This is still the most often used model in general-purpose CFD
codes. It relates the Reynolds stresses to the mean velocity gradients via the
eddy viscosity ν_t and expresses ν_t in terms of the turbulent kinetic k and the
dissipation rate e as the two parameters defining the velocity and length scale
of the turbulent motion. The distribution of k and ϵ in space and time is
determined from model transport equations. Standard model constants determined more than 25 years ago are usually employed [6]. For application
to vortex-shedding flow, the extension of the widely used steady model [6] is
straightforward in that all quantities appearing are now phase-averaged quantities, and a time-dependent term is added in each of the transport equations.

The standard $k - \epsilon$ model is based on an isotropic eddy-viscosity concept which is known to lead to unrealistically high production of k in stagnation regions which are present in the bluff-body application cases reported below. This is a consequence of the inability of this kind of model to simulate correctly the difference between normal stresses governing the production P_k in such regions. Kato and Launder [7] suggested as an ad hoc measure to replace the original production term $P_k = c_\mu \epsilon S^2$ by $P_k = c_\mu \epsilon S \Omega$, where

$$S = \frac{k}{\epsilon} \sqrt{\frac{1}{2} (U_{i,j} + U_{j,i})^2} \quad and \quad \Omega = \frac{k}{\epsilon} \sqrt{\frac{1}{2} (U_{i,j} - U_{j,i})^2}$$

denote respectively the strain and vorticity invariants. In simple shear flows, the behaviour remains unchanged as $\Omega \approx S$ while in stagnation regions $\Omega \approx 0$ so that the spurious turbulence production is eliminated. The application of this model version to the flow past a square cylinder and around a surface-mounted cube will be presented below.

Reynold-stress models (RSM). These models also known as second-order closure schemes do not employ the eddy-viscosity concept but solve model transport equations for the individual Reynolds stresses $\overline{u_i' u_j'}$. They are better suited for complex strain fields as well as for simulating transport and history and anisotropy effects of turbulence, and they automatically account for extra effects on turbulence such as due to streamline curvature, rotation, buoyancy and flow dilatation. Due to increased computing power and improved numerical techniques, RSM models have recently undergone extensive testing and are now applied to relatively complex flows and have been built into a number of commercial CFD codes. The RSM used for this purpose and in most of the applications reported below has been developed more than 20 years ago. It is the model of Launder, Reece and Rodi [8]. More advanced and also more complex Reynolds-stress models are under extensive development and testing these days [9] but have not been used in calculations that can be compared directly with LES calculations.

In an attempt to reduce the number of differential equations to be solved, algebraic stress models were developed by reducing the differential stress equations in Reynolds-stress models to algebraic equations through the introduction of simplifying assumptions about the convective and diffusive transport terms [10]. These algebraic stress equations are then used together with k- and ϵ-equations to form an extended two-equation model. However, in general situations the numerical solution of the coupled algebraic equation is not trivial and the trend has gone to use rather the original Reynolds-stress models. However, in recent years non-linear eddy-viscosity models have become popular as an alternative which are in some sense similar to algebraic stress models.

Near-wall treatment. In most practical calculations today, wall functions are still used in which the viscous sublayer is not resolved but the first grid point is located outside this layer. Basically, the quantities at this point are related to the friction velocity based on the assumption of the logarithmic velocity distribution and of local equilibrium of turbulence (production = dissipation). These assumptions are, however, questionable in separated flow regions. Recently, a variety of low-Reynolds-number versions of the $k - \epsilon$ model and the RSM models have been developed which can also be used in the viscous sublayer. Since these require, however, a rather fine numerical resolution an alternative are the so-called two-layer models in which the $k - \epsilon$ or RSM models are only used for the bulk of the flow away from walls and the viscosity-affected near-wall layer is resolved with a simpler model, e.g. a one-equation model employing only an equation for k together with a prescribed length-scale distribution very near the wall, e.g. a linear distribution. In the application examples to follow, mainly the Norris-Reynolds [11] one-equation model has been used in the context of two-layer model calculations.

4.   CALCULATION EXAMPLES

Several calculation examples are presented in this section, with an emphasis on more complex flow situations. It has been shown previously that the statistical quantities in the more simple plane-channel and boundary-layer flows can be simulated quite well with the LES method and with more advanced low-Reynolds-number $k - \epsilon$ models. Further, examples have been selected for which in general both LES and statistical model calculations are available so that the performance of the two methods can be compared.

4.1.   Developed Flow in Square Ducts

A characteristic feature of turbulent flow in square ducts is the existence of secondary motions which are driven by the turbulent fluctuations. In the average these motions are towards the corners (see Fig. 1) and are caused by differences in the turbulent normal stresses. These motions cannot be predicted with a statistical turbulence model employing an isotropic eddy viscosity like the standard $k - \epsilon$ model. Stress-equation or at least algebraic-stress models are required to reproduce this flow phenomenon. Breuer and Rodi [12] have calculated the flow in a square duct with the LES method at various Reynolds numbers. They first simulated the case at $Re = 4410$ for which DNS results of Gavrilakis [13] are available.

Fig. 1 shows on the right an instantaneous picture of the secondary velocities prevailing as turbulent vortices and on the left the average picture which is symmetrical with respect to the wall and corner bisectors. Both the Smagorinsky model and the dynamic subgrid-scale models have been used and the results for the mean velocity and the three normal stresses agree fairly well with the DNS calculations. A $41 \times 41 \times 62$ grid has been used. Breuer and Rodi then applied the LES method to calculate the flow at a much higher Reynolds number of 56690 (the Reynolds number in an experimental study of flow in a curved

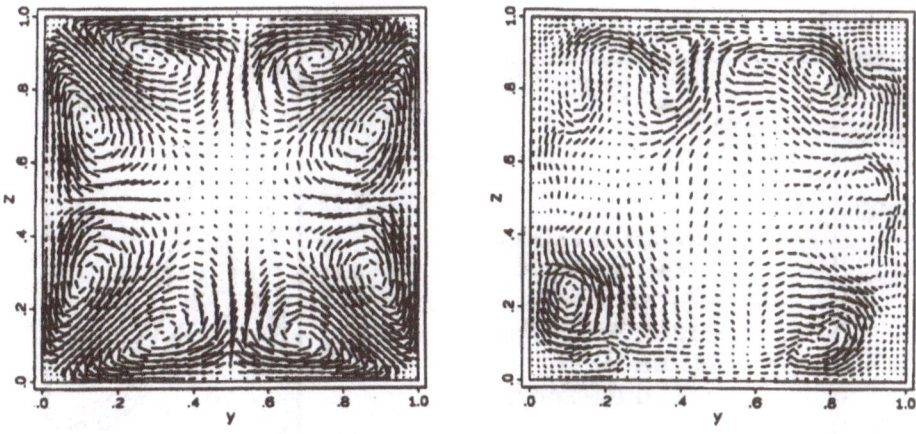

Fig. 1. — LES calculation [12] of secondary flow in square duct at $Re = 4410$ - left: time-averaged flow; right: velocities at one instant

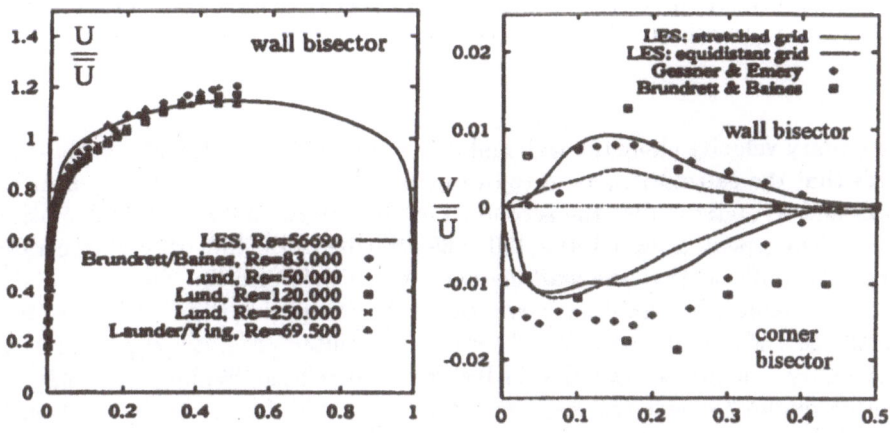

Fig. 2. — LES calculation [12] of developed square duct flow at $Re = 56690$; left: mean velocity distribution along wall bisector; right: distribution of secondary flow velocity

square duct which was also attempted with the LES method). Results were first obtained on a 101^3 grid using various wall functions.

Fig. 2 a compares the calculated velocity distribution along the wall bisector with various high-Reynolds number experimental results and Fig. 2b the

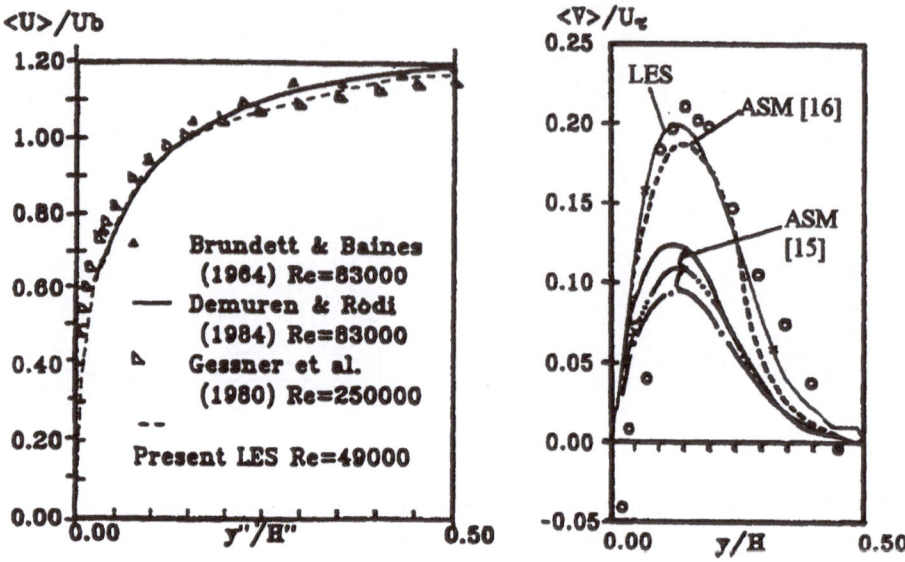

Fig. 3. — LES calculations by Su and Friedrich [14] for developed square duct flow
compared with ASM calculations; left: mean velocity along corner bisector; right:
mean secondary velocity along wall bisector

secondary velocity along the wall and corner bisectors. It is clear from these figures that the calculations tend to overpredict the streamwise velocity near the
wall and to underpredict the secondary motion considerably. The calculations
were then repeated on a $150 \times 150 \times 65$ stretched grid that was concentrated
near the walls so that the wall distance of the first grid line was $5 \times 10^{-4} D$;
in this case no-slip conditions were used. The computing time was rather high
with 120 CPU hours on a VP600 vector computer. Fig. 2b shows that the
secondary velocity along the wall bisector is increased significantly using the
better resolution near the wall, and the correct level of secondary motions is
now obtained. This shows that a good resolution of the corner region, where
the secondary motions are produced, is very important in this case. Somewhat
surprisingly, the velocity distribution along the wall bisector has not changed
significantly over the calculation with the coarser grid.

Su and Friedrich [14] have also calculated the square-duct flow for a Reynolds
number of $Re = 49000$ with their LES method employing Schumann's subgrid-
scale model and Schumann's wall function. They performed their calculations
on a $64 \times 64 \times 256$ grid and obtained quite good agreement with experiments
for the mean-velocity distribution and also the secondary velocity along the
wall bisector as shown in Fig. 3. It is at present not clear why their calculations appear to agree better with the experiments than the LES simulations of
Breuer and Rodi [12]. Much earlier, Demuren and Rodi [15] performed RANS

calculations with an algebraic-stress model and wall functions on a 20×20 grid. The results are also included in Fig. 3 and show reasonable agreement of the velocity distribution for the corner bisector but also a severe underprediction of the secondary velocity along the wall bisector. In even earlier calculations, Naot and Rodi [16] used a more empirical version of the algebraic-stress model in which aspects of the eddy-viscosity model were involved and obtained actually a better simulation of the secondary velocity level. This shows that by proper tuning of the model the correct results can be obtained. Also the calculations show that in principle the LES method without any tuning is capable of producing the complex phenomenon but that the results are sensitive to the details of the simulation.

4.2. Backward-Facing-Step and Other Separated Flows

Akselvoll and Moin [17] calculated with an LES method the backward-facing step flow that was simulated by direct numerical simulation by Le and Moin [18] and was measured by Jovic and Driver [19]. The Reynolds number (based on the step height and approach-flow velocity) was fairly low (5000). The dynamic subgrid-scale model was used in these calculations with various grids (finest: $230 \times 48 \times 32$). The LES calculations yielded fairly good agreement with the DNS results which themselves were in ageement with the measurements. The comparison with the DNS has also shown that comparable accuracy of quantities of practical interest can be achieved with LES on considerably coarser grids with much less computing effort (0.5%).

High Reynolds number ($Re = 1.65 \times 10^5$) backstep flow was calculated with LES by Arnal and Friedrich [20]. They simulated the experiments of Tropea [21] which has an expansion ratio of 2. They used Schumann's [4] subgrid-scale model and wall functions. As the approach flow was developed channel flow in the experiments, an LES was carried out first of developed channel flow which was then used for the inflow conditions. The calculations were carried out on various grids ranging from $80 \times 16 \times 16$ to $320 \times 64 \times 84$. The calculations on the finest grid required 240 CPU hours on a Cray XMP. The calculated reattachment length x_r was 7.7H on the coarsest grid and 7.2H on the finest grid, while the experimental value is 8.5H. This result is not better than the recirculation length of 7.52H calculated by Peric' et al. [22] with the standard $k - \epsilon$ model and wall functions. When comparing the profiles of the turbulent shear stress, the agreement with the measurements obtained with the finest grid is quite good but in terms of the mean velocity field, the LES calculations are no better than the $k - \epsilon$ model calculations.

The backward-facing step flow studied experimentally by Driver and Seegmiller [23] with a small expansion ratio of 1.125 and $Re = 3.46 \times 10^4$ was calculated with a variety of statistical turbulence models for the Collaborative Testing of Turbulence Models organised by Bradshaw et al. [24]. The opposite wall was either parallel to the bottom wall where the step was placed or inclined at an angle of $6°$ forming a diverging channel. $k - \epsilon$ models with wall functions underpredicted the length of the recirculation zone for the non-divergent channel by about 27% and for the diverging channel by about 32%. The calculations were improved both by resolving the viscosity-affected near-wall region with either a low-Re $k - \epsilon$ model or by a two-layer approach and also by switching to the stress-equation model, with wall functions. The recirculation lengths were, however, still short by approximately 20%. On the other hand, the $k - \omega$ model overpredicts the recirculation length. Similar trends were observed in calculations of the flow over a 2D model hill [25], but here the $k - \epsilon$-models resolving the near-wall region in one way or another and Reynolds-stress models actually yield the correct recirculation length.

LES and RANS calculations can be compared for the flow around a NACA 4412 airfoil near maximum lift ($a \approx 13°$). For this Coles and Wadcock [26] performed experiments with a flying hot-wire for a Reynolds number based on the chord length $Re = cU_0/\nu = 1.5 \times 10^6$, tripping the boundary layer on both the upper and lower side near the leading edge. LES calculations were carried out by Kaltenbach and Choi [27] with a dynamic subgrid-scale model on a $638 \times 79 \times 48$ C-mesh with 2.5 million nodes and a time step of $2 \times 10^{-4} c/U_\infty$. In these calculations, one time unit c/U_∞ took 90 CPU hours on a Cray C90 and several time units were necessary for calculating the statistics. In the experiment, a small separation zone was detected on the suction side near the trailing edge which could not be obtained in the LES calculations. On the other hand, the RANS calculations of Cordes et al. [28] yielded the small separation region and in general good agreement with the experiments. They used both the standard $k - \epsilon$ model with wall functions and the two-layer model and found that for this case with small separation region both models yield virtually the same results.

4.3.   Vortex-Shedding Flow Past a Square Cylinder

The flow past a square cylinder at $Re \approx 22000$ studied in detail experimentally by Lyn et al. [29] was posed as a test case at an LES workshop held in 1995 [30]. 9 groups contributed results of LES calculations for this test case. The same flow was also calculated with statistical turbulence models in a series of investigations in the author's research group. In these, ensemble- or phase-averaged two-dimensional time-dependent Navier-Stokes equations were solved to yield the periodic vortex-shedding motion and only the superimposed stochastic turbulent fluctuations were simulated by a turbulence model. A detailed comparison of the calculations obtained with the two methods is provided in [31] and this is only summarised here. A wide variety of methods was used for the LES calculations (different numerical schemes, Smagorinsky and dynamic subgrid-scale models, use of no-slip conditions and wall functions etc.), and the various results differ significantly as will be shown below. The RANS calculations were carried out with the standard $k - \epsilon$ model with wall functions, the Kato-Launder [7] modification to suppress the excessive turbulence production in stagnation regions, the two-layer version resolving the viscous sublayer, and with the standard Reynolds-stress model [8] both with wall functions and in a two-layer approach. Both LES and RANS methods yield the vortex-shedding motion in generally good qualitative agreement with the experiments. Also, in most calculations the non-dimensional shedding frequency, the Strouhal number is reproduced close to the experimental value of 0.13. There is more variation on the drag coefficient; here the LES calculations tend to overpredict this coefficient to various degrees while the RANS calculations range from an underprediction with the standard $k - \epsilon$ model and wall functions to an overprediction when the stress-equation model is used in the two-layer approach.

Fig. 4 displays the distribution of the time-mean velocity along the centre-line. Experimental data due to Lyn et al. [29] and Durao et al. [32] are included.

These agree fairly well in the near-cylinder region, but the approach to the free-stream velocity is quite different, perhaps due to different blockage and approach-flow turbulence. As can be seen, there are fairly large differences in the calculation results in the wake region. The standard $k-\epsilon$ model overpredicts the length of the separation zone considerably, introducing the Kato-Launder (KL) modification and the two-layer approach improves the calculations; in fact the combination of the two approaches gives the best agreement with measurements. The approach to the free-stream velocity is, however, faster than measured by Lyn et al. [29]. These RANS calculations are better than the RSM results [33] which yield too short a separation region. The LES results exhibit surprisingly large differences both in terms of the length of the separation region and the recovery behaviour. The recovery is generally predicted faster than that of Lyn et al. [29], but there is one calculation which is in fairly good accord and one even produces a slower recovery. The UKAHY2 results show an unrealistic slope of U at larger x-values; this is most likely caused by the relatively coarse grid in the downstream region as it goes away when a finer grid is used there [34].

Fig. 5 presents the distribution of the total (periodic plus turbulent) fluctuating kinetic energy along the centre-line. Here the various LES results show an even wider variation with an almost fourfold difference in the peak level of k_{tot}, but the picture is not entirely consistent. It can generally be observed that the total fluctuations are predicted too small when the drag coefficient and separation length are reasonable. Turning to the RANS calculations, it can be seen from Fig. 5 that the standard $k - \epsilon$ model with wall functions predicts the peak of the fluctuations considerably too far downstream and yields a much too small fluctuation level behind the cylinder. Switching to the KL modification moves the peak closer to the cylinder and raises ist value to roughly the correct levels and, when the combination of KL and two-layer approach is used, the location of the peak is roughly correct but the fluctuation level is now overpredicted.

Fig. 6 shows the corresponding distribution of the turbulent kinetic energy component of the fluctuations. Here only the UKAHY2 LES results are avail-

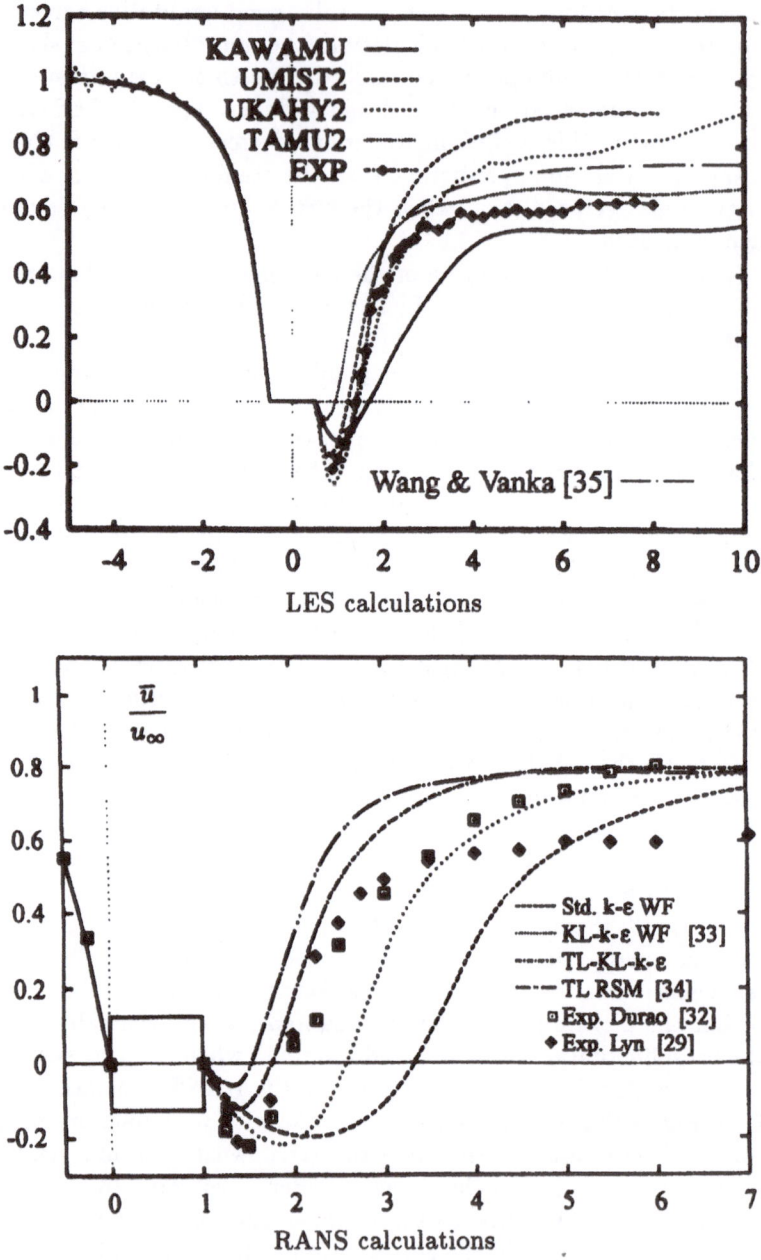

LES calculations

RANS calculations

Fig. 4. — Time-mean velocity along centre-line of square cylinder

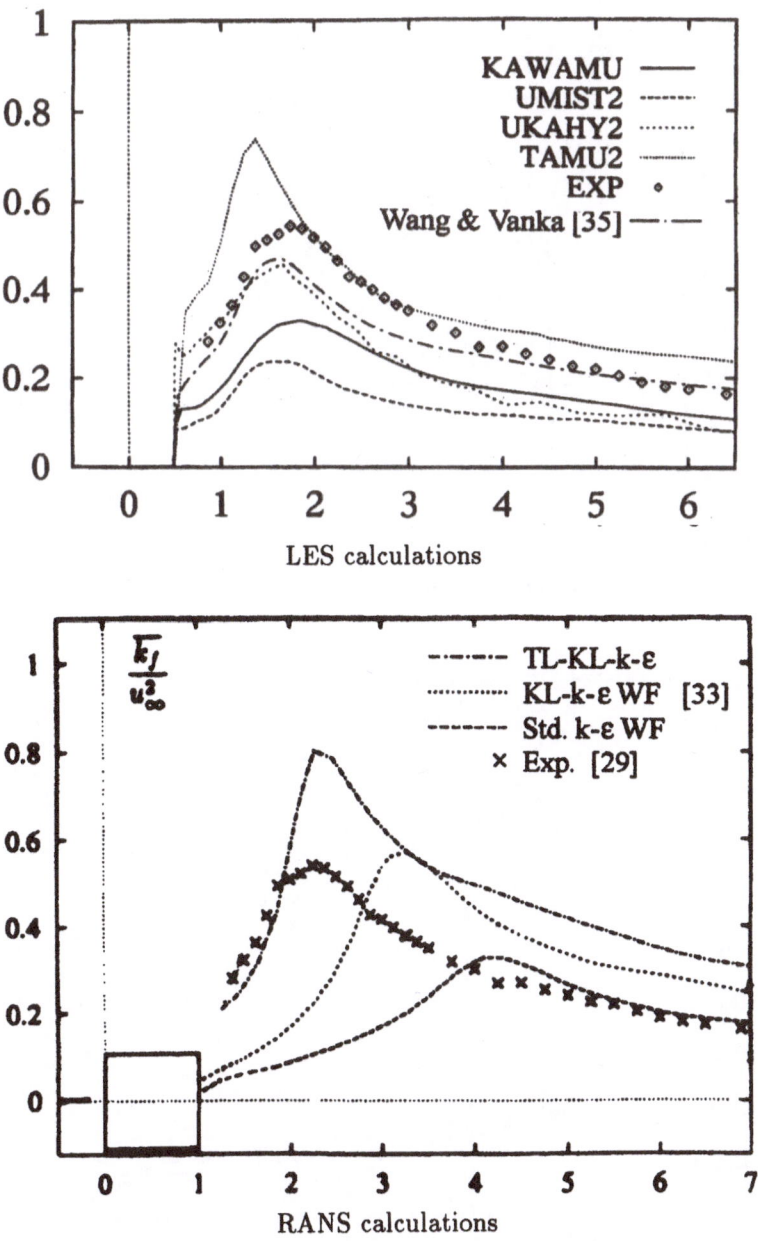

Fig. 5. — Total kinetic energy of fluctuations (periodic + turbulent) along centre-line of square cylinder

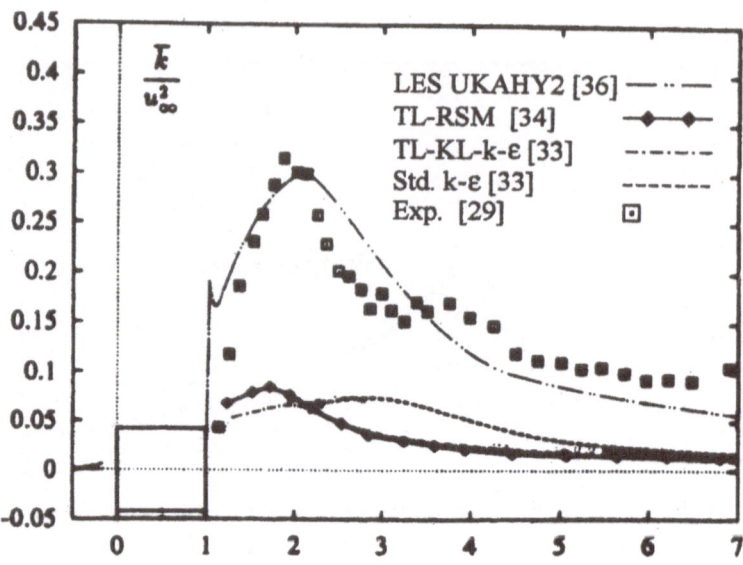

Fig. 6. — Turbulent component of kinetic energy along centre-line of square cylinder.

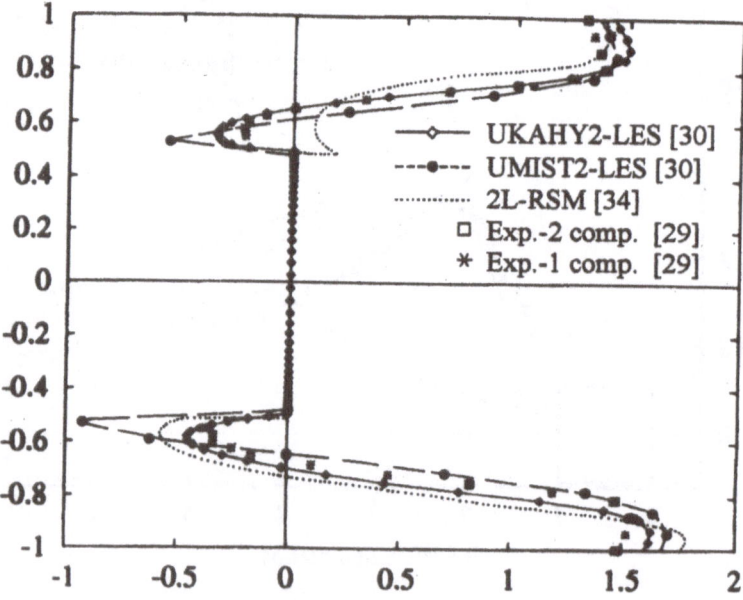

Fig. 7. — Profiles of phase-averaged velocity at x-location of rear cylinder wall for phase 1.

able. All RANS calculations are considerably too low while the LES results available are roughly in accord with the measurements. The very different behaviour of RANS and LES calculations is most likely due to the fact that the fairly high turbulent kinetic energy stems from contributions of low-frequency fluctuations modulating the shedding. In the experiments these originate from the 3D nature of the large-scale structures. The LES results can capture the 3D nature and count any low-frequency fluctuations originating from these as turbulence, while of course the 2D RANS calculations cannot.

The overall behaviour of the vortex-shedding flow is determined largely by the prediction of the evolution of the separated shear layer on the sides of the cylinder. Hence it is interesting to consider the U-velocity profile at the location of the rear face of the cylinder and this is done in Fig. 7 for the ensemble-averaged velocity at phase 1. In the experiments it was found that the separated shear layers do not reattach at any phase. From Fig. 7 it can be seen that on the upper side the RANS calculations predict unrealistically reattachment while on the lower side the shear layer remains detached in the calculations. Here, all LES calculations definitely are in closer agreement with the experiments; for two of them the predicted velocity distributions with the actual grid points are shown. This gives an impression of the resolution in this region: in the UKAHY2 calculations at least a few grid points are located in the region with reverse flow while in the UMIST2 calculations the first grid point is at the peak of the negative velocity. This points to a resolution problem in this area which is one of the main reasons for the disagreement with the experiments and also among the various LES calculations. On the other hand, the two-layer RSM calculations give poor results in spite of the much better resolution. The price to be paid for the generally improved simulation of the flow details by LES is a large increase in computing time: the UKAHY2 LES calculations ($165 \times 113 \times 17$ grid) took 73 hours on a SNI S600/20 vector computer while the RANS calculations using wall functions (100×76 grid) took 2 hours and the ones using the two-layer approach (170×170 grid) 8 hours on the same computer.

4.4. Flow Over a Surface-Mounted Cube

The flow over a cube placed in developed channel flow was also posed as test case at the 1995 LES workshop [30]. For a cube height H of half the channel height and $Re = U_B H/\nu = 40000$, flow visualisation studies and detailed LDA measurements were carried out by Martinuzzi and Tropea [35, 36]. From these, Martinuzzi [35] devised the flow picture given in Fig. 8, showing the complex nature of this flow in spite of the simple geometry. The flow separates in front of the cube with primary and secondary separation vortices; the main vortex is bent as horseshoe vortex around the cube into the wake where it has a typical converging-diverging behaviour. The flow separates at the front corners on the roof and side walls. In the mean it does not reattach on the roof. A large separation region develops behind the cube which interacts with the horseshoe

vortex. Originating from the ground plate an arch vortex develops behind the
cube. Predominant fluctuation frequencies were detected sideways behind the
cube, which were traced to some vortex shedding from the side walls. Finally,
bimodal behaviour of the flow separation and in particular of the vortices in
front and on the roof were observed.

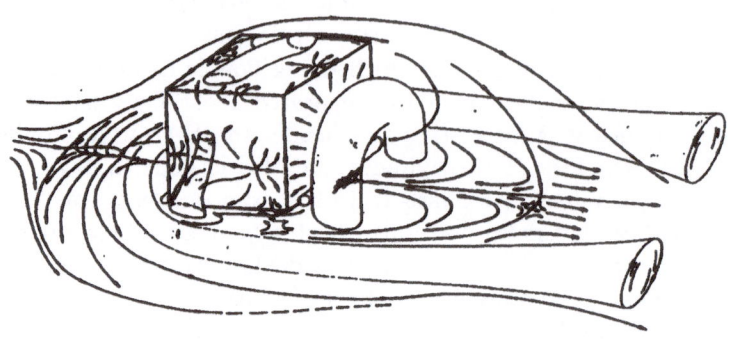

Fig. 8. — Flow around surface-mounted cube according to [35].

The three LES results submitted to the workshop which were based on sufficient averaging were in reasonably good agreement about most features of the
complex flow. Here only the results by Breuer and Rodi [12] obtained with the
dynamic subgrid-scale model are presented. Lakehal and Rodi [37] performed
RANS calculations of the same flow with the standard $k - \epsilon$ model using wall
functions, the Kato-Launder modification removing the excessive turbulence
production in the stagnation region and the two-layer version resolving the viscous sublayer. With each calculation method, the developed channel flow was
calculated first and the results were then used as inflow conditions. The LES
calculations were performed on a $165 \times 60 \times 97$ grid, the RANS calculations
with wall functions on a $110 \times 32 \times 32$ grid and the RANS two-layer calculations
on a $142 \times 84 \times 64$ grid. The height of the near-wall cells was 0.01H in the
RANS calculations with wall functions, 0.001H in the two-layer calculations
and 0.0125H in the LES calculations. In the latter, 160000 time steps were
necessary to obtain reasonably reliable statistics requiring 160 CPU hours on a
SNI S600/20 vector computer. On the same computer, the RANS calculations
using wall functions and the two-layer approach took 15 minutes and 6 hours,
respectively. Fig. 9 compares the streamlines in the plane of symmetry (left)
and near the channel floor (right). It is clear from this figure that on the whole
LES is able to simulate this complex flow very well. Only LES and the two-
layer version predict the location of separation in front of the cube correctly
and produce a small secondary separation in the corner. On the roof, the LES
calculations do not predict reattachment in the mean, as was also found in

the experiment, and the location and extent of the separation region are well reproduced. On the other hand, the $k - \epsilon$ model using wall functions leads to reattachment and to an underprediction of the separation region. Both the KL modification and the use of the two-layer model improve the flow prediction on the roof significantly, and in fact the best prediction there is obtained by combining the two-layer approach with the KL modification (not shown here).

All RANS models tested overpredict the extent of the separation region behind the cube. The standard $k - \epsilon$ model with wall functions already predicts this quantity 35% too long, and both the introduction of the KL modification and the two-layer approach increase the separation length further; a combination of both gives the most excessive length [37]. In the calculations with the KL modification, less turbulence is swept around the front corners and over the roof into the downstream region leading to lower eddy viscosity which explains the longer separation zone. Moving from wall functions to two-layer approach, the resulting larger separation zone on the roof also increases the separation behind the cube. LES clearly does a better job in the lee of the cube and predicts the separation length fairly well. In the experiments, some shedding from the side walls was observed which enhances the momentum exchange in the wake and can reduce significantly the length of the separation region behind obstacles. Even though there was no clear shedding detected in the LES results, the resolution of large-scale unsteady motions in these calculations seems to produce the correct effects. Steady RANS calculations can of course not account for such effects.

The complex behaviour of the surface streamlines near the channel floor as observed in the experimental oil-flow pictures is well reproduced in the LES calculations, including such details as the converging-diverging behaviour of the horseshoe vortex, the primary and secondary separation in front of the cube, the footprints of the arch vortices behind the cube and the reattachment line bordering the reverse-flow region. In the RANS approach, only the two-layer model can reproduce these details, but with a significantly too long reverse-flow region. Calculations obtained with wall functions yield a much simpler picture as the converging-diverging behaviour of the horseshoe vortex is absent and the whole separation region is basically filled by the arch vortices, which is in contrast to the experimental observation.

A comparison of the profiles of the streamwise velocity $\overline{U}$, shear stress $\overline{u'v'}$ and turbulent kinetic energy k at various downstream locations in the symmetry plane can be found in [31]. They show again that the LES produces much superior predictions in this case.

4.5. Flow in Baffled Stirred Tank Reactor

Baffled stirred tank reactors are very common in the chemical process industry, and reliable calculation methods of the flow in such reactors is particularly needed since the scale-up from model investigations if often problematic. Due to the impeller motion and the baffles, the flow is rather complex, and it is

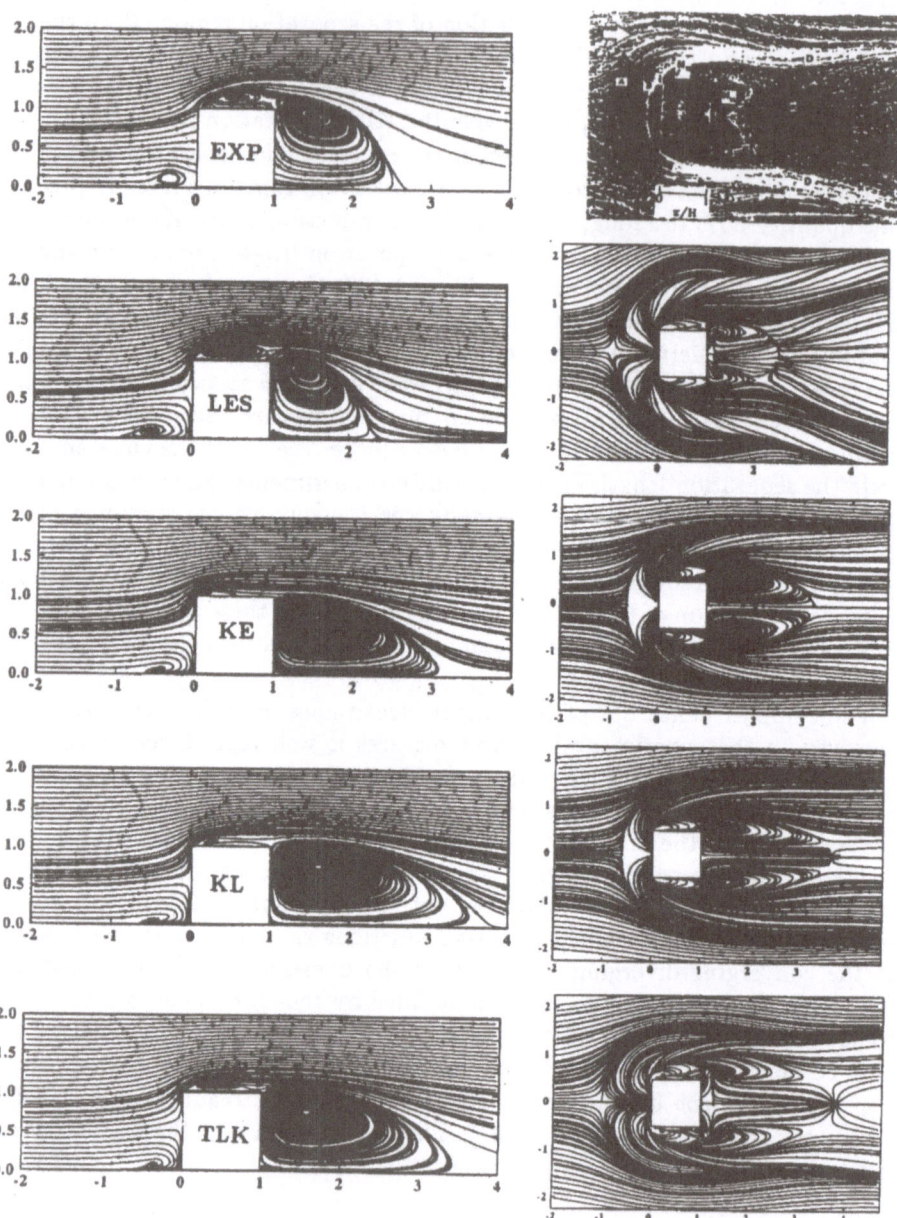

Fig. 9. — Streamlines in the symmetry plane (left) and near the channel floor (right) for flow around surface-mounted cube; LES [12], RANS [31], expts. [36].

important to predict the turbulence reliably in order to allow the simulation of the micromixing performance.

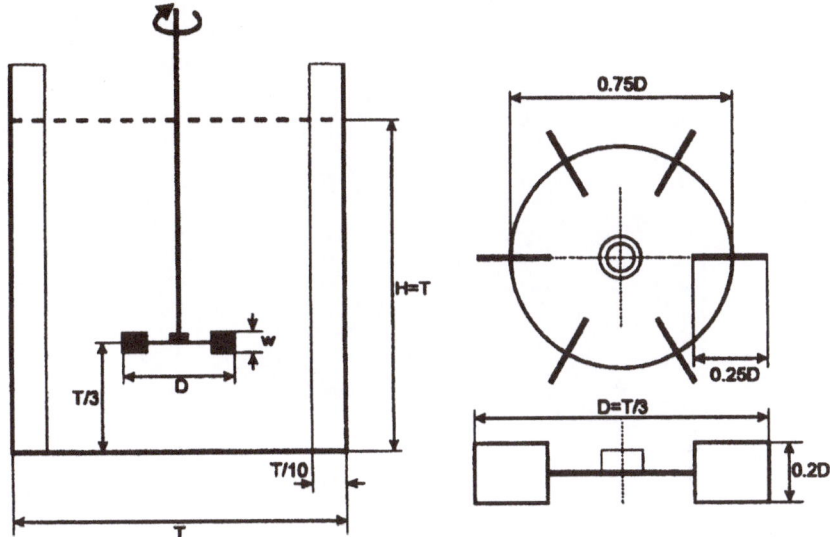

Fig. 10. — Baffled stirred tank reactor: geometry (reproduced from [38])

Eggels [38] of the Shell Company has carried out LES calculations of the flow in a baffled stirred tank reactor in which the flow and turbulence was induced by a six-blade disc impeller (see Fig. 10). The effect of this impeller was modelled by applying a force field to the momentum equations. Eggels used the Smagorinsky subgrid-scale model with C_S = 0.1 and no-slip conditions at the wall. The Reynolds number $Re = D^2/(T_o\nu)$ was 1×10^5 with T_o the impeller turn-around time and D the impeller diameter. The equations were solved with a Lattice-Boltzmann scheme. Simulations were carried out on a 120^3 grid with a sampling time of $30T_o$ and with a finer 240^3 grid (= 13.8×10^6 grid points) and a sampling time of $8T_o$. The coarser grid solution took 85 hours on an 8 node IBM SP2 and the fine grid solution nearly 9 hours for one turn-around time T_o on a 16 node IBM SP2 so that easily hundreds of CPU hours may be necessary for proper statistical averaging.

Fig. 11 shows both instantaneous and time-averaged velocity vector fields of the flow in the vertical plane through the centre of the tank reactor. As can be seen, the impeller induces a radial-jet motion which impinges on the tank wall, is diverted upwards and downwards there, turns back towards the centre of the tank and is sucked by entrainment to the jet, forming two large toroidal vortices, one above and one below the radial jet. The instantaneous flow field shows that there are indeed large-scale eddies both in the jet region and all over

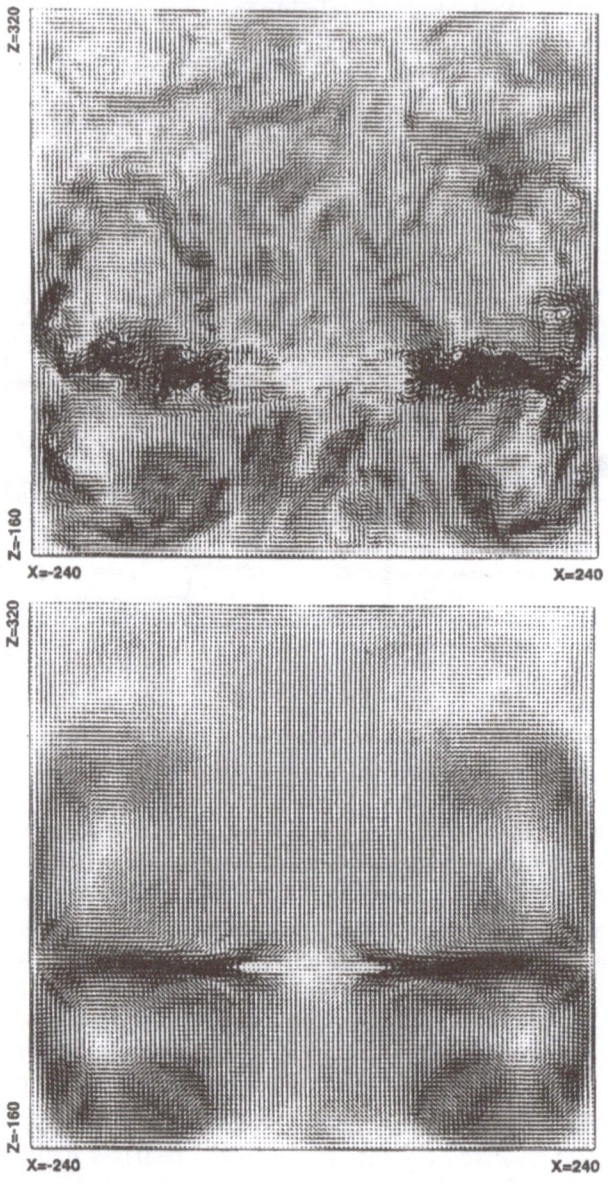

Fig. 11. — Instantaneous (top) and time-averaged (bottom) velocity vectors in vertical plane through symmetry axis of stirred tank reactor; LES calculations of [38]

the tank which dominate the mixing. In Fig. 12, a vertical profile from bottom to surface of the axial mean and RMS velocities is given and compared with measurements of Bakker [40]. The fine grid solution is shown which agrees quite well with the measurements, also for the turbulent fluctuations. The coarser grid solution does not agree so well with the measurements in the radial wall jet region because the impeller thickness is not well resolved by the numerical grid. In correspondence with the velocity vector field in Fig. 11, the mean velocity is upward (> 0) below the impeller level ($z = 0$) and downward (< 0) above the impeller level. The level of the turbulence fluctuations is fairly constant over most part of the tank but is considerably higher in the radial-jet region. All this is reproduced fairly well by the large-eddy simulations. According to Eggels [39], $k - \epsilon$ model calculations also performed at Shell using 40000 grid points (taking 80 to 100 CPU hours on a single-node SP2) did not give satisfactory predictions of the turbulence fluctuations. This study therefore shows the superiority of the LES method when large-scale structures are present in complex flows, and it shows also that LES can already be used as a tool to investigate turbulent flows in industrial applications.

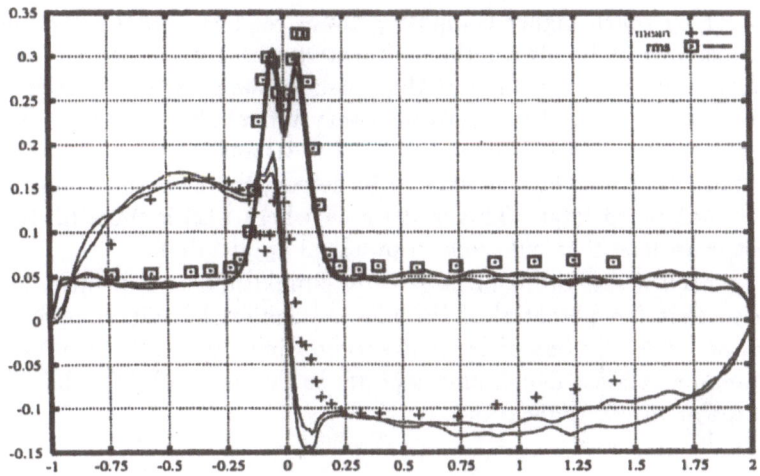

Fig. 12. — Stirred tank reactor: vertical profiles of axial mean and RMS velocities at r/R = 0.34; lines: LES, symbols: experiments (reproduced from [38])

5. CONCLUSIONS

The calculation examples presented have shown that LES is superior to the RANS method using statistical turbulence models when the flow is really complex and especially when large-scale structures dominate the turbulent transport and unsteady processes are involved. The details of the vortex-shedding flow past a square cylinder could be predicted more realistically, even though none of the LES calculations available is entirely satisfactory and further investigation s are necessary. The very complex flow over the geometrically simple cube with a variety of separation phenomena including some shedding from the side walls and fine-scale structure near the ground could be simulated significantly better with LES than with statistical models. It also seems clear that LES is required when pressure fluctuations causing dynamic loading of structures have to be determined and when the manipulation of turbulence by influencing directly the turbulent structures has to be investigated.

On the other hand, LES was not found to be clearly superior for more simple flows that are mainly of shear-layer type such as attached flow or flow with only small separation regions, and for predominantly two-dimensional flows. Here the RANS method with available statistical models, which sometimes have to be tuned and depend on the problem considered, often produce results which are sufficiently accurate for practical purposes; hence for these flows such methods will continue to be used in practice for many years to come.

The more realistic results obtained by LES for complex flows must generally be paid for by much higher computing times required - this is because LES calculations have to be always three-dimensional and unsteady; yet, it is often not a higher number of grid points that leads to the greatly increased computing times but the many time steps necessary for reliable statistical averaging. However, when the flow problem considered is 3D and unsteady anyway so that 3D unsteady RANS equations have to be solved, the differences in computing time may not be so large. This is the area where LES is most likely to take over first from RANS calculations in practical applications.

The numerical resolution in large-eddy simulations is still often not good enough so that the potential of the method cannot be fully exploited. This problem and also the one of excessive computing times will be alleviated by the continuing increase in computing power so that soon LES will be ready and feasible for practical applications.

References

[1] Tamura T., Ohta I., Kuwahara K.J., *Wind Eng. and Ind. Aerodyn.* **35** (1990) 275.

[2] Smagorinsky J.S., *Mon. Weather Rev.* **91** (1963) 99.

[3] Germano M., Piomelli U., Moin P., Cabot W.H., *Phys. Fluids A* **3** (1991) 1760.

[4] Schumann U.J., *Comp. Phys.* **18** (1975) 376.

[5] Werner A., Wengle H., Large-eddy simulation of turbulent flow over and around a cube in a plane channel, *Proc. 8th Symp. on Turbulent Shear Flows*, Munich, Germany (1991).

[6] Rodi W., Turbulence Models and their Application in Hydraulics, 3rd ed. (Balkema, Rotterdam, 1993)

[7] Kato M., Launder B.E., The modelling of turbulent flow around stationary and vibrating square cylinders, *Proc. 9th Symp. Turbulent Shear Flows*, Kyoto (1993).

[8] Launder B.E., Reece G.J., Rodi W., *J. Fluid Mech.* **86** (1975) 537.

[9] Launder B.E., Li S.-P., *Phys. Fluids A* **6** (1994) 999.

[10] Rodi W., *ZAMM* **56** (1976) T219.

[11] Norris H.L., Reynolds W.C., Turbulent channel flow with a moving wavy boundary, *Dept. Mech. Eng. Rept. FM/10*, Stanford University (1975).

[12] Breuer M., Rodi W., Large-eddy simulation of complex turbulent flows of practical interest, in *Flow Simulation with High-Performance Computers II*, ed. E.H. Hirschel, Notes on Numerical Fluid Mechanics 53, Vieweg Verlag (1996) pp. 258-274.

[13] Gavrilakis S., *J. Fluid Mech.* **244** (1992) 101.

[14] Su M.D., Friedrich R., *ZAMM* **73** (1993) T563.

[15] Demuren A.O., Rodi W., *J. Fluid Mech.* **140** (1984) 189.

[16] Naot D., Rodi W., *ASCE J. Hydraulics Div.* **108** (1982) 948.

[17] Akselvoll K., Moin P., Large-eddy simulation of turbulent confined coannular jets and turbulent flow over a backward-facing step, *Rept. No. TF-63*, Dept. of Mech. Eng., Stanford University (1995).

[18] Le H., Moin P., Direct numerical simuation of flow over a backward-facing step, *Rept. No. TF-58*, Dept. of Mech. Eng., Stanford University (1994).

[19] Jovic S., Driver D.M., Backward-facing step measurement at low Reynolds number, $Re = 5000$, *NASA Technical Memorandum, No. 108807* (1994).

[20] Arnal M., Friedrich R., Investigation of the pressure and velocity fields in a turbulent separated flow using the LES technique, *Paper AIAA 91-0251* (1991).

[21] Tropea C., Die turbulente Strömung in Flachkanälen und offenen Gerinnen, *Dissertation*, University of Karlsruhe, 1982.

[22] Peric' M., Rüger M., Scheuerer G., A finite-volume multigrid method for calculating turbulent flows, *Proc. 7th Symp. Turbulent Shear Flows*, Stanford University (1989).

[23] Driver D.M., Segmiller H.J., *AIAA J.* **23** (1985) 163.

[24] Bradshaw P., Launder B.E., Lumley J.L., *J. Fluids Eng.* **118** (1996) 243.

[25] Bonnin J.-Ch., Buchal T., Rodi W., Data bases and testing of calculation methods for turbulent flows, *ERCOFTAC Bulletin No. 28* (1996).

[26] Coles D., Wadcock A.J., *AIAA J.* **17** (1979) 321.

[27] Kaltenbach H.-K., Choi H., Large-eddy simulation of flow around an airfoil on a structured mesh, *Annual Research Briefs 1995*, Center for Turbulence Research, Stanford (1995) pp. 51-60.

[28] Cordes J., Rodi W., Cho. N.-H., Calculation of separated flows with a two-layer turbulence model, in *Notes on Numerical Fluid Mechanics 40*, Vieweg (1993) pp. 27-36.

[29] Lyn D.A., Einav S., Rodi W., Park J.-H., *J. Fluid Mech.* **304** (1995) 285.

[30] Rodi W., Ferziger J.H., Breuer M., Pourquié M., Status of large-eddy simulation: Results of a workshop, to appear in *J. Fluids Eng.*

[31] Rodi W., Comparison of LES and RANS calculations of the flow around bluff bodies, to appear in *J. Wind Eng. and Ind. Aerodyn.*

[32] Durao D.F.G., Heitor M.V., Pereira J.C.F., *Experiments in Fluids 6* (1988) 298.

[33] Franke R., Rodi W., Calculation of vortex shedding past a square cylinder with various turbulence models, in *Turbulent Shear Flows 8*, eds. U. Schumann et al., Springer Verlag (1993).

[34] Pourquié M., private communication (1996).

[35] Martinuzzi R., Experimentelle Untersuchung der Umströmung wandgebundener rechteckiger prismatischer Hindernisse, *Dissertation*, University Erlangen-Nürnberg (1992).

[36] Martinuzzi R., Tropea C., *J. Fluid Eng.* **115** (1993) 85.

[37] Lakehal D., Rodi W., Calculation of the flow past a surface-mounted cube with two-layer turbulence models, to appear in *J. Wind Eng. and Ind. Aerodyn.*

[38] Eggels J.G.M., *Int. J. Heat and Fluid Flow* **17** (1996) 307.

[39] Eggels J.G.M., private communication (1996).

[40] Bakker R.A., Laser-doppler measurements in a baffled stirred tank reactor with a disc turbine, *Ph. D. Thesis*, Delft University of Technology (1996).

Modeling Compressibility Effects on Turbulence

T. B. Gatski

NASA Langley Research Center
Hampton, Virginia 23681
USA

1. INTRODUCTION

The goal of this lecture is to present a cohesive overview of compressible turbulence research that is related to the *numerical simulation* and the *predictive modeling* of such flows.

The approach is to identify topical areas of research in numerical simulations of the Navier-Stokes (NS) equations and in predictive modeling of turbulent flows and to provide some background for the focus of the work. The interested reader is encouraged to review the cited references for more detail on the various topics.

Simulations of compressible turbulent flows have only begun in earnest within the last decade, and these simulations have focused mainly on homogeneous and temporally developing flows. Only recently have inhomogeneous flows been attempted. In contrast, numerical solutions of the compressible (Reynolds) averaged Navier-Stokes (RANS) equations with suitable correlation closure models, have been available for more than two decades.

In the first half of this lecture, the recent results of the numerical simulations, from either direct numerical simulation (DNS) or large eddy simulation (LES), are discussed. These approaches were primarily intended to provide an information basis for model development and, as a secondary benefit, to provide an unobstructed view of flow dynamics for simpler flows. The second half of the lecture is devoted to compressible turbulence modeling, with a particular focus on the correlations that are unique to the compressible regime. The space here to identify all of the compressible engineering-type flows that have been computed. Suffice it to say that many have been computed with varied success. Detailing these flows would only detract from the more global view desired here. However, this lecture is intended to provide the reader with a better understanding of the model development and, therefore, a better awareness of turbulence model limitations.

Before the numerical simulations are discussed, the governing equations are examined, including the compressible Navier-Stokes equations, which are relevant to the DNS; the filtered compressible Navier-Stokes equations, which are relevant to the LES; and the averaged Navier-Stokes equations, which are relevant to the RANS approach. Because the DNS provides the necessary information about the mean and turbulent flow field variables required to model the unknown turbulent correlations that appear in the LES and RANS formulations, it is advantageous to present the averaged (filtered) equations so that these correlations can be identified. Then, during the discussion of the simulations, the relevance of the terms can be quickly assessed.

The starting point is the mass, momentum, and energy conservation equations which are given by

$$\frac{\partial \rho}{\partial t} + \frac{\partial}{\partial x_j}(\rho u_j) = 0, \tag{1}$$

$$\frac{\partial(\rho u_i)}{\partial t} + \frac{\partial(\rho u_i u_j)}{\partial x_j} = -\frac{\partial p}{\partial x_i} + \frac{\partial \sigma_{ij}}{\partial x_j}, \tag{2}$$

$$\sigma_{ij} = 2\mu\left(S_{ij} - \frac{1}{3}S_{kk}\delta_{ij}\right), \tag{2a}$$

and

$$\frac{\partial}{\partial t}\left[\rho\left(e + \frac{u_i u_i}{2}\right)\right] + \frac{\partial}{\partial x_j}\left[u_j\rho\left(e + \frac{u_i u_i}{2} + \frac{p}{\rho}\right)\right] = \frac{\partial}{\partial x_j}(u_i\sigma_{ij}) - \frac{\partial q_j}{\partial x_j}, \tag{3}$$

where μ is the molecular viscosity, $e = c_v T$ (c_v is the specific heat at constant volume), and $q_j = -k_T \partial T/\partial x_j$ (k_T is the thermal conductivity). For the present purposes, the equation of state is $p = \rho RT$.

In section 1.1, these equations are filtered to yield a set of equations for the large-scale motions of the flow; in section 1.2, these equations are averaged to yield a set of equations for the statistical mean motion of the flow. In the compressible formulation, Favre, or mass-weighted variables [1, 2], are generally used. For a dependent variable f, the Favre "average" is defined as

$$\tilde{f} = \frac{\overline{\rho f}}{\overline{\rho}}, \tag{4}$$

where the overbar will represent either a filtering process, used in the LES approach or an averaging process, used in the RANS approach.

1.1. Filtered Conservation Equations

In the LES approach, the dependent variables are filtered to yield a set of equations for the resolved large-scale motions. This filtering process is defined as

$$\overline{f}(\mathbf{x}, t) = \int_D G(\mathbf{x} - \mathbf{x}^*) f(\mathbf{x}^*, t) d\mathbf{x}^* \tag{5}$$

where G is a spatial filter with compact support. Several types of filters can be applied; some of these filters are described in Ref. [3]. When Eqs (1), (2), and (3) are filtered, the resulting equations in Favre-variables for the large-scale motions are as follows.

1.1.1. Mass

$$\frac{\partial \overline{\rho}}{\partial t} + \frac{\partial}{\partial x_j}(\overline{\rho}\tilde{u}_j) = 0 \tag{6}$$

1.1.2. Momentum

$$\frac{\partial(\overline{\rho}\tilde{u}_i)}{\partial t} + \frac{\partial}{\partial x_j}(\tilde{u}_j\overline{\rho}\tilde{u}_i) = -\frac{\partial \overline{p}}{\partial x_i} + \frac{\partial \overline{\sigma}_{ij}}{\partial x_j} - \frac{\partial(\overline{\rho}\tau_{ij})}{\partial x_j} \tag{7}$$

where the resolved-scale stress tensor is

$$\overline{\sigma}_{ij} = \overline{2\mu\left(S_{ij} - \frac{1}{3}S_{kk}\delta_{ij}\right)} \simeq 2\overline{\mu}\left(\tilde{S}_{ij} - \frac{1}{3}\tilde{S}_{kk}\delta_{ij}\right) \tag{7a}$$

with $\overline{\mu}$ the large-scale molecular viscosity, and τ_{ij} is the subgrid scale (SGS) stress tensor.

1.1.3. Resolved Total Energy

In the LES formulation, several different approaches can be used to close the governing equation set. The solution of the total energy equation introduces several new SGS terms. An alternative is to use a resolved or pseudo total energy equation that does not introduce new SGS terms. A nonconservative formulation has also been used, which involves either internal energy or some other thermodynamic variable such as pressure. For illustrative purposes, the resolved total energy equation is presented, and the other formulations can be deduced from this equation. Let the resolved total energy be given by

$$\tilde{\mathcal{E}} = e + \frac{\tilde{u}_i\tilde{u}_i}{2}, \tag{8}$$

and

$$\frac{\partial(\overline{\rho}\tilde{\mathcal{E}})}{\partial t} + \frac{\partial}{\partial x_j}\left(\tilde{u}_j\overline{\rho}\tilde{\mathcal{H}}\right) = \frac{\partial(\tilde{u}_i\overline{\sigma}_{ij})}{\partial x_j} - \tilde{u}_i\frac{\partial \tau_{ij}}{\partial x_j} - \frac{\partial}{\partial x_j}\left(\overline{q}_j + \overline{\rho}\widetilde{u_j e}\right); \tag{9}$$

where

$$\tilde{\mathcal{H}} = \tilde{\mathcal{E}} + \frac{\overline{p}}{\overline{\rho}}, \tag{9a}$$

$$\bar{q}_j = -k_T \overline{\frac{\partial T}{\partial x_j}} \simeq -\overline{k}_T \frac{\partial \tilde{T}}{\partial x_j}, \tag{9b}$$

$e = c_v \tilde{T}$, and $\overline{k}_T$ is the large-scale thermal conductivity. The perfect gas equation of state is given by $\bar{p} = \bar{\rho} R \tilde{T}$.

In the present formulation, this filtering operation has introduced both an SGS stress tensor and an SGS heat flux vector:

$$\tau_{ij} = \bar{\rho} \left(\widetilde{u_i u_j} - \tilde{u}_i \tilde{u}_j \right), \tag{10}$$

and

$$\overline{\rho u_j e} = \bar{\rho} c_v \left(\widetilde{u_j T} - \tilde{u}_j \tilde{T} \right), \tag{11}$$

which require modeling.

1.2. Mean Conservation Equations

In the RANS approach, the dependent variables are partitioned into mean and fluctuating parts given by

$$f = \tilde{f} + f'' \tag{12}$$

where the averaging process given in Eq. (4) is now defined in terms of some statistical average. If the flow is statistically steady, then a long-time average of f is taken; if the flow is statistically homogeneous, then a volume average is taken. If the flow is neither *stationary* nor *homogeneous* and the average characteristics vary with time or space, then an *ensemble mean* over all realizations (samples) is required. For example, if the turbulence is stationary, then the appropriate average is

$$\overline{f(\mathbf{x})} = \lim_{\tau \to \infty} \frac{1}{\tau} \int_{t_0}^{t_0 + \tau} f(\mathbf{x}, t) \, dt, \tag{12a}$$

In the compressible formulation, equations for the mean conservation of mass and the mean conservation of (total) energy are needed, as well as the momentum equation.

1.2.1. Mass

$$\frac{\partial \bar{\rho}}{\partial t} + \frac{\partial}{\partial x_j} (\bar{\rho} \tilde{u}_j) = 0, \tag{13}$$

1.2.2. Momentum

$$\frac{\partial (\bar{\rho} \tilde{u}_i)}{\partial t} + \frac{\partial}{\partial x_j} (\tilde{u}_j \bar{\rho} \tilde{u}_i) = -\frac{\partial \bar{p}}{\partial x_i} + \frac{\partial \bar{\sigma}_{ij}}{\partial x_j} - \frac{\partial (\bar{\rho} \tau_{ij})}{\partial x_j}, \tag{14}$$

where

$$\bar{\sigma}_{ij} = \overline{2\mu \left(S_{ij} - \frac{1}{3} S_{kk} \delta_{ij} \right)} \simeq 2\bar{\mu} \left(\tilde{S}_{ij} - \frac{1}{3} \tilde{S}_{kk} \delta_{ij} \right), \tag{14a}$$

is the viscous stress tensor, $\bar{\mu}$ the mean molecular viscosity, and $\tau_{ij} \equiv \widetilde{u_i'' u_j''}$ is the Favre-averaged correlation tensor. Equation (14a) neglects contributions from μ', and assumes that $\bar{u} \approx \tilde{u}$.

1.2.3. Total Energy

A mean total energy equation of the form

$$\frac{\partial(\bar{\rho}\tilde{E})}{\partial t} + \frac{\partial}{\partial x_j}\left(\tilde{u}_j\bar{\rho}\tilde{H}\right) = \frac{\partial}{\partial x_j}\overline{\Sigma}_j - \frac{\partial}{\partial x_j}\left(\overline{q}_j + \overline{\rho E'' u_j''}\right),\qquad(15)$$

is used, where

$$\tilde{H} = \tilde{E} + \frac{\bar{p}}{\bar{\rho}},\qquad(15a)$$

$$\overline{\Sigma}_j = \overline{\sigma}_{ij}\tilde{u}_i + \overline{\sigma}_{ij}\overline{u_i''} + \overline{\sigma_{ij}'u_i'},\qquad(15b)$$

$$\overline{q}_j = -\overline{k_T T}_{,j} \simeq -\overline{k}_T \tilde{T}_{,j},\qquad(15c)$$

and

$$\overline{\rho E'' u_j''} = \overline{\rho c_p u_j'' T''} + \bar{\rho}\tilde{u}_i \tau_{ij} + \frac{\overline{\rho u_i'' u_i'' u_j''}}{2},\qquad(15d)$$

Above, the specific total energy is

$$\tilde{E} = c_v\tilde{T} + \frac{\tilde{u}_i\tilde{u}_i}{2} + \frac{\widetilde{u_i''u_i''}}{2}\qquad(16)$$

In Eq. (15c), fluctuations in the thermal conductivity are neglected, and the Favre-averaged and Reynolds-averaged mean temperatures are taken as approximately equal. An equation of state is also required and for a perfect gas is given by $\bar{p} = \bar{\rho}R\tilde{T}$:

$$\bar{p} = (\gamma - 1)\left[\bar{\rho}\tilde{E} - \frac{1}{2}\bar{\rho}\left(\tilde{u}^2 + \tilde{v}^2 + \tilde{w}^2\right) - \bar{\rho}K\right]\qquad(17)$$

where γ is the ratio of specific heats (c_p/c_v). The presence of the turbulent kinetic energy term K suggests a strong coupling between the mean equations and the turbulent transport equations. Because application to flows with shocks is more common in the RANS than the LES approach, the total energy is used to effectively employ shock-capturing techniques in the numerical formulation.

Just in the formulation of the mean conservation equations, the following unknown correlations that require closure have appeared:

- Favre-averaged correlation τ_{ij}

- Turbulent heat flux $\overline{\rho}c_p\widetilde{u_j''T''}$

- Turbulent mass flux $\overline{u_i''} = -\overline{\rho'u'}/\bar{\rho}$

- Turbulent transport or diffusion $\overline{\rho u_i''u_i''u_j''}$.

1.3. Turbulent Stress Transport Equations

Several similarities exist between the incompressible and compressible turbulent stress transport formulations [4]. These similarities suggest that simple variable-density extensions of existing incompressible models can be justified. The remaining terms that arise solely as a result of the compressibility of the flow can then be easily identified.

The resulting Reynolds-averaged equation in Favre variables for the turbulent stress tensor $\bar{\rho}\tau_{ij}$ is given by

$$\frac{\partial \bar{\rho}\tau_{ij}}{\partial t} + \frac{\partial}{\partial x_k}\left(\tilde{u}_k \bar{\rho}\tau_{ij}\right) = \bar{\rho}\tilde{P}_{ij} + \bar{\rho}\Pi_{ij}^d + \bar{\rho}\Pi_{ij}^{dl} + M_{ij} - \bar{\rho}\epsilon_{ij}$$

$$+ \frac{\partial \bar{\rho}\tilde{D}_{ijk}^t}{\partial x_k} + \frac{\partial D_{ijk}^v}{\partial x_k} \tag{18}$$

where the right-hand side represents the rate of change of $\bar{\rho}\tau_{ij}$ that is produced by the turbulent production $\bar{\rho}\tilde{P}_{ij}$, the deviatoric part of the pressure strain-rate correlation $\bar{\rho}\Pi_{ij}^d$, the pressure dilatation $\bar{\rho}\Pi_{ij}^{dl}$, the mass flux variation M_{ij}, the turbulent dissipation rate $\bar{\rho}\epsilon_{ij}$, the turbulent diffusion $\bar{\rho}\tilde{D}_{ijk}^t$, and the viscous diffusion D_{ijk}^v. These terms are given by

$$\bar{\rho}\tilde{P}_{ij} = -\bar{\rho}\tau_{ik}\frac{\partial \tilde{u}_j}{\partial x_k} - \bar{\rho}\tau_{jk}\frac{\partial \tilde{u}_i}{\partial x_k}, \tag{18a}$$

$$\bar{\rho}\Pi_{ij}^d = \overline{p'\left(\frac{\partial u_i'}{\partial x_j} + \frac{\partial u_j'}{\partial x_i}\right)} - \frac{2}{3}\overline{p'\frac{\partial u_k'}{\partial x_k}}\delta_{ij}, \tag{18b}$$

$$\bar{\rho}\Pi_{ij}^{dl} = \frac{2}{3}\overline{p'\frac{\partial u_k'}{\partial x_k}}\delta_{ij}, \tag{18c}$$

$$M_{ij} = \overline{u_i''}\left(\frac{\partial \bar{\sigma}_{jk}}{\partial x_k} - \frac{\partial \bar{p}}{\partial x_j}\right) + \overline{u_j''}\left(\frac{\partial \bar{\sigma}_{ik}}{\partial x_k} - \frac{\partial \bar{p}}{\partial x_i}\right), \tag{18d}$$

$$\bar{\rho}\tilde{D}_{ijk}^t = -[\bar{\rho}\widetilde{u_i'' u_j'' u_k''} + \overline{p'(u_i'\delta_{jk} + u_j'\delta_{ik})}], \tag{18e}$$

$$\bar{\rho}\epsilon_{ij} = \frac{2}{3}\epsilon\delta_{ij} + {}_D\epsilon_{ij} = \overline{\sigma_{ik}'\frac{\partial u_j'}{\partial x_k}} + \overline{\sigma_{jk}'\frac{\partial u_i'}{\partial x_k}}, \tag{18f}$$

and

$$D_{ijk}^v = \overline{(\sigma_{ik}'u_j' + \sigma_{jk}'u_i')} \approx \bar{\mu}\left(\frac{\partial \tau_{jk}}{\partial x_i} + \frac{\partial \tau_{ki}}{\partial x_j} + \frac{\partial \tau_{ij}}{\partial x_k}\right). \tag{18g}$$

In the definition of the pressure-strain correlation (Eq. (18b)), the trace (the pressure dilatation) is subtracted out. This partitioning is intended to associate the extra compressibility effects with the modeling of the pressure-dilatation term $\bar{\rho}\Pi_{ij}^{dl}$. The deviatoric part Π_{ij}^d can be modeled by a variable-density extension of the incompressible models, with the restriction that the turbulent Mach number $M_t(= \sqrt{2K}/\bar{c}) < 0.3$. (See section 3.1.2.)

The compressible turbulent kinetic energy equation is easily extracted from Eq. (18) by taking the trace

$$\frac{\partial(\bar{\rho}K)}{\partial t} + \frac{\partial}{\partial x_j}(\tilde{u}_j\bar{\rho}K) = \bar{\rho}\tilde{P} + \bar{\rho}\Pi^{dl} + \mathcal{M} + \frac{\partial}{\partial x_j}\tilde{\mathcal{D}}_j^t - \bar{\rho}\epsilon + \frac{\partial}{\partial x_j}\left(\bar{\mu}\frac{\partial K}{\partial x_j}\right) \quad (19)$$

where the right-hand side represents the turbulent kinetic energy transport produced by the turbulent production $\bar{\rho}\tilde{P} \equiv \bar{\rho}\tilde{P}_{ii}/2$, the pressure dilatation $\bar{\rho}\Pi^{dl} \equiv \bar{\rho}\Pi_{ii}^{dl}/2$, the mass flux variation $\mathcal{M} \equiv M_{ii}/2$, the turbulent diffusion $\tilde{\mathcal{D}}_j^t \equiv \tilde{D}_{iij}^t/2$, the isotropic turbulent dissipation rate $\epsilon \equiv \epsilon_{kk}/2$, and the viscous diffusion.

The deviatoric part of the dissipation rate, which is required for the turbulent stress transport equation, is usually absorbed into the pressure-strain correlation to account for any anisotropic dissipation rate effects; $\bar{\rho}\Pi_{ij}^d = \bar{\rho}\Pi_{ij}^d - \bar{\rho}_D\epsilon_{ij}$ is modeled. This structuring is not unique and an alternative is to develop models for the deviatoric part of the dissipation rate.

As shown in the next section, the isotropic dissipation rate is generally partitioned [5, 6] as

$$\epsilon = \varepsilon + \varepsilon^d \quad (20)$$

where ε is the solenoidal (incompressible) dissipation and ε^d is the dilatation (compressible) dissipation. The DNS results have been able to quantify the effect of this partitioning, as well provide the basis for new dilatation models.

2.   SIMULATION OF COMPRESSIBLE TURBULENT FLOWS

Both DNS and LES of a wide variety of compressible flows have been performed. In the case of DNS, where the motivation has been to understand the underlying physics, the studies have evolved through a hierarchy of flows; in the case of LES, the motivation has focused on SGS model validation in the simple homogeneous flows, but primarily on the prediction of complex flow fields.

2.1.   DNS

In this section, compressible flow simulations are presented that range from simple homogeneous, isotropic turbulence to inhomogeneous, wall-bounded flows. With increasing flow complexity, information about some of the unknown correlations appearing in the compressible conservation and transport equations (section 1) can be obtained.

2.1.1.   Homogeneous Turbulence

Homogeneous flows make up the simplest class of turbulent flows that can be studied and are the most amenable to theoretical analysis and verification. An analysis of the linearized, fluctuating equations of motion reveals three types of fluctuation modes [7]: the vorticity, the acoustic, and the entropy modes.

Each of these three modes has a linearized governing equation associated with it.

These modes can be either coupled to or decoupled from one another, depending on the mean flow field. In a shear-free isotropic flow, the three modes are decoupled in the inviscid case but are coupled in the viscous case. (This coupling can be removed by a repartitioning; however, then the modes no longer consist of single fields.) In a shear flow, the entropy mode is decoupled from the vorticity and acoustic modes, but the vorticity and dilatation fields are coupled. The resultant energy exchange between modes is important because it implies that the modes become independent of initial conditions. In the isotropic flow case, the decoupling between modes implies a strong dependence on the initial conditions. Erroneous or misleading closure model calibration is less likely if the flows used for such purposes are free from any initial condition or intermediate transient bias.

When the turbulence is compressed by either shocks (one dimension) or spherical compressions such as in internal combustion engines (three dimensions), the flow is amenable to analysis through rapid distortion theory (RDT). The theory is applicable to flows in which the time scale of the turbulence is long in comparison with the time scale of the mean deformation. In this case, the turbulence does not have time to interact with itself.

As discussed in the next few subsections, an extensive database of DNS results exists that has provided insight into some of these basic building-block flows.

2.1.1.1. <u>Isotropic Turbulence</u>. This turbulent flow is the simplest homogeneous turbulence to simulate and has provided extensive insight into the modeling of key dynamical features of compressible flows. In these simulations, the compressibility level is usually quantified through either the turbulent Mach number M_t given by

$$M_t = \frac{\sqrt{2K}}{\bar{c}} \tag{21}$$

where $K = \widetilde{u_i'' u_i''}/2$ is the turbulent kinetic energy and $\bar{c}$ $(= \sqrt{\gamma R \tilde{T}})$ is the mean speed of sound, or the *rms* Mach number M_{rms} given by

$$M_{rms} = \sqrt{\overline{\left(\frac{u_i''}{c}\right)^2}} \tag{22}$$

Based on the appearance of eddy shocklets in two-dimensional [8] temporally decaying, isotropic turbulence, Zeman [5] analyzed the flow in the vicinity of a shocklet (Fig. 1) and developed a model for the dilatation dissipation. The model is given by

$$\varepsilon^d \propto \mathcal{F}(M_t, \mathrm{K})\varepsilon \tag{23}$$

where K is the kurtosis $(\overline{u'^4}/(\overline{u'^2})^2)$. The model has been calibrated against a variety of flows; its final form is discussed in section 3.2.1. Blaisdell *et al.* [9],

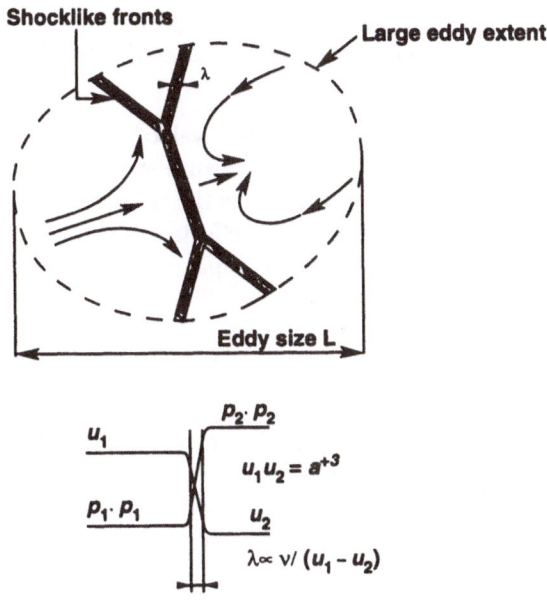

Fig. 1. — Sketch of eddy shocklet and normal shock relations [5].

who studied three-dimensional decaying turbulence over a range of initial M_{rms} from 0.3 to 0.7, also found such structures, although the effect of these structures diminished as the turbulence evolved.

Another theoretical approach has been used in the development of a model for the dilatation dissipation. Sarkar *et al.* [6] simulated low M_t turbulence in order to apply linear acoustics to the prediction of such turbulent flows. Because the three modes (*i.e.*, vorticity, acoustic, and entropy) are decoupled, the acoustic mode can be effectively isolated and a relevant theory can be applied. The asymptotic theory of Sarkar *et al.* identified a nondimensional parameter F, which represented the partition between the internal and the kinetic energy in the acoustic mode. This parameter evolved to a quasi-equilibrium phase, which Sarkar *et al.* termed an acoustic equilibrium, given by $F \simeq 1$. This equipartition of energy was confirmed by the DNS of decaying, isotropic, compressible turbulence up to $M_t = 0.5$. In the acoustic equilibrium limit, the pressure-dilatation is zero, and the compressible dissipation is nonzero. From the theoretical and simulation results, a model for the compressible dissipation of the form

$$\varepsilon^d \propto \mathcal{F}(M_t)\varepsilon \qquad (24)$$

was obtained (see section 3.2.1).

Although this flow was able to provide enough information in regard to the dilatation dissipation to yield closure models, the flow was not well-suited for the pressure-dilatation term because the term is essentially zero in the mean in decaying isotropic turbulence. An examination of the more complex case of homogeneous shear flow is needed to assess the behavior of the pressure-dilatation.

2.1.1.2. Sheared Turbulence. The next level of complexity within homogeneous flows is sheared turbulence. The mean flow shearing introduces anisotropies into the turbulence which complicates the interaction dynamics of the flow field. In these flows, the compressibility effects are characterized by the turbulent or *rms* Mach number, as in the isotropic case, but also by a new measure, the gradient Mach number [10] M_g:

$$M_g = \frac{Sl}{\bar{c}}, \tag{25}$$

where S is the mean shear rate and l is a representative length scale of the turbulence in the direction of shear.

Unlike the isotropic case, the shear rate acts to couple the vorticity, acoustic, and entropy modes within the compressible turbulence. This coupling allows the compressible turbulence to evolve to equilibrium states that are independent of the initial state of the turbulence and that are parameterized by both M_t and M_g.

In the isotropic case, the pressure-dilatation $\bar{p}\Pi_{ij}^{dl}$ can be neglected at equilibrium; however, in homogeneous shear flow, it cannot be neglected. In the simulations [11, 9], the temporal evolution of the pressure-dilatation is oscillatory with a tendency toward a negative (temporal) mean value.

Zeman [12] proposed a model based on the fact that the pressure-dilatation can be related directly to the evolution of the pressure variance for $M_t < 1$, so that

$$\overline{p'\frac{\partial u'_k}{\partial x_k}} = -\frac{1}{2\gamma\bar{p}}\frac{d}{dt}\overline{p'^2}, \tag{26}$$

where γ is the ratio of specific heats and $\bar{p}$ is the mean pressure. Additional modifications were later made for the application to wall-bounded flows. These modifications are discussed later in section 3.2.2, along with a more usable approximation to Eq. (26).

Sarkar [11] further analyzed the DNS of sheared turbulence by partitioning the pressure into compressible p'^C and incompressible parts p'^I and then evaluated the corresponding pressure-dilatation terms $\overline{p'^C u'_{k,k}}$ and $\overline{p'^I u'_{k,k}}$. The compressible pressure component was oscillatory with a zero mean, and the incompressible component was less oscillatory with a nonzero (negative) mean. Thus, only the incompressible pressure contribution to the dilatation term $\overline{p'^I u'_{k,k}}$ was significant. A further partitioning of the incompressible pressure into rapid and slow parts was utilized to develop models for the two contributions to the pressure-dilatation (see section 3.2.2):

$$\overline{p'^R\frac{\partial u'_k}{\partial x_k}} = \mathcal{F}(\bar{p}, \tilde{\mathcal{P}}, M_t, \chi), \qquad \overline{p'^S\frac{\partial u'_k}{\partial x_k}} \propto \bar{p}M_t^2\varepsilon. \tag{27}$$

As mentioned, in addition to the parameterization of the flow by M_t, homogeneous shear flow can also be parameterized by the gradient Mach number M_g. In a combined theoretical and DNS study, Sarkar [10] varied both M_t and M_g separately in order to identify the effect of the compressibility on the turbulence. By isolating the effect of M_g, the turbulent energy growth rate was shown to decrease as M_g increased. This effect on the turbulent production mechanism, rather than on direct dilatational effects, shows the stabilizing [10] effect of compressibility on the turbulence. Because $M_{g[m-l]} \gg M_{g[b-l]}$, Sarkar's analysis led to an explanation for the reduced importance of compressibility corrections in boundary-layer flows relative to free-shear flows such as mixing layers.

2.1.1.3. <u>Compressed Turbulence.</u> The final level of complexity to be discussed in this section is compressed, compressible turbulence. Simulations of one-dimensional and spherically compressed, isotropic turbulence [13, 14] have been performed. The results of these simulations have led to a better understanding of important physical processes in practical flows, such as compression ramps (one dimension) and internal combustion engines (three dimensions). Rapid compressions were simulated so that RDT could be used to analyze the data. Use of RDT requires that the eddy turnover time of the turbulence K/ε to be much greater than the time scale of the compression, which is related to the mean dilatation rate.

For axial compressions (one dimension) [14, 15], the RDT analysis yielded a parameter Δm, which is defined as the ratio of the mean deformation rate to the inverse sonic time scale. The parameter can be interpreted as the change of the mean flow Mach number across the integral scale of the turbulence. For $\Delta m \ll 1$ [15], the coupling between the acoustic and vortical modes can be neglected. In this limit, a model for the pressure-dilatation can be derived from considerations of the form of the pressure-variance (section 3.2.2). For $\Delta m \gg 1$ at finite values of the turbulent Mach number, the turbulence is strongly dilatational with no damping effect from the pressure because correlations that involve the fluctuating pressure and velocity are negligible (i.e., a "pressure released" solution [16]).

For spherical compressions, the solenoidal pressure fluctuations are zero, whereas the dilatational pressure fluctuations are unaltered by an increase in the dimensionality of the turbulence. Both the turbulent kinetic energy and the pressure variance are easily determined and are relatively insensitive to variations in turbulent Mach number [13].

Although these combined simulation and theoretical studies yield insight into the dynamics of multidimensional, compressed turbulence, they also provide extensive information for the development of closure models for the pressure dilatation and even for the pressure-strain correlation. These models are discussed in sections 3.2.2 and 3.1.2, respectively.

2.1.2. Inhomogeneous Turbulence

Studies of homogeneous flows just discussed provide useful information about the turbulence dynamics, and the results have contributed to the development of closure models (section 3). In addition, these flows are also amenable to theoretical analysis. Relevant compressible engineering flows are more complex. Embedded and impinging shocks in wall-bounded flows are not uncommon and have not been extensively explored. Free-shear flows are also important and, to date, have provided the strongest motivation for incorporating compressibility corrections into turbulent closure models, although a recent analysis of temporal mixing-layer data has raised some contradictory evidence [17]. (See section 3.2.1.) In this section, shock-turbulence interactions, isolated from the presence of walls, are discussed, as well as recent simulations and initiatives in accounting for the presence of walls in supersonic flows.

2.1.2.1. Shock-Turbulence Interactions.

In turbulent flows with shocks, the dual effect of turbulence modification by the shock and shock-wave modification by the turbulence is present. The strong mean flow compression in the direction normal to the shock leads to vorticity generation (shock curvature effect) and turbulent kinetic energy generation (caused by unsteady shock movement). Linear theories such as RDT and compressible linear interaction analysis can be used to predict these effects. The RDT accounts for the mean flow compression; the linear interaction analysis includes both vorticity and turbulent kinetic energy generation, as well as the mean flow compression. To study these effects, simulations of the interaction of isotropic turbulence with a normal shock were studied by Lee *et al.* [18] and Hannappel and Friedrich [19]. Lee *et al.* [18] considered the case of quasi-incompressible turbulence interacting with a weak normal shock (with an upstream Mach number $M^U \sim 1.1$), and Hannappel and Friedrich [19] considered the case of compressible turbulence interacting with a normal shock (with an upstream Mach number $M^U \sim 2.0$). Although resolution requirements dictated the need for low Reynolds numbers (*i.e.*, microscale Reynolds number of $Re_\lambda \sim 18$ [18] and $Re_\lambda \sim 6$ [19]), the dynamics of the turbulence could be studied in a more detailed manner than would be possible in more realistic flows.

In these flows, the normal turbulent stress components were amplified across the shock, and the shear stress levels were unaffected and remained small because of the flow symmetry. On the other hand, the transverse components of the vorticity fluctuations were enhanced, and the streamwise component was unaffected. This enhanced level of vorticity fluctuations implies an increase in the turbulent dissipation rate levels downstream of the shock as well. In the cases studied, the time scale associated with the return to isotropy was much larger than the decay rate time scale.

An important finding for modeling purposes was the conclusion that the pressure work term was the main contributor to the increase in turbulent kinetic energy. A decomposition of this term into a pressure-dilatation term $\overline{p}\Pi_{ij}^{dl}$ and a pressure transport term $\overline{p'(u'_i\delta_{jk} + u'_j\delta_{ik})}_{,k}$ showed that the latter was the main contributor to the pressure work downstream of the shock. The pressure-dilatation term acted to convert the mean internal energy into turbulent kinetic energy. Although no models have yet been proposed based on these results, the data should provide a key basis for such development. Another result with modeling implications was the determination that Morkovin's hypothesis did not hold in such shock-turbulence interactions. The requirement that the density fluctuations be associated with the entropy fluctuations was not satisfied.

Because the shock wave was modified by the turbulence, this caused the rms value of the peak compression varied significantly across the transverse plane. The instantaneous non-uniformities associated with the turbulence caused both the location and thickness of the shock to vary in the transverse direction. For weak upstream turbulent intensities ($M_t < M^U - 1$), well-defined fronts were present with a single mean compression peak. For strong upstream turbulent intensities ($M_t > M^U - 1$), shock-front distortions occurred which caused multiple compression peaks inside the shock and no well-defined fronts in the transverse direction.

By increasing the compressibility of the incoming turbulence [19], the same qualitative results were obtained as were found in the case of incompressible or quasi-compressible turbulence; however, weaker length-scale reductions, reduced amplification of thermodynamic variables across the shock, and reduced effects of the pressure work outside the shock region were realized. Clearly, additional studies on compressibility effects are warranted under more realistic conditions and with improved accuracy in the immediate vicinity of the shock.

2.1.2.2. <u>Wall-Bounded Flows</u>. With the improvement in computer capability, simulations of compressible wall-bounded flows are being attempted over a modest range of conditions. Recently, a temporal simulation of an isothermal, supersonic channel flow has been performed, as well as a spatial simulation of a supersonic boundary layer with an adiabatic wall.

In the channel flow simulation [20, 21], the cold-wall case was studied with centerline-to-wall temperature ratios of 1.38 and 2.49, and Mach numbers of 1.5 and 3, respectively. The Reynolds numbers based on bulk density and velocity were 3000 and 4880, which correspond to the Mach 1.5 and Mach 3 cases, respectively. Over most of the flow, the mass flux variations were small, so that the Favre-averaged and Reynolds-averaged quantities were essentially the same; only near the wall did some deviation occur. In this cold-wall case, unlike the shock-turbulence interaction discussed earlier, the major effect of Mach number was attributable to mean density variations which reaffirms Morkovin's hypothesis. Both dilatation dissipation and pressure-dilatation effects were negligible, although currently proposed models (sections 3.2.1 and 3.2.2) based on turbulent Mach number over-predicted their effect.

In the boundary-layer simulation [22], the flow was initiated upstream of a transition point that was forced with sinusoidal suction and blowing at the wall at the most unstable mode of the laminar layer. The free-stream momentum thickness Reynolds number was ≈ 5300. Morkovin's hypothesis held for both the mean and turbulent quantities, and the dilatation dissipation and pressure-dilatation effects were negligible. In the log layer, the Van-Driest scaling also held.

Although the simulations just discussed were important first steps in under-standing compressible, supersonic flows, these simulations did not shown the strong dilatational effects displayed by the shock-turbulence interaction prob-lem. For this reason, the obvious next step is to study wall-bounded flows with shocks. Two such studies that deal with shock–boundary-layer interactions are underway. Adams and Shariff [23] have recently examined some of the numeri-cal issues that are associated with the application of a hybrid compact upwind scheme and a high-order essentially non-oscillatory (ENO) scheme for the so-lution of a compression ramp flow. A related problem, which is presently being initiated (Gatski and Grasso, private communication), deals with an oblique shock that impinges on a turbulent boundary layer. In that study, a hybrid compact difference scheme with a weighted ENO scheme is proposed. A sketch of the computational domain is shown in Fig. 2; in Fig. 3, an example of the ability of the scheme to predict the correct behavior of a shock–entropy-wave interaction is shown. Recently, Hannappel *et al.* [24] performed such a compar-ison on ENO and total variation diminishing (TVD) schemes for the compu-tation of shock-turbulence interaction. Although these results are relevant to the impinging-shock problems mentioned in this section, they are also relevant to the shock-turbulence interaction problem discussed in section 2.1.2.1.

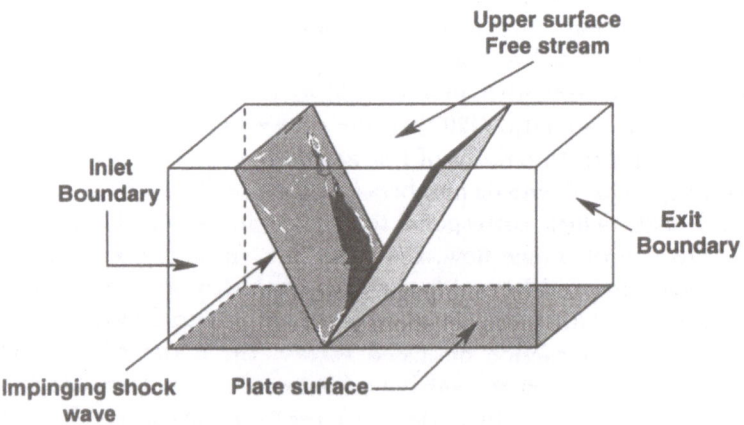

Fig. 2. — Schematic of computational region (not to scale).

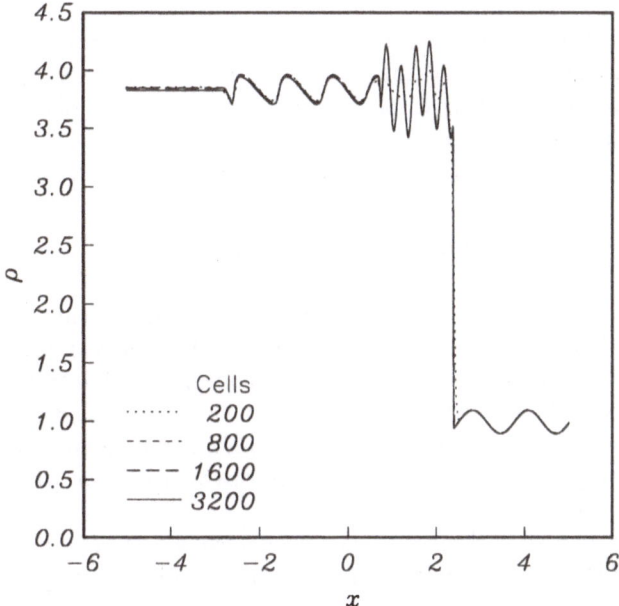

Fig. 3. — Shock–Entropy-Wave Interaction (amplitude = 0.1).

2.2. LES

Because a significant portion of this book is devoted to a variety of LES applications and strategies for the development of SGS models, a comprehensive review of all of the compressible simulations that have been performed is not necessary here. Unfortunately, an extensive set of LES studies of homogeneous flows does not exist as for DNS, and those studies that have been performed have focused primarily on isotropic decay. Some recent work has been done on homogeneous shear flow [25] which will be discussed shortly.

Yoshizawa[26] appears to be the first to develop a Smagorinsky-type model for weakly compressible flows that uses direct interaction approximation (DIA) theory. Speziale *et al.* [28] also developed a Smagorinsky-type model but used Favre variables. The model was not confined to flows with small density fluctuations and was validated against a DNS of compressible decaying turbulence.

The behavior of the dynamic SGS model in decaying turbulence was studied by Spyropoulos and Blaisdell [27] who examined the performance of the dynamic models of both Moin [29] and Lilly [30] in correctly capturing compressibility effects and in assessing related implementation issues. Spyropoulos and Blaisdell compared their results to the DNS results that were discussed previously (section 2.1.1.1). Within the compressibility and Reynolds number ranges studied, the dynamic model of Lilly performed the best and was able to capture not only the rms density fluctuations but other statistical quantities, including spectra, as well.

An examination of implementation related issues [27, 31] provides some insight into the level of inaccuracy that can arise in such simulations. These examinations found that aliasing errors could be reduced when the convective terms were modified to skew-symmetric form. For example, in the momentum equation (Eq. (7)) the convection term should be rewritten in the form

$$\frac{\partial}{\partial x_j}(\tilde{u}_j \bar{\rho} \tilde{u}_i) = \frac{1}{2}\frac{\partial}{\partial x_j}(\tilde{u}_j \bar{\rho} \tilde{u}_i) + \frac{1}{2}\tilde{u}_j \frac{\partial}{\partial x_j}(\bar{\rho} \tilde{u}_i) + \frac{1}{2}\bar{\rho} \tilde{u}_i \frac{\partial}{\partial x_j}(\tilde{u}_j), \qquad (28)$$

in order to preclude a spectral buildup at high wave numbers that could lead to instabilities. This same buildup occurred when the resolved total energy equation (Eq. (9)) was used instead of the (nonconservative) internal energy equation. These results show that even though the isotropic decay case is simple, it does provide an unobstructed (*i.e.*, no inhomogeneities) view of some fundamental accuracy problems, which are less obvious in more complicated flows but no less important.

In homogeneous sheared turbulence [25], an LES using a dynamic SGS model qualitatively reproduced the reduction of turbulent kinetic energy growth with compressibility, although the reduction was overestimated.

For the simple inhomogeneous flows, supersonic flat-plate boundary-layer flows have been simulated, and these have focused on transition-to-turbulence effects [32, 33] rather than the fully turbulent regime. These studies have used either the dynamic or the structure function [34, 35] (SF) (or the filtered structure function (FSF)) SGS models. The results have been useful in understanding the transition process. For the fully turbulent case, Spyropoulos and Blaisdell [36] have attempted to simulate a flat-plate flow ($M_\infty = 2.25$) with adiabatic wall conditions. Because this flow was spatially evolving, a finite-difference scheme was used. Spyropoulos and Blaisdell found that a high-order scheme was necessary to capture the smaller resolved scales and also proposed that the SGS model be applied to scales that were not properly resolved by the numerical scheme, as well as by the subgrid scales.

3.   COMPRESSIBLE TURBULENCE CLOSURE MODELS

In the RANS formulation presented in sections 1.2 and 1.3, the mean conservation equations (mass, momentum, and energy) and the mean transport equation for the turbulent stresses were presented, along with the unknown correlations that are associated with this set of equations. These correlations require closure, and the DNS results presented in the last section yield significant information about many of these terms. These simulation results have led to the direct formulation of models for terms such as the dilatation dissipation ε^d and the pressure-dilatation $\bar{p}\Pi_{ij}^{dl}$, as well as to possible modifications of existing closures for the solenoidal dissipation rate and the pressure-strain rate correlation (both of which are usually closed through variable-density extensions of their incompressible forms). In this section, the models developed based on these simulation results are presented.

3.1. Compressibility Extensions to Incompressible Closures

Although the compressible regime requires the development of entirely new models for closure, some models can be easily extended to compressible flows through either variable-density extensions or modifications that are dictated by results from simulations such as those just presented. In this subsection, these alterations are discussed for the solenoidal dissipation rate, the pressure-strain rate correlation, and the turbulent triple-velocity correlation.

3.1.1. Solenoidal Dissipation Rate

The high Reynolds number form of the variable-density extension to the isotropic solenoidal dissipation rate equation is

$$\frac{\partial \overline{\rho}\varepsilon}{\partial t} + \frac{\partial}{\partial x_j}\left(\tilde{u}_j \overline{\rho}\varepsilon\right) = \overline{\rho}\tilde{P}_\varepsilon - \frac{4}{3}\overline{\rho}\varepsilon\frac{\partial \tilde{u}_k}{\partial x_k} - \overline{\rho}\tilde{D}_\varepsilon + \frac{\partial}{\partial x_j}\overline{\rho}\tilde{D}_{\varepsilon j}^t + \frac{\partial}{\partial x_j}\left(\overline{\mu}\frac{\partial \varepsilon}{\partial x_j}\right), \quad (29)$$

where

$$\tilde{P}_\varepsilon = -C_{\varepsilon 1}\frac{\varepsilon}{K}\tau_{ij}\left(\frac{\partial \tilde{u}_i}{\partial x_j} - \frac{1}{3}\frac{\partial \tilde{u}_k}{\partial x_k}\delta_{ij}\right), \quad (29a)$$

$$\tilde{D}_\varepsilon = C_{\varepsilon 2}\frac{\varepsilon^2}{K}, \quad (29b)$$

$$\overline{\rho}\tilde{D}_{\varepsilon j}^t = \overline{\rho}C_\varepsilon\left(\tau_{ij}\frac{K}{\varepsilon}\frac{\partial \varepsilon}{\partial x_i}\right), \quad (29c)$$

and $C_{\varepsilon 1}$, $C_{\varepsilon 2}$, and C_ε are assumed to take on their incompressible values. For a two-equation $K - \varepsilon$ formulation, the only change is in the turbulent transport model with

$$\overline{\rho}\tilde{D}_{\varepsilon j}^t = \frac{\overline{\mu}_t}{\sigma_\varepsilon}\frac{\partial \varepsilon}{\partial x_j}, \quad (30)$$

where

$$\overline{\mu}_t = \overline{\rho}C_\mu\frac{K^2}{\varepsilon}, \quad (C_\mu \approx 0.09), \quad (30a)$$

and

$$\sigma_\varepsilon = \frac{\kappa^2}{(C_{\varepsilon 2} - C_{\varepsilon 1})\sqrt{C_\mu}}, \quad (\kappa = 0.41). \quad (30b)$$

Other than the variable-density extension, this form differs from the incompressible form because of the addition of the mean dilatation term and the traceless production-of-dissipation term.

Results from RDT and from DNS of spherically compressed turbulence (section 2.1.1.3) have shown that the omission of a mean dilatation term causes the model to incorrectly predict both the decrease in the integral length scale for isotropic expansion and the increase in the length scale for isotropic compression [37]. A further modification proposed by Coleman and Mansour [38] also accounts for variations of mean kinematic viscosity. Their proposal, however, simply alters the coefficient in front of the mean dilatation term.

The dissipation rate equation is inappropriate in the log layer of a wall-bounded flow because it incorrectly accounts for the mean density variation near the wall [39, 40]s. This incorrect accounting causes the slope of the Van Driest velocity in the log layer to be different than the inverse of the Von Karman constant κ^{-1}.

3.1.2. Pressure-Strain Rate Correlation

The pressure-strain rate correlation term has received the greatest amount of attention from developers of higher order models in the incompressible regime. In the absence of dilatation effects, this term is the same order as the production term and acts as a redistribution term between the Reynolds stress components. Thus, the term acts on the anisotropy of the stress field to diminish the difference between the normal stress components.

Models for $\overline{p}\Pi_{ij}^d$ are classified (linear, quadratic, or cubic) by their dependency on the turbulent stress anisotropy tensor

$$b_{ij} = \frac{\overline{\rho}\tau_{ij}}{2\overline{\rho}K} - \frac{1}{3}\delta_{ij} \tag{31}$$

and have taken the general form [41, 42, 43] of

$$\overline{p}\Pi_{ij}^d = \overline{\rho}\varepsilon\mathcal{A}_{ij}(\mathbf{b}) + \overline{\rho}K\mathcal{M}_{ijkl}(\mathbf{b})\frac{\partial\overline{u}_k}{\partial x_l}, \tag{32}$$

where $\mathcal{A}_{ij}(\mathbf{b})$ and $\mathcal{M}_{ijkl}(\mathbf{b})$ are related to integrals over the flow volume derived from a Poisson equation for the pressure. The $\mathcal{A}_{ij}(\mathbf{b})$ term is usually associated with the "slow" relaxation of the turbulence toward isotropy, and the $\mathcal{M}_{ijkl}(\mathbf{b})$ term is usually associated with the "rapid" response of the turbulence to imposed mean velocity gradients. The interpretation of each of these terms is exact in homogeneous turbulent flows. This partitioning has its origins in the splitting of the turbulent pressure field p' into slow $p'^{(S)}$ and rapid $p'^{(R)}$ parts. The $p'^{(S)}$ is the solution of a Poisson equation that involves gradients of the turbulent velocity field, and the $p'^{(R)}$ part is the solution of a Poisson equation that involves the mean velocity gradients. Several reviews of these models have been conducted (*e.g.*, [44, 4, 45]); the interested reader can find the further details in these references.

From the DNS, indications that simple variable-density extensions of the (deviatoric) pressure-strain rate correlation $\overline{p}\Pi_{ij}^d$ may not be adequate. For the case of homogeneous shear flow (section 2.1.1.2), Speziale *et al.* [46] showed that inclusion of the dilatational effects alone does not properly account for the stress anisotropy levels. They concluded that the model extensions for $\overline{p}\Pi_{ij}^d$ may not be adequate and that some additional effects may be required in cases for which $M_t > 0.3$.

An RDT analysis and a subsequent comparison with DNS results for rapidly compressed axial turbulence [14] (section 2.1.1.3) has suggested that increased compressibility (quantified through Δm) reduces the effects of this redistribution term. By incorporating a decaying exponential factor into the pressure-strain and pressure-dilatation terms, the effect of these terms is removed in the limit of large Δm. For the DNS cases studied, this damping correction yields very good agreement with the (temporal) evolution of the axial turbulent stress component.

3.1.3. Turbulent Diffusion Models

The pressure-velocity correlation, or pressure transport, and the triple-velocity correlation term are usually combined and modeled by using a gradient diffusion assumption in the incompressible case. For the most part, variable-density extensions of these incompressible models are also used in the compressible case. Thus, a gradient transport hypothesis is invoked for the model of the turbulent diffusion term $\bar{D}^t_{ijk}$. The most common models used in conjunction with the second-moment closures (e.g., [44, 4]) fall into the general functional form

$$\bar{\rho}\bar{D}^t_{ijk} = d_{ijklmn}(\bar{\rho}\tau_{pq})\frac{\partial\tau_{lm}}{\partial x_n}. \tag{33}$$

Other, more complicated models have also been proposed. Lumley [41] proposed a model that contains both a model of the form of Eq. (33) as a tensorial basis, and an explicit model for the pressure transport term. As might be expected, the form of the model is quite complex and has not been used extensively. Magnaudet [47] has proposed a model that is consistent with the two-dimensional turbulence encountered near a wall and at a free surface.

Although most modeling has not explicitly accounted for the effects of pressure transport $\overline{p'(u'_i\delta_{jk} + u'_j\delta_{ik})}$ in compressible modeling, the results of section 2.1.2.1 have shown that this term can have a significant effect on the behavior of the turbulence in the vicinity of the shock. Because terms of the form $\overline{p'u'_i}$ are related to other compressible correlations through the relations

$$\overline{p'u'_i} \equiv \overline{p'u''_i}, \tag{34}$$

$$p' = R(\rho T'' + \rho'\tilde{T}), \tag{34a}$$

and

$$\overline{p'u''_i} = R(\bar{\rho}\widetilde{u''_iT''} - \bar{\rho}\tilde{T}\overline{u''_i}), \tag{35}$$

a consistent pressure-transport model can be developed in a straightforward manner. However, because models for the mass and the heat flux are generally neglected in compressible modeling, no attempt has been made to incorporate a pressure transport model into a compressible formulation. In weakly compressible flows, such omissions probably have little impact on the results of the model predictions; however, in strongly compressible flows with shocks or strong temperature (density) gradients, these terms are important, and their effect should be properly taken into account.

3.1.4. Algebraic Stress Models

Before the dilatation and scalar flux models are discussed in the next section, some comments on compressible algebraic stress models are presented. These models are extensions of the low-order Boussinesq approximation, which is easily extended to compressible applications through

$$\bar{\rho}\tau_{ij} = \frac{2}{3}\bar{\rho}K\delta_{ij} - 2\mu_t \left(\tilde{S}_{ij} - \frac{1}{3}\tilde{S}_{kk}\delta_{ij} \right). \tag{36}$$

The algebraic stress models, or the more recently developed explicit algebraic stress models [48], which yield an algebraic relationship between the turbulent Reynolds stresses and the mean velocity field, are an effective compromise between the full second-moment closure and the two-equation models. These models can bee derived from equilibrium hypotheses imposed on both the convective and diffusive terms, which can be readily carried over to the compressible regime. The convective equilibrium hypothesis is

$$\bar{\rho}K\frac{Db_{ij}}{Dt} \equiv \frac{D\bar{\rho}\tau_{ij}}{Dt} - \frac{\tau_{ij}}{K}\frac{D\bar{\rho}K}{Dt} = 0, \tag{37}$$

and the diffusive equilibrium hypothesis is

$$\frac{\partial}{\partial x_k}\left(\bar{\rho}\tilde{D}^t_{ijk} + \bar{\rho}D^v_{ijk}\right) = 0. \tag{38}$$

By using these relations, Eq. (37) can be expanded as

$$\bar{\rho}(\tilde{\mathcal{P}} - \varepsilon)b_{ij} = -\frac{2}{3}\bar{\rho}K\tilde{S}_{ij} \quad -\bar{\rho}K\left(b_{ik}\tilde{S}_{jk} + b_{jk}\tilde{S}_{ik} - \frac{2}{3}b_{mn}\tilde{S}_{mn}\delta_{ij}\right)$$
$$- \bar{\rho}K(b_{ik}\tilde{W}_{jk} + b_{jk}\tilde{W}_{ik}) + \frac{1}{2}\bar{\rho}\Pi^d_{ij} \tag{39}$$

Equation (39) is the traditional starting point for the algebraic stress models. This equation is an implicit relationship that is dependent only on the (deviatoric) part of the pressure-strain rate correlation.

An explicit, quadratic constitutive relationship for compressible flows can then be written as

$$\bar{\rho}\tau_{ij} = \frac{2}{3}\bar{\rho}K\delta_{ij} - 2\mu_t^* \left[\left(\tilde{S}_{ij} - \frac{1}{3}\tilde{S}_{kk}\delta_{ij} \right) + \alpha_2\frac{K}{\varepsilon}\left(\tilde{S}_{ik}\tilde{W}_{kj} + \tilde{S}_{jk}\tilde{W}_{ki} \right)\right.$$
$$\left. - 2\alpha_3\frac{K}{\varepsilon}\left(\tilde{S}_{ik}\tilde{S}_{kj} - \frac{1}{3}\tilde{S}_{kl}\tilde{S}_{kl}\delta_{ij} \right)\right], \tag{40}$$

where

$$\mu_t^* = \bar{\rho}C_\mu^*\frac{K^2}{\varepsilon} \tag{41}$$

and

$$C_\mu^* = C_\mu^* \left(II_{\bar{S}}, II_{\bar{W}} \right), \quad \alpha_2 = \alpha_2 \left(II_{\bar{S}}, II_{\bar{W}} \right), \quad \alpha_3 = \alpha_3 \left(II_{\bar{S}}, II_{\bar{W}} \right) \quad (41a)$$

are functions of the invariants of the mean strain rate and rotation rate tensors. Further details on the development and application of these models are given in Refs. [49] and [50].

3.2. Dilatation and Scalar Flux Closure Models

As shown in section 1, the compressible formulation of the mean conservation equations and the turbulent transport equations introduced several higher order correlations that are directly related to dilatational effects. These correlations include the dilatation dissipation ε^d, the pressure-dilatation $\overline{p} \Pi_{ij}^{dl}$, and the mass flux $\overline{u_i''}$ ($= \overline{\rho' u_i'}/\overline{\rho}$). These terms were of central importance in the DNS studies discussed in section 2 because their role was important to the understanding of the dynamics of the flows. In this subsection, the models derived for these dilatation correlations are discussed, as well as the model development for the compressible turbulent heat flux (which would be significant in flows with strong temperature gradients, such as heated or cooled wall-bounded flows).

3.2.1. Dilatation Dissipation

In section 1.3, the isotropic dissipation rate was partitioned into a solenoidal and a dilatational part. The solenoidal part was modeled by way of a transport equation, which was discussed in section 3.1.1. Here, the more common closure models for the dilatation dissipation are presented.

The models for the dilatation dissipation take the general form

$$\varepsilon^d = \alpha^* \mathcal{F}(M_t)\varepsilon. \quad (42)$$

The Zeman form [5, 54] is given by

$$\mathcal{F}(M_t) = 1 - \exp{-[0.5(\gamma + 1)(M_t - M_{t0})^2/\Lambda^2]}\mathcal{H}(M_t - M_{t0}), \quad (43)$$

where $\alpha^* = 0.75$, $\mathcal{H}(M_t)$ is the Heaviside step function, $M_{t0} = 0.10[2/(\gamma + 1)]^{1/2}(\Lambda = 0.6)$ for mixing layers and $M_{t0} = 0.25[2/(\gamma + 1)]^{1/2}(\Lambda = 0.66$ for boundary layers. As mentioned in section 2.1.1.1, Zeman deduced the form of his model from an analysis of the flow near an eddy shocklet, which introduced a dependency on the turbulent kurtosis. This explicit dependency has been removed so that the final model has dependence only on M_t.

The form proposed by Sarkar et al. [6] is given by

$$\mathcal{F}(M_t) = M_t^2, \quad (44)$$

with $\alpha^* = 0.6$.

Wilcox [39] also developed a model with the same general functional form as shown above:

$$\mathcal{F}(M_t) = \left[M_t^2 - M_{t0}^2\right]\mathcal{H}\left(M_t - M_{t0}\right) \quad (45)$$

with $M_{t0} = 0.25$ and $\alpha^* = 1.5$. The model is intended for use in conjunction with a $K - \omega$ two-equation model.

In the compressible mixing layer, these models have a significant effect on the spreading rate (Fig. 4). The inclusion of these models increases the mixing-layer spreading and brings the turbulent model predictions into close agreement with experimental results. A recent study by Vreman *et al.* [53] critically reviewed

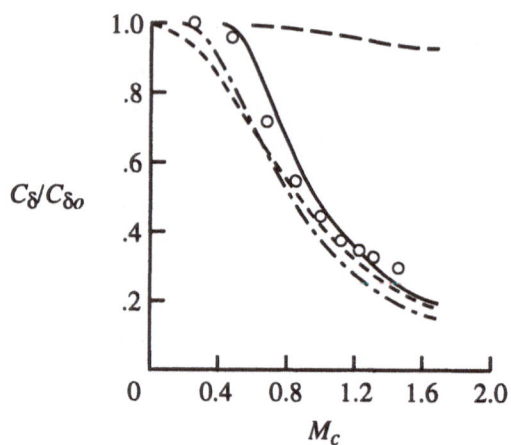

Fig. 4. — Comparison of computed and measured spreading rate for compressible mixing layer [51].   — — Unmodified $K - \omega$;   —— Wilcox model;   – – – Sarkar model;   — – — Zeman model;   o Langley curve [52].

the data from a temporal DNS of a compressible mixing layer. Vreman *et al.* concluded that although these dilatation corrections yielded, for example, improved growth-rate predictions, the actual dilatation effects were much less important in the simulated flow field. This result held even when eddy shocklets were present.

In flat-plate boundary layers, however, the log law results can be adversely affected when the Zeman or Sarkar models are used. Zeman [54] alleviated this problem by accounting for density effects in the specification of the pressure-dilatation model which is discussed in the next section. The Wilcox model, coupled with the $K - \omega$ formulation, is not as adversely affected.

3.2.2. *Pressure-Dilatation*

Because the velocity gradient tensor is not traceless in compressible flow, both the compressible stress transport equation Eq. (18) and the turbulent kinetic energy equation Eq. (19) explicitly account for a pressure dilatation term. With the formulation of the equations discussed in section 1, the compressibility effects that result from the pressure-strain correlation are more directly attributable to the pressure-dilatation term. The remaining deviatoric part can then be modeled in much the same way as its incompressible counterpart by using a variable-density extension. As shown in the discussion in section 3.1.2, this use of the variable-density extension has been the general trend, although some DNS results have suggested that such procedures may not be generally valid.

Zeman [12, 54] argued that for small, turbulent Mach numbers the pressure-dilatation $\overline{\rho}\Pi^{dl}$ is solely dependent on the transport equation for the pressure variance. Originally, Zeman [12] proposed a pressure relaxation model based on comparisons with decaying turbulence and later applied to rapid spherical and axial compression [13, 15]. (See section 2.1.1.3.) For application to boundary layers in quasi-equilibrium, a more thorough analysis of the pressure variance equation yielded a relationship with the mean density gradient and the pressure transport term $\overline{p'u_i'}$. As shown in section 3.1.3, this required models for both the mass and heat flux, which were approximated as

$$
\overline{p'\frac{\partial u_k'}{\partial x_k}} = 2f_\rho(M_t)\overline{a}^2\tau_{ij}\frac{K}{\varepsilon}\frac{1}{\overline{\rho}}\frac{\partial \overline{\rho}}{\partial x_i}\frac{\partial \overline{\rho}}{\partial x_j}, \tag{46}
$$

where $f_\rho(M_t) = 0.02[1 - \exp(-M_t^2/0.2)]$ and $\overline{a} = \sqrt{\gamma R\tilde{T}}$ is the sound speed. The model was able to preserve the Van Driest compressible law of the wall, even when used with a dilatation dissipation model.

Sarkar [11] proposed a model obtained from a formal solution for homogeneous turbulence that is applicable to free shear flows. The pressure field was split into an incompressible and compressible part, and DNS data were utilized to show that only the contribution from the incompressible pressure field p'^I to the pressure dilatation correlation was significant. Because the incompressible pressure component is most relevant, some ideas from incompressible homogeneous shear flow can be implemented. In that case, the incompressible pressure p'^I is normally partitioned into a rapid part p'^R and a slow part p'^S. (See section 3.1.2.) The form of the model for homogeneous shear flow is

$$
\overline{p'\frac{\partial u_k'}{\partial x_k}} = 0.15\overline{\rho}\tilde{P}M_t + 0.20\overline{\rho}\varepsilon M_t^2. \tag{47}
$$

Taulbee and VanOsdol [55] proposed a model for the combination $\overline{\rho}\Pi^{dl} - \overline{\rho}\epsilon$. The model was dependent on the density variance and mass flux. Modeled transport equations for the density variance and average mass fluctuating velocity were then included in the governing equation set. The full set of equations is rather extensive; however, the general form of the average mass-fluctuating velocity will be presented in the next section.

3.2.3. Mass Flux

The average mass-fluctuating velocity $\overline{u_i''}$ is related to the mass flux $\overline{\rho' u_i'}$ by

$$\overline{u_i''} \equiv -\frac{\overline{\rho' u_i'}}{\overline{\rho}}. \tag{48}$$

Recent attempts have been made to model this correlation, although none of the models have been extensively tested. Some of these models are presented here.

Rubesin [56] assumed that the density and pressure fluctuations were related through a polytropic gas law:

$$\overline{u_i''} = -c_e \tau_{ij} \frac{K}{\varepsilon} \frac{1}{\tilde{T}} \frac{\partial \tilde{T}}{\partial x_j}, \tag{49}$$

where c_e is a closure coefficient. Huang et al. [21] found in comparisons with compressible channel flow simulations (section 2.1.2.2) that no unique value for c_e could be found to satisfy both velocity components. For the streamwise component, $c_e = 1.1$; for the transverse component, $c_e = 0.21$.

In the discussion of the pressure-dilatation, Taulbee and VanOsdol [55] require the solution of a modeled transport equation for the mass flux:

$$\frac{D\overline{\rho' u_i'}}{Dt} = -C_{u_2} \frac{\varepsilon}{K} \overline{\rho' u_i'} + \mathcal{P}_{mf} + \mathcal{D}_{mf}^t + \mathcal{D}_{mf}^v, \tag{50}$$

where $C_{u_2} = 5.3$ is a model parameter determined from comparison with experimental data. The first term on the right represents the relaxation of $\overline{u_i''}$. The contributions to the production $\mathcal{P}_{mf}$ are from both mean velocity and mean density gradients. The diffusion terms $\mathcal{D}_{mf}^t$ and $\mathcal{D}_{mf}^v$ are closed by gradient diffusion hypotheses. In comparisons with experiments on an adiabatic flat-plate boundary layer and a compressible free shear flow, agreement is generally favorable.

Zeman and Coleman [13] derived a transport equation for the mass flux in order to study the response of turbulence to a shock (section 2.1.2.1):

$$\frac{D\overline{\rho' u_i'}}{Dt} \simeq -\frac{\overline{\rho' u_i'}}{\tau_a} - \tau_{ij} \frac{\partial \overline{\rho}}{\partial x_j} - \overline{\rho' u_j'} \frac{\partial \tilde{u}_i}{\partial x_j}, \tag{51}$$

where τ_a is an acoustic timescale determined from the DNS data. The transport equation consists of a relaxation term that is needed to represent the decay of the turbulence after the shock and two production-type terms based on mean velocity and mean density gradients.

3.2.4. Turbulent Heat Flux

Although the heat flux can be an important term in the mean energy equation (Eq. (15)), this term is also a contributor to the pressure transport term $\overline{p' u_i'}$.

In the latter role, the term has not been extensively utilized, nor has it been consistently implemented, even when the heat flux term has been accounted for elsewhere. Gradient diffusion models have been the most popular and simplest closures used for the turbulent heat fluxes. Sommer et al. [57] proposed a gradient diffusion model that also displayed the correct asymptotic consistency in the near-wall region (see section 3.4):

$$\widetilde{u_i''T''} = -\kappa_T \frac{\partial \tilde{T}}{\partial x_i}, \tag{52}$$

with turbulent heat diffusivity $(Pr_t = \bar{\nu}_t/\kappa_T)$

$$\kappa_T = C_\lambda f_\lambda K \left(\frac{K\widetilde{T''^2}}{\varepsilon \varepsilon_T} \right)^{1/2} \tag{52a}$$

where $C_\lambda = 0.11$ is the model constant, f_λ is a near-wall damping function that is a function of the distance from the wall and the turbulent Reynolds number, and ε_T is the dissipation rate of the temperature variance. As can be seen, this formulation required transport equations for the temperature variance and temperature-variance dissipation rate. The tests on incompressible and compressible boundary layers with adiabatic and cooled-wall boundary conditions done by Sommer et al. showed that Morkovin's hypothesis was valid for high-Mach-number flows (consistent with the DNS results in Refs. [20] and [21]; furthermore in highly cooled-wall flows the Prandtl number was not constant, and the assumption of dynamic similarity between momentum and heat transport was not applicable.

Huang and Coakley [58] have extracted an algebraic relationship from a transport equation for the heat flux by using the assumptions associated with the development of algebraic stress models:

$$-\widetilde{u_k''T''} \frac{\partial \tilde{u}_i}{\partial x_k} - c_{1T} \widetilde{u_i''T''} \frac{\varepsilon}{K} + c_{2T} \widetilde{u_k''T''} \frac{\partial \tilde{u}_i}{\partial x_k} = \tau_{ki} \frac{\partial \tilde{T}}{\partial x_k}, \tag{53}$$

where $c_{1T} = 3$ and $c_{2T} = 0.5$ are model coefficients. The equations were solved by direct inversion, and tests were run over a range of Mach numbers and wall conditions, including a case of hypersonic flow with a shock–boundary-layer interaction. In addition, a simple gradient transport model proposed by Ha Minh et al. [59] was also tested:

$$\widetilde{u_i''T''} = -c_T \frac{K}{\varepsilon} \tau_{ik} \frac{\partial \tilde{T}}{\partial x_k}, \tag{54}$$

where $c_T = 0.313$ was obtained by optimization to free shear flow predictions. The results showed a somewhat limited predictive capability of the combined stress closure and heat flux models (with little difference between the two models).

Finally, El Baz and Launder [60] used transport equations for the heat flux, temperature variance, and variance dissipation rate in conjunction with a full second-moment closure to compute the compressible mixing layer. Other proposals have been put forth for turbulent heat flux modeling; these models are briefly discussed in the review by Cousteix and Aupoix [61].

3.3. Wall Functions

The high Reynolds number solutions must be patched to a solution field that is valid in the vicinity of the wall. This patching is done in two ways: the method presented in this section, which uses a prescribed functional distribution, and the method described in the next section, which uses alternative closure models that are valid directly to the wall.

The wall function approach to be described requires a functional distribution for the mean field and the turbulent field to which the high Reynolds number solution can be matched. For illustrative purposes, only a zero pressure-gradient (ZPG) flow in Cartesian coordinates will be considered. The starting point is the mean momentum equation (14) and total energy equation (15):

$$\bar{\sigma}_{tot} = \bar{\sigma}_{xy} - \bar{\rho}\tau_{xy} = \bar{\sigma}_w \tag{55}$$

and

$$\bar{q}_{tot} = \bar{q}_y + c_p \bar{\rho}\widetilde{v''T''} = \bar{q}_y\big|_w + \bar{\sigma}_{xy}\tilde{u} - \bar{\rho}\tilde{u}\tau_{xy} = \bar{q}_w + \tilde{u}\bar{\sigma}_{tot}, \tag{56}$$

where the mass flux and turbulent transport are neglected. These equations, coupled with the well-known functional distributions in the log layer of the flow, yield the necessary temperature and velocity wall distributions.

3.3.1. Temperature and Velocity Law of the Wall

To explicitly obtain the compressible law of the wall, a relation between the velocity and the temperature is needed. In the log-law region,

$$\frac{\partial \tilde{u}}{\partial y} = \frac{(\bar{\sigma}_w/\bar{\rho})^{\frac{1}{2}}}{\kappa y} \tag{57}$$

and

$$\frac{\partial \tilde{T}}{\partial y} = -\mathrm{Pr}_t \frac{(\bar{q}_{tot}/\bar{\rho}c_p)}{(\bar{\sigma}_w/\bar{\rho})^{\frac{1}{2}}\kappa y}. \tag{58}$$

The heat flux at the wall, $\bar{q}_w$, is then obtained by using Eqs. (56), (57), and (58) as follows:

$$\bar{q}_w = -c_p \frac{\bar{\mu}_t}{\mathrm{Pr}_t} \frac{\partial \tilde{T}}{\partial y} - \tilde{u}\bar{\sigma}_w, \tag{59}$$

where

$$\bar{\mu}_t = \bar{\sigma}_w \left(\frac{\partial \tilde{u}}{\partial y}\right)^{-1} = \bar{\rho}C_\mu \frac{K^2}{\varepsilon}. \tag{59a}$$

From Eq. (58), the temperature distribution in the inner layer is given by

$$\tilde{T} \simeq \tilde{T}_w - \frac{\text{Pr}_t}{c_p}\left(\frac{\overline{q}_w \tilde{u}}{\overline{\sigma}_w} + \frac{\tilde{u}^2}{2}\right). \tag{60}$$

Equation (60) is only an approximate relationship because the integration limit on Eq. (58) extends directly to the wall and the expression is not valid in the viscous sublayer.

Equations (57) and (60) can be combined to obtain the Van Driest velocity because density is inversely proportional to temperature (pressure is assumed to be constant within the layer):

$$\tilde{u}_c \simeq \left(\frac{2c_p \tilde{T}_w}{\text{Pr}_t}\right)^{\frac{1}{2}}\left[\arcsin\left(\frac{\overline{q}_w + \tilde{u}\overline{\sigma}_w}{\overline{\sigma}_w D}\right) - \arcsin\left(\frac{\overline{q}_w}{\overline{\sigma}_w D}\right)\right], \tag{61}$$

where

$$D = \left(\frac{\overline{q}_w^2}{\overline{\sigma}_w^2} + \frac{2c_p \tilde{T}_w}{\text{Pr}_t}\right)^{\frac{1}{2}}. \tag{61a}$$

The Van Driest velocity satisfies the compressible law of the wall given by

$$\tilde{u}_c^+ \equiv \frac{\tilde{u}_c}{\tilde{u}_\tau} = \frac{1}{\kappa_c}\ln y^+ + B_c, \tag{62}$$

where

$$\tilde{u}_\tau = \sqrt{\overline{\sigma}_w/\overline{\rho}_w}, \quad y^+ = \overline{\rho}_w \tilde{u}_\tau y/\overline{\mu}_w, \tag{62a}$$

$$\kappa_c = \kappa_c(M_\tau, B_q, \text{Pr}_t), \quad B_c = B_c(M_\tau, B_q, \text{Pr}_t), \tag{62b}$$

and

$$B_q = \overline{q}_w/\overline{\rho}_w c_p \tilde{u}_\tau \tilde{T}_w. \tag{62c}$$

These dependencies generally have a minimal effect in the low supersonic regime. More complex, multilayer wall-function formulations [62] can be used, but these are cumbersome to apply to a Reynolds stress formulation.

3.3.2. Wall Function Implementation

The main steps for implementing the wall functions for compressible flows can be outlined as follows:

(a) Set an initial guess for the wall shear stress, with either $\overline{\sigma}_w = 0.3\overline{\rho}_w K_{(2)}$ or $\overline{\sigma}_w$ from the previous iteration.

(b) *Isothermal Condition:*
Compute the heat flux at the wall, $\overline{q}_w$, from Eq. (60), evaluated at the first interior point (2).
Adiabatic or Specified $\overline{q}_w$ Condition:
Compute the temperature at the wall, $\tilde{T}_w$, from Eq. (60), evaluated at the first interior point (2).

(c) Given the temperature and velocity profiles from the previous iteration (time) step, calculate $\tilde{u}_{c(2)}$ from Eq. (61).

(d) Update the wall shear stress $\overline{\sigma}_w$ from Eq. (62).

(e) Repeat steps (b), (c), and (d) until convergence.

(f) Update the turbulent kinetic energy $K_{(2)}$ from $\tilde{u}_\tau^2/\sqrt{C_\mu}$ or the turbulent stresses $\tau_{ij(2)}$ from their respective transport equations.

(g) Update the dissipation rate $\varepsilon_{(2)}$ from $K^{\frac{3}{2}}/2.5y$.

(h) Return to step (a) for the next iteration (time) step.

Care should be exercised when applying the compressible wall functions in the hypersonic regime. As the Mach number increases, the viscous sublayer becomes thicker, so that in order to keep the first interior point inside the turbulent region ($y^+ \approx 30$) the physical distance from the first interior point to the wall should be increased with the Mach number. Computations of hypersonic flows with wall functions have difficulty capturing flow separation because the reverse flow in the viscous layer is not resolved.

3.4. Near-Wall Models

An alternative to wall functions or multi-layer models is the development of near-wall closure models, which allow for direct integration to the wall without recourse to the specified wall distributions just described. The advantages to such a process are that a single set of equations is solved for the entire flow field and that the model is sensitized to low Reynolds number effects as well as to surface proximity effects. The disadvantages are the need for improved grid resolution near the wall and the complexity of the near-wall models.

3.4.1. Second-Moment Models

The models that have been developed have been based on asymptotic consistency of turbulent transport equations near the wall. Each turbulent variable is assumed to be a polynomial function of wall distance; from a balance of the respective transport equations, a set of polynomial coefficients that balance the equations can be obtained. These models have been developed for incompressible flows; variable-density extensions to the incompressible pressure-strain correlation and the (solenoidal) dissipation rate equation are used. Other than the work of Zeman [54], who proposed a model that is capable of handling wall-bounded flows (section 3.2.2), no account has been taken of near-wall effects on dilatation dissipation, pressure-dilatation, mass flux, or heat flux.

In the development of near-wall models for the pressure-strain rate and (solenoidal) dissipation rate correlations that have been proposed, the fundamental problem has been the introduction of wall normals; for example, terms such as the following have usually appeared:

$$\frac{\tau_{ij} + \tau_{ik}n_k n_j + \tau_{jk}n_k n_i + n_i n_j \tau_{kl}n_k n_l}{1 + 3\tau_{kl}n_k n_l/2K}. \tag{63}$$

As this term suggests, the coding aspects in complex geometries become burdensome. Zhang *et al.* [63] have developed such closures based on a variable-density extension of the Lai and So [64] model. Other incompressible approaches, such as Durbin's [65] elliptic relaxation model, which requires the solution of a Helmholtz-type equation for the *relaxed* pressure-strain, could also be extended to the compressible regime.

3.4.2. *Two-Equation Models*

The development of near-wall two-equation models is straightforward in that damping-coefficient factors are introduced into various terms in the turbulent dissipation rate equations. A damping function is also included in the defining relation for the turbulent eddy viscosity ν_t in order to obtain the correct peak in the K levels, and a damping function is included in the destruction of dissipation ε^2/K term because of the singularity that occurs at the wall. This function is also chosen to ensure the correct asymptotic distribution of the dissipation rate at the wall. Additional damping functions, factored into the production term, have also been included to improve the model performance in flows for which the effects of cooled or heated walls need to be taken into account.

4. SUMMARY

The intent of this lecture is to provide a cohesive overview of the recent advances in compressible turbulence research that involve both numerical simulations and closure-model development. In the direct numerical simulation approach, results have focused on both homogeneous and inhomogeneous flows, with a particular emphasis on the relationship between the simulation statistics and the models for the compressible correlations that appear in the mean conservation and turbulent transport equations. New insights on compressibility effects in turbulent flows, obtained from the simulations, have also been highlighted. In the large-eddy simulation approach, the subgrid scale model development has primarily focused on variable-density extensions. Most of the large-eddy simulation studies have focused on more complex inhomogeneous flow fields, where effects of walls and other external forces play a role. Only recently has some fundamental work been performed on homogeneous flows in order to examine the level of both numerical and subgrid-scale model accuracy.

Acknowledgments

Financial support to attend the session has been provided by the DRET and the Ministère des Affaires Étrangères.

References

[1] Favre A., *J. Mécanique* **4**(3) (1965) 361.

[2] Favre A., *In* "Studies in Turbulence" T. B. Gatski, S. Sarkar, C. G. Speziale, Eds., (Springer-Verlag, New York, 1991), pp. 324–341.

[3] Ferziger J. H., *In* "Simulation and Modeling of Turbulent Flows" T. B. Gatski, M. Y. Hussaini, and J. L. Lumley, Eds., (Oxford University Press, New York, 1996), pp. 109–154.

[4] Gatski T. B., *In* "Handbook of Computational Fluid Dynamics" R. Peyret, Ed., (Academic Press Ltd., London, 1996), pp. 339–415.

[5] Zeman O., *Phys. Fluids A* **2**(2) (1990) 178.

[6] Sarkar S., Erlebacher G., Hussaini M. Y., Kreiss H. O., *J. Fluid Mech.* **227** (1991) 473.

[7] Kovasznay L. S. G., *J. Aero. Sci.* **20** (1953) 657.

[8] Passot T., Pouquet A., *J. Fluid Mech.* **181** (1987) 441.

[9] Blaisdell G.A., Mansour N.N., Reynolds W.C., *J. Fluid Mech.* **256** (1993) 443.

[10] Sarkar S., *J. Fluid Mech.* **282** (1995) 163.

[11] Sarkar S., *Phys. Fluids A* **4**(12) (1992) 2674.

[12] Zeman O., *Phys. Fluids A* **3**(5) (1990) 951.

[13] Zeman O., Coleman G.N., *In* "Proceedings of the Eighth Symposium on Turbulent Shear Flows" F. Durst, R. Friedrich, B. E. Launder, F. W. Schmidt, U. Schumann, and J. H. Whitelaw, Eds., (Springer-Verlag, Heidelberg, 1993), pp. 283–296.

[14] Cambon C., Coleman G.N., Mansour N.N., *J. Fluid Mech.* **257** (1993) 641.

[15] Durbin P.A., Zeman O., *J. Fluid Mech.* **242** (1992) 349.

[16] Jacquin L., Cambon C., Blin E., *Phys. Fluids A* **5**(10) (1993) 2539.

[17] Vreman B., Kuerton H., Geurts B., *Phys. Fluids* **7**(9) (1995) 2105.

[18] Lee S., Lele S.K., Moin P., *J. Fluid Mech.* **251** (1993) 533.

[19] Hannappel R., Friedrich R., *Appl. Sci. Res.* **54** (1995) 205.

[20] Coleman G.N., Kim J., Moser R.D., *J. Fluid Mech.* **305** (1995) 159.

[21] Huang P.G., Coleman G.N., Bradshaw P., *J. Fluid Mech.* **305** (1995) 185.

[22] Rai M.M., Gatski T.B., Erlebacher G., *AIAA Paper No. 95-0583* January 1995.

[23] Adams N.A., Shariff K., *J. Comp. Phys.* **127** (1996) 27.

[24] Hannappel R., Hauser T., Friedrich R., *J. Comp. Phys.* **121** (1995) 176.

[25] Spyropoulos E.T., Ph.D. Dissertation, Purdue University (1996).

[26] Yoshizawa A., *Phys. Fluids* **29**(7) (1986) 2152.

[27] Spyropoulos E.T., Blaisdell G.A., *AIAA J.* **34**(5) (1996) 990.

[28] Speziale C.G., Erlebacher G., Zang T.A., Hussaini M.Y., *Phys. Fluids* **31**(4) (1988) 940.

[29] Moin P., Squires K., Cabot W., Lee S., *Phys. Fluids A* **3**(11) (1991) 2746.

[30] Lilly D.K., *Phys. Fluids A* **4**(3) (1992) 633.

[31] Blaisdell G.A., Spyropoulos E.T., Qin J.H., *Appl. Num. Math.* **20** (1996) 1.

[32] El-Hady N.M., Zang T.A., *Theor. Comput. Fluid Dyn.* **7** (1995) 217.

[33] Ducros F., Comte P., Lesieur, M., *J. Fluid Mech.* **326** (1996) 1.

[34] Métais O., Lesieur M., *J. Fluid Mech.* **239** (1992) 157.

[35] Lesieur M., Métais O., Annu. Rev. Fluid Mech. 28 (Annual Reviews Inc., Palo Alto, 1996) p. 45.

[36] Spyropoulos E.T., Blaisdell G.A., *AIAA Paper No. 97-0429* January 1997.

[37] Speziale C.G., Sarkar S., *AIAA Paper No. 91-0217* January 1991.

[38] Coleman G.N., Mansour N.N., *Phys. Fluids A* **3**(9) (1991) 2255.

[39] Wilcox D.C., *AIAA J.* **30**(11) (1992) 2639.

[40] Huang P.G., Bradshaw P., Coakley T.J., *AIAA J.* **32**(4) (1994) 735.

[41] Lumley J.L., *In* "Advances in Applied Mechanics" **18**, C-S. Yih, Ed., (Academic Press, New York, 1978), pp. 123–176.

[42] Reynolds W.C., "Lecture Notes – Von Karman Institute AGARD Lecture Series 86," 1987.

[43] Speziale C.G., Gatski T.B., Sarkar S., *Phys. Fluids A* **4**(12) (1992) 2887.

[44] Speziale C.G., Annu. Rev. Fluid Mech. 23 (Annual Reviews Inc., Palo Alto, 1991) p. 107.

[45] Gatski T.B., Hussaini M.Y., Lumley J.L. (Eds.), Simulation and Modeling of Turbulent Flows (Oxford University Press, New York, 1996).

[46] Speziale C.G., Abid R., Mansour N.N., *ZAMP* **46** (1995) S717.

[47] Magnaudet J., *Appl. Sci. Res.* **51** (1993) 525.

[48] Gatski T.B., Speziale C.G., *J. Fluid Mech.* **254** (1993) 59.

[49] Abid R., Rumsey C., Gatski T.B., *AIAA J.* **33**(11) (1995) 2026.

[50] Abid R., Morrison J., Gatski T.B., *AIAA J.* (1996), to appear.

[51] Wilcox D.C., Turbulence Modeling for CFD (DCW Industries, Inc., La Cañada, California, 1993) pp. 183–189.

[52] Kline S.J., Cantwell B.J., Lilley G.M., (Eds.), "The 1980-81 AFOSR-HTTM-STANFORD on Complex Turbulent Flows: Comparison of Computation and Experiment," (Thermosciences Division, Mechanical Engineering Department, Stanford University, 1982) pp. 364–368.

[53] Vreman A.W., Sandham N.D., Luo K.H., *J. Fluid Mech.* **320** (1996) 235.

[54] Zeman O., *AIAA Paper No. 93-0897* January 1993.

[55] Taulbee D.B., VanOsdol J., *AIAA Paper No. 91-0524* January 1991.

[56] Rubesin M.W., NASA Contractor Report 177556, 1990.

[57] Sommer T.P., So R.M.C., Zhang H.S., *AIAA J.* **31**(1) (1993) 27.

[58] Huang P.G., Coakley T.J., *In* "Near-Wall Turbulent Flows" R. M. C. So, C. G. Speziale, and B. E. Launder, Eds., (Elsevier Science Publishers, Amsterdam, 1993), pp. 199–208.

[59] Ha Minh H., Rubesin M.W., Vandromme D., Viegas J.R., "International Symposium on Computational Fluid Dynamics," September 12–16, Tokyo, Japan, 1985.

[60] El Baz A.M., Launder B.E., *In* "Engineering Turbulence Modelling and Experiments 2" W. Rodi and F. Martelli, Eds., (Elsevier Science Publishers, Amsterdam, 1993), pp. 63–72.

[61] Cousteix J., Aupoix B., AGARD Report No. 764, 1990.

[62] Viegas J.R., Rubesin M.W., Horstman C.C., *AIAA Paper No. 85-0180* January 1985.

[63] Zhang H.S., So R.M.C., Gatski T.B., Speziale C.G., *In* "Near-Wall Turbulent Flows" R. M. C. So, C. G. Speziale, and B. E. Launder, Eds., (Elsevier Science Publishers, Amsterdam, 1993), pp. 209–218.

[64] Lai Y.G., So R.M.C., *J. Fluid Mech.* **221** (1990) 641.

[65] Durbin P.A., *J. Fluid Mech.* **249** (1993) 465.

Reynolds Averaged and Large Eddy Simulation Modeling for Turbulent Combustion

D. Veynante([1]) and T. Poinsot([2])

([1]) *Laboratoire E.M2.C.*
CNRS et Ecole Centrale Paris
Grande Voie des Vignes
92295 Chatenay - Malabry cedex
France
([2]) *IMFT et CERFACS*
Av. Camille Soula
31400 Toulouse cedex
France

1. INTRODUCTION

Combustion is a widely used technique in energy transformation and is encountered in many practical systems such as heaters, domestic or industrial furnaces, thermal power plants, automotive and aeronautic engines, rocket engines,... In most applications, combustion occurs in turbulent gaseous flows. Accordingly, the interaction between turbulence and combustion has to be described. Combustion phenomena may be characterized by:

- **a strong and irreversible heat release**. Heat release occurs in very thin zones (typical flame thicknesses δ_L are about 0.1 to 1 mm) and induces strong temperature gradients (temperature ratio between burnt and unburnt gases, T_b/T_u, are about 5 to 7) leading to strong heat transfers and large density variations.

- **a stiff highly non linear reaction rate** (Arrhenius law).

Various coupling mechanisms occur in combusting flow fields. **Chemical reaction schemes** have to be described to estimate the consumption rate of fuel, the formation of combustion products and pollutant species and are required to predict ignition, stabilization or extinction. **Mass transfers** of chemical species by molecular diffusion, convection and turbulent transport are also an important ingredient of combustion processes. The heat release due to chemical reactions induces strong conductive, convective or radiative **heat transfers** inside the flow field and with the surroundings (walls for example). This thermal energy is then used for heating or is converted into mechanical energy in a gas turbine or in a piston engine. Of course, gaseous combustion requires the description of the flow field (**fluid mechanics**). In some systems, other aspects may be also involved. Two (liquid fuel) and three (solid fuel) phase systems may be encountered. Phenomena such as spray formation and vaporization, droplet combustion,... have to be taken into account. Soot formation leads to carbon particles which will be transported by the flow motions.

The main difficulty in turbulent combustion modeling comes from the interaction between turbulent flow field and heat release. A large range of time and length scales is encountered: turbulent scales such as turbulent kinetic energy k, integral, l_t, and Kolmogorov, l_k, length scales,... and chemical scales (flame thickness, chemical time scales...). For example, in most pratical situations, the fuel oxidation reaction occurs in short times compared to turbulence times but other reactions such as pollutant formations have longer characteristic times.

The aim of this lecture is to propose a review of turbulent combustion modeling approaches. A first part (section 2) will be devoted to a brief description of laminar flames and how combustion may be affected by turbulent motions. The next section (3) is devoted to a summary of classical modeling of the mean flow field. A promising method to study turbulent combustion is based on large eddy simulation (LES) techniques: corresponding studies have started recently and will be described in Section 4. Direct numerical simulations (DNS) are also a powerful tool for combustion studies, both to analyze combustion phenomena and to develop and validate models. Two recent review papers [1] [2] and a book [3] have been devoted to this subject which will not be discussed here.

2. PREMIXED AND NON PREMIXED FLAMES

2.1. Introduction

Two limiting cases depending on the introduction of reactants in the combustion zone have been identified: perfectly premixed combustion and non premixed flames. In **premixed combustion**, the reactants, fuel and oxidizer are perfectly mixed before entering the reaction zone (Fig. 1a). This situation is favorable in terms of burning efficiency but a flame is able to propagate in premixed reactants upstream of the burner, leading to safety problems. In

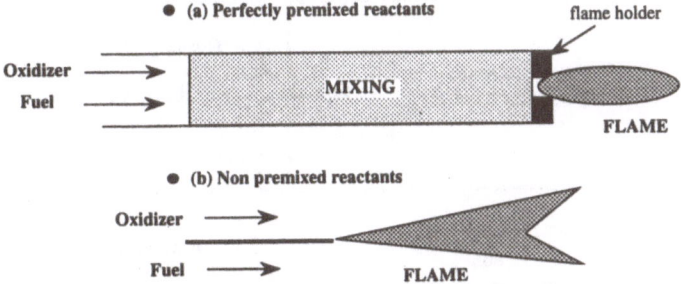

Fig. 1. — Two limiting reacting cases depending if reactants are initially premixed (a) or not (b)

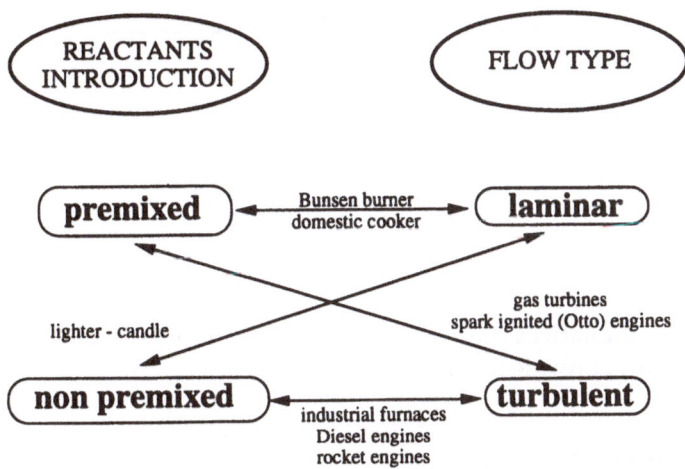

Fig. 2. — Classification of reacting flows depending on the introduction of reactants (premixed or not) and on the flow characteristics (laminar or turbulent). Corresponding practical systems are also indicated.

non premixed combustion, the reactants are introduced separately in the reaction zone. The simplest geometry of a non premixed (or diffusion) flame is a mixing layer (Fig. 1b). In this situation, the flame is unable to propagate upstream but the reaction rate may be limited by the mixing of the reactants, controlled by convection, molecular diffusion and turbulent transport.

A simple classification of combustion regimes is presented in Fig. 2. This description identifies four limiting cases (premixed, non-premixed, laminar, turbulent) and some corresponding practical systems are indicated.

Main characteristics of premixed and non-premixed laminar flames will be

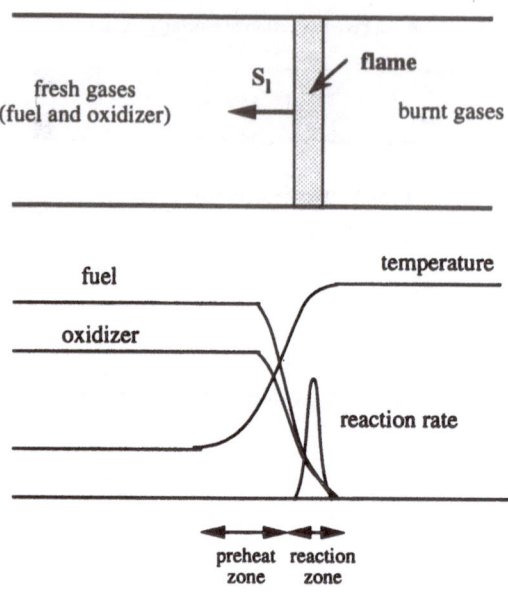

Fig. 3. — Structure of a laminar plane premixed flame.

now briefly summarized and some experimental results about corresponding turbulent flames provided.

2.2.   Premixed Flames

The structure of a laminar premixed flame is displayed on Fig. 3. Fresh gases (fuel and oxidizer) and burnt gases (combustion products) are separated by a thin reaction zone. A strong temperature gradient is observed (typical ratios between burnt and fresh gases temperatures are about 5 to 7). The main characteristic of a premixed flame is its ability to propagate towards the fresh gases. Because of the temperature gradient and the corresponding thermal fluxes, fresh gases are preheated and then start to burn. The propagation speed S_L of a laminar flame depends on various parameters (fuel and oxidizer compositions, fresh gases temperature,...) and is about 0.1 to 1 m/s.

With turbulent fresh gases, a propagating premixed flame is also observed but the turbulent propagation speed S_T is larger than the corresponding laminar flame speed S_L ($S_T \gg S_L$). The thickness of the turbulent flame brush, δ_T is greater than the laminar flame thickness δ_L ($\delta_T \gg \delta_L$). The following relation is generally proposed from experimental data [4] [5] or theoretical arguments [6]:

$$\frac{S_T}{S_L} = 1 + \alpha \left(\frac{u'}{S_L} \right)^n \tag{1}$$

where α and n are two constants of the order of 1.

The turbulent flame speed S_T depends both on the laminar flame speed S_L and the rms velocity u'. S_T is found to increase with the turbulence level, at least up to the observation of a small decrease just before the flame extinction (see [4] [5]). Unfortunately, S_T is not a well defined quantity [7] and depends on various parameters (chemistry characteristics, flow geometry,...).

As observed by Damköhler [8], the turbulent flame speed S_T is linked to the flame front wrinkling:

$$\frac{S_T}{S_L} = \frac{A_T}{A_L} \tag{2}$$

where A_T/A_L is the ratio of the surfaces of turbulent (A_T) and laminar (A_L) flames and corresponds to the wrinkling of the flame front due to turbulent motions.

2.3. Non Premixed Flames

In a non premixed (or diffusion) flame, reactants (fuel and oxidizer) are separated by the flame front. Species and temperature profiles are displayed on Fig. 4. The reaction rate is mainly controlled by molecular diffusion processes and the flame structure depends on the Damköhler number (ratio of chemical and molecular diffusion time scales). Such flames are unable to propagate because combustion may only occur where fuel and oxidizer coexist.

Let us now consider a simple chemical reaction between fuel (F) and oxidizer (O):

$$F + sO \rightarrow (1 + s)P \tag{3}$$

where s is the mass stoechiometric ratio. The balance equation for the fuel (Y_F) and oxidizer (O) mass fractions are respectively:

$$\frac{\partial \rho Y_F}{\partial t} + \frac{\partial \rho u_i Y_F}{\partial x_i} = \frac{\partial}{\partial x_i} \left(\rho D_F \frac{\partial Y_F}{\partial x_i} \right) + \dot{\omega}_F \tag{4}$$

$$\frac{\partial \rho Y_O}{\partial t} + \frac{\partial \rho u_i Y_O}{\partial x_i} = \frac{\partial}{\partial x_i} \left(\rho D_O \frac{\partial Y_O}{\partial x_i} \right) + \dot{\omega}_O \tag{5}$$

where the species molecular diffusions are described using the Fick's law. $\dot{\omega}_F$ and $\dot{\omega}_O$ are respectively the fuel and oxidizer reaction rates.

A new parameter, Z, may be defined as:

$$Z = \frac{\frac{\phi Y_F}{Y_F^\infty} - \frac{Y_O}{Y_O^\infty} + 1}{\phi + 1} \tag{6}$$

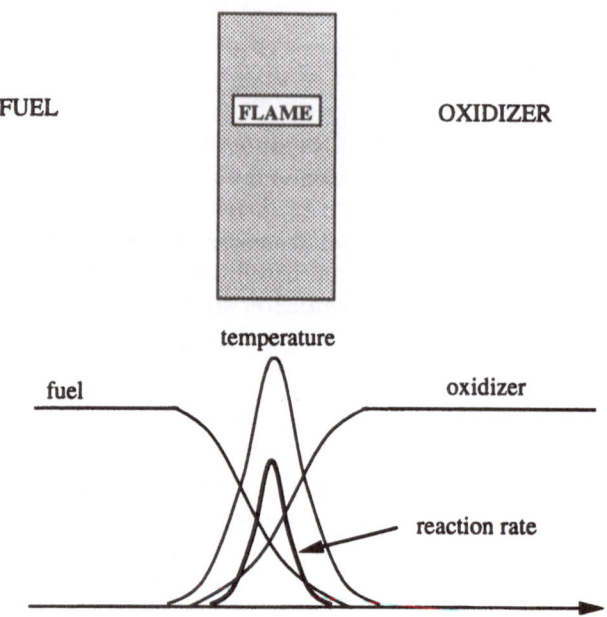

FUEL   FLAME   OXIDIZER

Fig. 4. — Structure of a non premixed laminar flame.

where Y_F^∞ is the fuel mass fraction in the pure fuel flow and Y_O^∞ the oxidizer mass fraction in the oxidizer flow. ϕ is the chemical equivalence ratio, defined as:

$$\phi = \frac{sY_F^\infty}{Y_O^\infty} \tag{7}$$

Assuming that the molecular diffusion are the same for the fuel and the oxidizer (i.e. $\rho D_F = \rho D_O = \rho D$), Z is a **passive scalar** (also called Schwab-Zeldovitch variable) and follows a transport equation without source terms:

$$\frac{\partial \rho Z}{\partial t} + \frac{\partial \rho u_i Z}{\partial x_i} = \frac{\partial}{\partial x_i}\left(\rho D \frac{\partial Z}{\partial x_i}\right) \tag{8}$$

The passive scalar Z has the following properties:

- $Z = 1$ in the pure fuel flow.
- $Z = 0$ in the pure oxidizer flow
- Under the assumption of a fast chemistry (fuel and oxidizer cannot coexist at the flame front where $Y_F = Y_O = 0$), the flame is located on the $Z_{st} = 1/(\phi + 1)$-isolevel surface.

Species and temperature transport equations may be recast in a new reference frame where Z is chosen as an independant variable (see [9] for details).

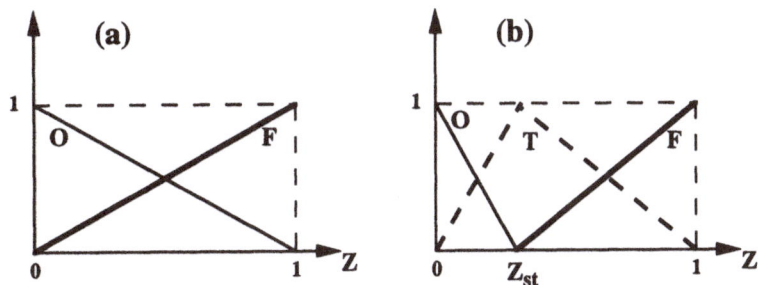

Fig. 5. — Reduced species mass fraction (F: fuel; O: oxidizer) and temperature (T) as a function of the passive scalar Z. (a) mixing of inert reactant. (b) infinitely fast chemistry (Burke and Schumann solution).

Assuming that only gradients normal to the Z iso-surface are non negligible, Eq. (4) and (5) become:

$$\rho \frac{\partial Y_k}{\partial t} = \chi \frac{\partial^2 Y_k}{\partial Z^2} + \dot{\omega}_k \tag{9}$$

where the scalar dissipation χ of the passive scalar Z is introduced:

$$\chi = \rho D \left(\frac{\partial Z}{\partial x_i} \right)^2 = \rho D \, |\nabla Z|^2 \tag{10}$$

This analysis shows that a diffusion flame may be described using a passive scalar Z following a transport equation without source terms. Species mass fractions and temperature are then related to the Z field and its scalar dissipation χ through Eq. (9). Under steady state operations, in two particular cases (mixing of inert reactants and combustion with infinitely fast chemistry), species mass fractions and temperature, solutions of Eq. (9), are the linear functions of Z displayed on Fig. 5.

The influence of a turbulent flow on a diffusion flame may be analyzed from the "jet-flame". A fuel jet (diameter d_0, velocity u_0) is discharged in the atmosphere. Experimental data show that the flame length L_f depends on the jet Reynolds number $Re = u_0 d_0 / \nu$ (where ν is the jet flow viscosity) as displayed on Fig. 6.

When the flow is laminar, the flame length increases and is proportional to the Reynolds number Re. Then, for a sufficiently high Reynolds number, a transition is observed and the flame tip becomes turbulent. As Re continues to increase, the flame length L_f decreases and reaches a constant value. The transition point between laminar and turbulent flame moves upstream, closer to the jet exit. When the Reynolds number increases, the fuel mass flow rate also increases. The constant value of the turbulent flame length L_f when Re increases clearly shows that the reaction rate is enhanced by turbulent motions.

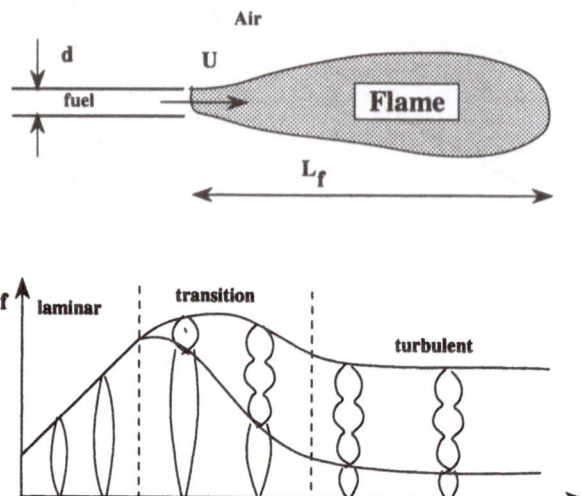

Fig. 6. — Flame length L_f of a jet flame as a function on the jet Reynolds number Re.

3.   MODELING OF AVERAGED TRANSPORT EQUATIONS

3.1.   Introduction

Numerical simulations of practical systems are generally based on averaged balance equations for momentum, species mass fractions and energy. The objective of modeling is to propose phenomenological closures for these equations. Compared to non-reacting constant density flows, additional difficulties occur in combustion because of the strong heat release (leading to high temperature and density gradients) and the stiff non-linear reaction rate.

3.1.1.   Favre (Mass Weighted) Averaging

Splitting any quantity Q in averaged $\overline{Q}$ and fluctuating Q' parts, $Q = \overline{Q} + Q'$ (Reynolds averaging), the averaged continuity equation may be written:

$$\frac{\partial \overline{\rho}}{\partial t} + \frac{\partial}{\partial x_i}\left(\overline{\rho}\,\overline{u_i} + \overline{\rho' u_i'}\right) = 0 \tag{11}$$

where velocity and density fluctuations correlations $\overline{\rho' u_i'}$ needs to be described. To avoid the explicit modeling of such correlations, a Favre (mass weighted) average $\tilde{Q}$ is introduced and any quantity is then splitted as $Q = \tilde{Q} + Q''$:

$$\tilde{Q} = \frac{\overline{\rho Q}}{\overline{\rho}} \quad ; \quad \widetilde{Q''} = \frac{\overline{\rho\left(Q - \tilde{Q}\right)}}{\overline{\rho}} = 0 \tag{12}$$

Then, the Favre averaged continuity equation:

$$\frac{\partial \overline{\rho}}{\partial t} + \frac{\partial \overline{\rho}\tilde{u}_i}{\partial x_i} = 0 \tag{13}$$

is formally identical to the Reynolds averaged continuity equation for constant density flows. This result is true for any balance equations (momentum, energy, mass fractions...). Nevertheless, Favre averaging is **only a mathematical formalism** leading to the following comments:

- There is no simple relation between Favre, $\tilde{Q}$, and Reynolds, $\overline{Q}$, averages. A relation between $\tilde{Q}$ and $\overline{Q}$ requires a description of density fluctuations correlations $\overline{\rho' Q'}$ which remain hidden in Favre averaged quantities:

$$\overline{\rho}\tilde{Q} = \overline{\rho}\,\overline{Q} + \overline{\rho' Q'} \tag{14}$$

- Comparisons between numerical simulations, providing Favre averaged quantities $\tilde{Q}$ with experimental results are not obvious. Most experimental techniques lead to Reynolds averaged data $\overline{Q}$ and differences between $\tilde{Q}$ and $\overline{Q}$ may be significant.

- Favre turbulent fluxes such as $\widetilde{u_i'' u_j''}$, $\widetilde{u_i'' Y_k}$, ... remain to be modeled taking into account heat release and density effects.

3.1.2. *Balance Equations for Mass Fractions*

The averaged conservation equation for the mass fraction Y_k of the species k may be written :

$$\frac{\partial \overline{\rho}\widetilde{Y_k}}{\partial t} + \frac{\partial \overline{\rho}\tilde{u}_i\widetilde{Y_k}}{\partial x_i} + \frac{\partial \overline{\rho}\widetilde{u_i'' Y_k''}}{\partial x_i} = \overline{\frac{\partial \mathcal{J}_{k,j}}{\partial x_j}} + \overline{\dot{\omega}}_k \tag{15}$$

where ρ is the mass density, u_i the flow velocity components. $\mathcal{J}_k$ is the molecular diffusion flux and $\dot{\omega}_k$ is the volumetric production rate of the chemical reaction of species k. The three terms in the left-hand side correspond respectively to unsteady effects, convection by the mean flow, turbulent transport of the species k. The two terms in the right hand side correspond respectively to molecular diffusion and reaction rate of the species k.

The contribution of molecular diffusion $\mathcal{J}_k$ is usually neglected for high Reynolds number flows. In Eq.(15), two terms need to be modeled: the mean reaction rate $\overline{\dot{\omega}}_k$ and the turbulent transport $\overline{\rho}\widetilde{u_i'' Y_k''}$ terms. The first term has received considerable attention in recent years and various models, briefly described in the following, have been derived and incorporated into practical codes for turbulent combustion. The second term, however, has received considerably less attention and is generally described with a simple classical gradient eddy-viscosity model:

$$\overline{\rho u_i'' Y_k''} = \bar{\rho} \widetilde{u_i'' Y_k''} = -\frac{\mu_t}{\sigma_k} \frac{\partial \widetilde{Y}_k}{\partial x_i} \tag{16}$$

where μ_t denotes the turbulent dynamic viscosity and σ_k a turbulent Schmidt number.

From theoretical and experimental research [10] [11], it is known that counter-gradient transport takes place in some turbulent premixed flames: flames where the turbulent flux $\overline{\rho u_i'' Y_k''}$ and the $\widetilde{Y}_k$ gradient, $\partial \widetilde{Y}_k / \partial x_i$, have the same sign in certain regions and cannot be described with Eq. (16). This is generally due to the differential effect of pressure gradients on cold reactants and hot products. Recent studies based on direct numerical simulations of turbulent premixed flames without [12] and with [13] externally imposed pressure gradients have confirmed that counter-gradient diffusion was found in simulations, but that classical gradient diffusion was also possible. A criterion to delineate between gradient and counter-gradient diffusion situations was derived from these studies, showing that counter gradient may occur in practical situations. Counter-gradient transport is also analyzed in terms of coherent structure motions in [19]. Without any externally imposed pressure gradient, counter-gradient turbulent diffusion has to be expected when the Bray number N_B is higher than unity:

$$N_B = \frac{\tau}{2\alpha \dfrac{u'}{S_L}} \leq 1 \tag{17}$$

where τ is the heat release factor ($\tau = T_b/T_u - 1$ where T_u and T_b are respectively the temperature of fresh and burnt gases) and u'/S_L the velocity ratio between the RMS velocity u' and the laminar flame speed S_L. α is an efficiency function taking into account the fact that smaller turbulent length scales are unable to affect the flame front. α is of order unity for large values of the length scale ratio l_t/δ_L and decreases towards 0 for low length scale ratios [12]. Counter-gradient turbulent transport is promoted by high values of the heat release factor τ and by favorable externally imposed pressure gradients (i.e. pressure decreasing from fresh gases to burnt gases) [13] whereas high turbulence levels and adverse pressure gradients promote classical gradient turbulent diffusion.

Turbulent transport plays an important role in the propagation of turbulent premixed flames [14]. Accordingly, a more refined description of turbulent fluxes than proposed in Eq. (16) is required. It is generally based on the closure of exact balance equations for $\bar{\rho} \widetilde{u_i'' Y_k''}$ [10] [15] [16]. Few attempts of such an approach are up to now available in practical situations [17] [18].

3.2. A First Simplified Analysis: Taylor's Expansion

We consider a simple irreversible reaction between fuel (F) and oxidizer (O): $F + sO \rightarrow (1+s)P$ where the fuel mass reaction rate $\dot{\omega}_F$ is expressed from the Arrhenius law as:

$$\dot{\omega}_F = -A\rho^2 T^b Y_F Y_O \exp\left(-\frac{E}{RT}\right) \tag{18}$$

where Y_F and Y_O are respectively the fuel and the oxidizer mass fractions, A the pre-exponential constant, E the activation energy and R the perfect gases constant.

As the reaction rate is highly non-linear, the averaged reaction rate $\overline{\dot{\omega}}_F$ cannot be easily expressed as a function of the mean mass fractions $\widetilde{Y}_F$ and $\widetilde{Y}_O$, the mean density $\overline{\rho}$ and the mean temperature $\widetilde{T}$. The first idea is to write the mean reaction rate $\overline{\dot{\omega}}_F$ as a serie expansion:

$$\overline{\dot{\omega}}_F = -A\overline{\rho}^2 \widetilde{T}^b \widetilde{Y}_F \widetilde{Y}_O \exp\left(-\frac{E}{R\widetilde{T}}\right) \left[1 + \frac{\widetilde{Y_F'' Y_O''}}{\widetilde{Y}_F \widetilde{Y}_O} + (P_1 + Q_1)\left(\frac{\widetilde{Y_F'' T''}}{\widetilde{Y}_F \widetilde{T}} + \frac{\widetilde{Y_O'' T''}}{\widetilde{Y}_O \widetilde{T}}\right) \right.$$
$$\left. + (P_2 + Q_2 + P_1 Q_1)\left(\frac{\widetilde{Y_F'' T''^2}}{\widetilde{Y}_F \widetilde{T}^2} + \frac{\widetilde{Y_O'' T''^2}}{\widetilde{Y}_O \widetilde{T}^2}\right) + \dots \right] \tag{19}$$

where

$$P_n = \sum_{k=1}^{n} (-1)^{n-k} \frac{(n-1)!}{(n-k)! \left[(k-1)!\right]^2 k} \left(\frac{E}{R\widetilde{T}}\right)^k \tag{20}$$

$$Q_n = \frac{b(b+1)\dots(b+n-1)}{n!} \tag{21}$$

Equation (19) leads to various difficulties. First, new quantities such as $\widetilde{Y_k'' T''^n}$ have to be closed, using algebraic expressions or transport equations. Because of non linearities, large truncature errors are introduced when only few terms of the series expansion are taken into account. Expression (19) is quite complicated but is only valid for a simple irreversible reaction and cannot be easily extended to realistic chemistry schemes (at least 9 species and 19 reactions for hydrogen combustion, several hundred species and several thousand reactions for hydrocarbon combustion). For these reasons, reaction rate closures in turbulent combustion are not based on Eq. (19) but are derived from physical analysis as described in the next section.

Nevertheless, this approach is used in some practical simulations, for example in supersonic reacting flow fields, where chemical times cannot be neglected against mechanical times [20] or to describe reaction in atmospheric boundary layer [21]. In these situations, only the first two terms in the series expansion are taken into account and the unknown quantity $I_s = \widetilde{Y_F'' Y_O''}/\widetilde{Y}_F \widetilde{Y}_O$, called segregation factor, is postulated or is provided from a balance equation (see [22] in a large eddy simulation context).

3.3.   Physical Analysis

As the mean reaction rate $\bar{\dot{\omega}}_k$ cannot be found from an averaging of Arrhenius laws, a physical approach is required to develop turbulent combustion models. As already mentioned, turbulent combustion involves various lengths, velocity and time scales describing turbulent flow field and chemical reactions. The Damköhler number compares the turbulent time scale τ_t and the chemical time scale τ_c :

$$Da = \frac{\tau_t}{\tau_c} \tag{22}$$

In the limit of high Damköhler numbers ($Da \gg 1$), the chemical time is short compared to the turbulent time, corresponding to a thin reaction zone distorted and convected by the flow field. The internal structure of the flame is not affected by turbulence and may be described as a laminar flame element called *flamelet*. The turbulent structures wrinkle and strain the flame surface. On the other hand, a low Damköhler ($Da \ll 1$) corresponds to a slow chemical reaction. Reactants and products are mixed by turbulent structures before reaction. In this *perfectly stirred reactor* limit, the mean reaction rate may be expressed from Arrhenius laws using mean mass fractions and temperature. Most practical situations correspond to high to medium values of the Damköhler numbers. It is worth noting that various chemical time scales may be encountered: fuel oxidation generally corresponds to short chemical time scales ($Da \gg 1$) whereas pollutant production or destruction such as CO oxidation or NO formation are slower ($Da \approx 1$).

As previously described, two limiting cases, depending on the introduction of reactants in the combustion chamber are also distinguished in the physical description of turbulent combustion: premixed and non-premixed situations (see Fig. 2).

In **turbulent premixed flames**, the chemical time scale τ_c may be estimated as the ratio between the flame thickness δ_L and the laminar flame speed S_L. The turbulent time scale corresponds to the integral length scale l_t and is estimated as $\tau_t = l_t/u'$ where u' is the rms velocity (or the square root of the turbulent kinetic energy). Then, the Damköhler number becomes:

$$Da = \frac{\tau_t}{\tau_c} = \frac{l_t}{\delta_l} \frac{S_L}{u'} \tag{23}$$

where velocity (u'/S_L) and length scale (l_t/δ_L) ratios are identified. For large values of the Damköhler number ($Da \gg 1$), the flame front is thin and its structure is not affected by turbulence motions. This regime is called *flamelet regime* or *thin wrinkled flame regime* (Fig. 7a). As the turbulence level increases, turbulent structures become able to affect the preheat zone of the flame (Fig. 7b), corresponding to the *thickened-wrinkled flame regime*. If the turbulence level continues to increase (and the Damköhler number to decrease), turbulent motions becomes sufficiently strong to affect the whole flame structure. It is the *thickened flame regime* (Fig. 7c). These regimes may be identified on a combustion diagram (Fig. 8). This analysis is due to Borghi and Destriau [23] but similar classification may be found in the litterature [9] [24] [25] [26]. Such classifications are only qualitative and cannot be used to determine a combustion regime from the knowledge of turbulent (l_t and u') and chemical (δ_L and S_L) scales. Various assumptions are implicitly made to derive combustion diagrams. For example, turbulence is supposed to be homogeneous, isotropic and unaffected by heat release. Both direct numerical simulations [27] and experiments [28] show that the flamelet regime is larger than predicted by these qualitative analysis. An essential mechanism revealed by DNS is that small turbulent scales, which are supposed in classical theories to have the strongest effects on flames, have small lifetimes because of viscous dissipation and therefore only limited effects on combustion. Fig. 8 shows the flamelet regime limit obtained from Poinsot *et al.* [27] using DNS and the criterion delineating between gradient and counter gradient turbulent diffusion (§ 3.1.2) proposed by Veynante *et al.* [12].

Identification of combustion regimes is more complicated in non premixed turbulent flames because the estimate of a chemical time τ_c cannot be correlated to a flame propagation speed S_L and the flame thickness δ_L is not well defined. Nevertheless some classifications have been proposed [23] [29] but will not be discussed here.

3.4. Premixed Combustion

Various models have been proposed to describe the mean reaction rate $\overline{\dot{\omega}}_F$ in turbulent premixed combustion. A brief summary of the main approaches is proposed below.

3.4.1. Turbulent Flame Speed

Turbulent flames are often described in terms of "turbulent flame speed" S_T using Eq. (1). Unfortunately, the turbulent flame speed S_T is not a well defined quantity and depends on various parameters (chemistry characteristics, flow geometry, ...). While this global approach is not well suited to close Favre averaged transport equations, it constitutes an attractive description for Large Eddy Simulations to describe the flame front displacement (see section 4).

3.4.2. Algebraic Models: Eddy-Break-Up Model

The Eddy-Break-Up (EBU) model, devised by Spalding [30], is used in many industrial applications because of its simplicity. When oxidizer is in excess, the fuel reaction rate is expressed as :

$$\overline{\dot{\omega}}_F = C_{EBU}\bar{\rho}\frac{\varepsilon}{k}\frac{\tilde{Y}_F}{Y_F^0}\left(1 - \frac{\tilde{Y}_F}{Y_F^0}\right) \tag{24}$$

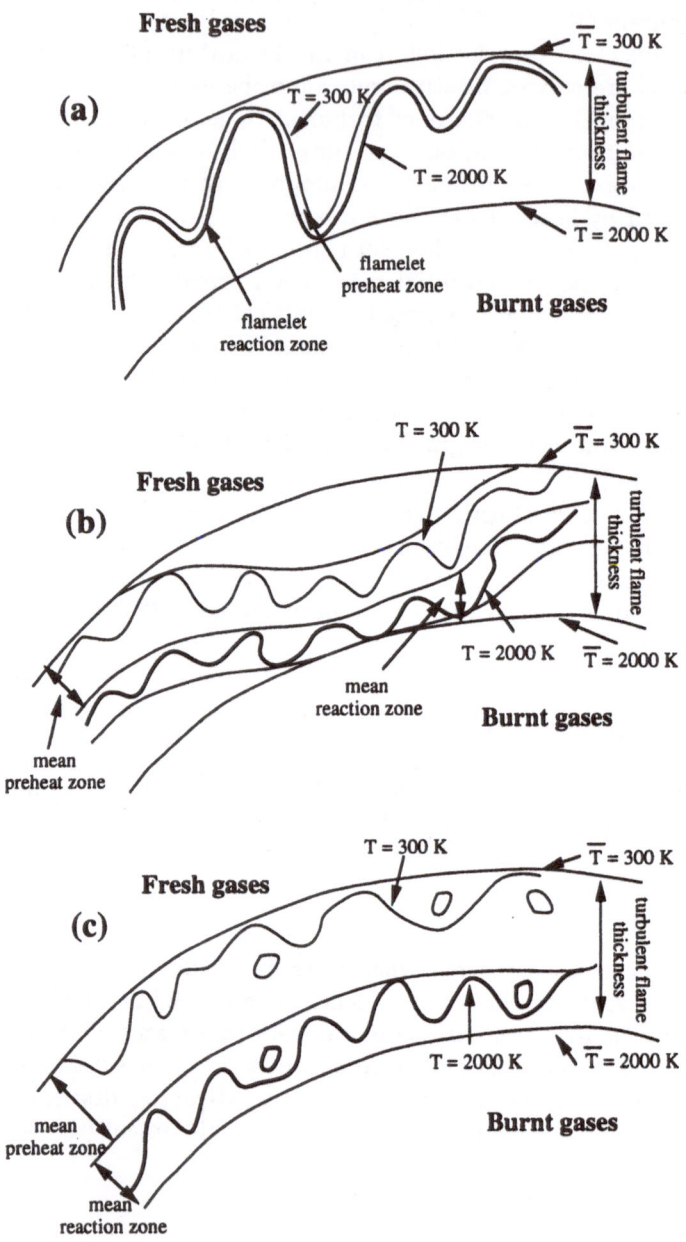

Fig. 7. — Turbulent premixed combustion regimes identified by Borghi and Destriau (1995). (a) Thin wrinkled flame (flamelet) regime. (b) thickened-wrinkled flame regime. (c) thickened flame regime.

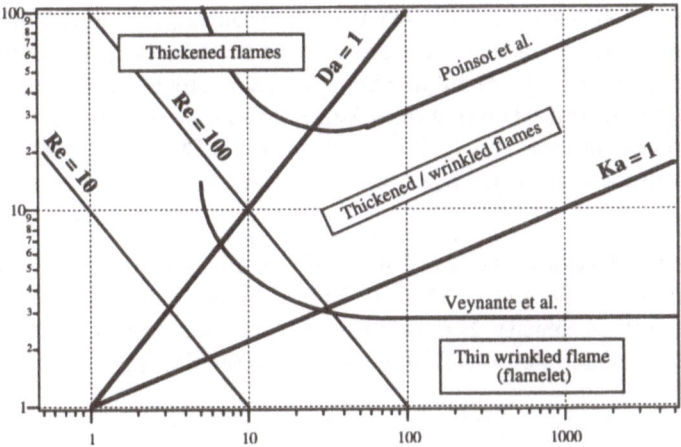

Fig. 8. — Turbulent premixed combustion diagram (Borghi and Destriau, 1995). Flame regimes are identified as a function of the length scale (l_t/δ_L) and velocity (u'/S_L) ratios. The Klimov-Williams criterion ($Ka = 1$) corresponds to a flame thickness δ_L equal to the Kolmogorov turbulent length scale l_k. Below this line, the flame is thinner than any turbulent length scales. The flamelet limit proposed by Poinsot *et al.* (1991) is also plotted. The criterion proposed by Veynante *et al.* to delineate between gradient (above) and counter-gradient (below) turbulent diffusion is displayed for a heat release factor $\tau = T_b/T_u - 1 = 6$ where T_u and T_b are respectively the fresh and burnt gases temperature.

where Y_F^0 is the fuel mass fraction in fresh gases, k and ε are respectively the turbulent kinetic energy and its dissipation rate. C_{EBU} is a model constant. In this approach, the reaction rate $\overline{\dot{\omega}}_F$ is assumed to be proportional to the variance $\widetilde{Y_F''^2}$ of the fuel mass fraction and inversely proportional to a turbulent time $\tau_t = k/\varepsilon$.

This model is attractive because the reaction rate is simply written as a function of known quantities without any additional transport equation. Nevertheless, this reaction rate does not depend on chemical characteristics. The Eddy-Break-Up model also tends to overestimate the reaction rate, especially in highly strained regions (large values of ε/k).

3.4.3. Flame Surface Density Approaches

In the limit of high Damköhler numbers, the reaction zone may be viewed as a collection of thin reaction layers having the structure of laminar flames ("flamelets"). Following this analysis, the mean reaction rate of the species k may be modeled as :

$$\overline{\dot{\omega}}_k = Q_k \Sigma \tag{25}$$

where Σ is the flame surface density (i.e. flame surface per unit volume) and Q_k the reaction rate of the species k per unit of flame area. The main advantage of this formulation is to decouple the chemical problem, described by the local reaction rate Q_k and estimated from laminar flame studies, from the interaction between flame surface and turbulent flow field. The flame surface density Σ may be estimated from an algebraic expression or as a solution of a balance equation.

3.4.3.1. **Algebraic closures.** Two kinds of closures have been proposed for the flame surface density Σ. In the Bray - Moss - Libby (BML) model [31], the flame surface density Σ is derived from the analysis of the intermittency between fresh reactants and fully burnt gases, leading to:

$$\Sigma = g\frac{\bar{c}(1-\bar{c})}{\sigma_y L_y} = \frac{g}{\sigma_y L_y}\frac{1+\tau}{(1+\tau\tilde{c})^2}\tilde{c}(1-\tilde{c}) \tag{26}$$

where c is a reduced mass fraction (or temperature), with $c = 0$ in fresh reactants and $c = 1$ in fully burnt ones. The parameter σ_y is an orientation factor assumed to be constant ($\sigma_y \approx 0.6$), g a model constant of the order of unity and L_y the wrinkling length scale of the flame front. $\tau = T_b/T_u - 1$ is the heat release factor, where T_u and T_b are respectively the fresh and burnt gases temperatures. A submodel is required for L_y, generally assumed to be proportional to the turbulent integral length scale l_t:

$$L_y = C_l l_t \left(\frac{S_L}{u'}\right)^n \tag{27}$$

where C_l and n are two constants of the order of unity.

One has to note that the BML approach proposes a simple algebraic closure for Σ but the attention is focussed on a precise description of the turbulent fluxes $\overline{\rho u_i'' Y_k''}$ to predict the occurence of counter-gradient turbulent transport.

In a second approach, the flame front is described as a fractal surface [32]:

$$\Sigma = \frac{1}{L_{outer}}\left(\frac{L_{outer}}{L_{inner}}\right)^{D-2} \tag{28}$$

where L_{inner} and L_{outer} are respectively the inner and outer cut-off scales and D the fractal dimension of the flame front. The cut-off scales are generally respectively related to integral turbulence length scale l_t and Kolmogorov length scale l_k. In a recent review paper, Gülder and Smallwood [33] discuss scaling for flame inner cut-off.

3.4.3.2. **A transport equation for Σ.** First devised by Marble and Broadwell [34] for non-premixed turbulent combustion, this approach is now widely used to model turbulent premixed combustion. Recent theoretical studies have lead to an exact transport equation for the density of a propagating surface [35] [36] [37] [38] [39]:

$$\frac{\partial \Sigma}{\partial t} + \nabla.(\tilde{u}\Sigma) + \nabla.(\langle u'' \rangle_s \Sigma) = (\nabla.\tilde{u} - \langle nn \rangle_s : \nabla\tilde{u}) \Sigma$$
$$+ \langle \nabla.u'' - nn : \nabla u'' \rangle_s \Sigma$$
$$- \nabla.(\langle wn \rangle_s \Sigma) + \langle w\nabla.n \rangle_s \Sigma \qquad (29)$$

where n is the vector normal to the flame front pointing towards the fresh gases. The flame front curvature, $\nabla.n$, is chosen positive convex towards the fresh gases. $\langle \rangle_s$ denotes the average along the flame front surface as defined by Pope [35]. The two last terms on the left-hand side of Eq.(29) represent convection by the mean flow and turbulent diffusion. The terms on the right-hand side correspond respectively to the tangential strain rate acting on the flame surface and due to the mean flow field, turbulent strain rate, flame propagation and combined effects of flame curvature and propagation. The propagation speed w of the flame surface depends on laminar flame characteristics and includes complex stretch and curvature effects. It may be quite different from the laminar flame speed S_L [40]. A summary and a comparison of various closure schemes for the Σ-equation (29) may be found in [14]. Recent developments are discussed in several papers [19] [38] [39] [41].

3.4.4. Probability Density Functions

A statistical description of the flow field may be founded on probability density functions (PDF). Defining the probability density function $P(\Psi_1, \Psi_2 \ldots, \Psi_N)$ as:

$$\text{Probability } (\Psi_1 \leq Y_1 < \Psi_1 + d\Psi_1, \ldots, \Psi_N \leq Y_N < \Psi_N + d\Psi_N) =$$
$$P(\Psi_1, \ldots, \Psi_N) \, d\Psi_1 \ldots d\Psi_N \qquad (30)$$

for a given location, with the normalization relation:

$$\int_{\Psi_1, \Psi_2, \ldots \Psi_N} P(\Psi_1, \Psi_2 \ldots, \Psi_N) \, d\Psi_1 \, d\Psi_2 \ldots d\Psi_N = 1 \qquad (31)$$

where $Y_1, Y_2, \ldots, Y_N$ are the thermochemical variables (mass fractions, temperature,...), any averaged quantity $\overline{Q}$, for example the species mass fractions or the mean reaction rate $\overline{\dot{\omega}}_k$, may be determined as:

$$\overline{Q} = \int_{\Psi_1, \Psi_2, \ldots \Psi_N} Q(\Psi_1, \ldots, \Psi_N) \, P(\Psi_1, \Psi_2 \ldots, \Psi_N) \, d\Psi_1 \, d\Psi_2 \ldots d\Psi_N \qquad (32)$$

This stochastic description has many theoretical advantages. Probability density functions may be defined in any turbulent reacting flow field without any hypothesis. These functions may also be extracted from experimental data or direct numerical simulations. Two main paths are available to determine $P(\Psi_1, \Psi_2 \ldots, \Psi_N)$:

3.4.4.1. Presumed probability density functions. In this approach, the shape of the PDF is a priori assumed [24]. The probability density function $p(c)$, where c is the progress variable, is generally supposed to depend only on the mean progress variable $\tilde{c}$ and its variance $\widetilde{c''^2}$ as a β-function:

$$p(c) = \frac{1}{B(a,b)} c^{a-1} (1-c)^{b-1} \tag{33}$$

where $B(a,b)$ is a normalization factor:

$$B(a,b) = \int_0^1 c^{a-1} (1-c)^{b-1} \, dc \tag{34}$$

The PDF parameters a and b are determined from $\tilde{c}$ and $\widetilde{c''^2}$ as:

$$a = \tilde{c}\left[\frac{\tilde{c}(1-\tilde{c})}{\widetilde{c''^2}} - 1\right] \quad ; \quad b = \frac{a}{\tilde{c}} - a \tag{35}$$

This simplified approach is used in some industrial simulations. Additional balance equations for fluctuating quantities $\widetilde{c''^2}$ have to be modeled and solved.

3.4.4.2. Transport equation for the probability density function P. An exact transport equation for PDF may be derived [42] [43]:

$$\bar{\rho}\frac{\partial \tilde{P}}{\partial t} + \bar{\rho}\tilde{U}_k \frac{\partial \tilde{P}}{\partial x_k} = \quad - \quad \underbrace{\frac{\partial}{\partial x_k}\left[\bar{\rho}\,\overline{(U_i''|\underline{Y} = \underline{\Psi})}\,\tilde{P}\right]}_{\text{Turbulent convection}}$$

$$- \quad \underbrace{\bar{\rho}\sum_{i=1}^{N}\sum_{j=1}^{N}\frac{\partial}{\partial \Psi_i}\frac{\partial}{\partial \Psi_j}\left[\overline{D\left(\frac{\partial Y_i}{\partial x_k}\frac{\partial Y_j}{\partial x_k}|\underline{Y} = \underline{\Psi}\right)}\,\tilde{P}\right]}_{\text{Molecular mixing}}$$

$$- \quad \underbrace{\bar{\rho}\sum_{i=1}^{N}\frac{\partial}{\partial \Psi_i}\left(\dot{\omega}_i(\Psi_1, \Psi_2, \ldots, \Psi_N)\,\tilde{P}\right)}_{\text{Chemical reaction}} \tag{36}$$

where $\overline{Q|\underline{Y} = \underline{\Psi}}$ correspond to conditional averaging of quantity Q for the sampling values Ψ_i.

The reaction term of this equation is closed and does not require any modeling but the modeling of the unclosed mixing diffusion term is not obvious. The numerical solution of the PDF-equation is generally based on Monte-Carlo algorithms and is very time consuming. The above PDF balance equation (36) involves only thermochemical quantities. In a more refined approach, joint velocity-thermochemical variables probability density functions may be considered leading to a more complex transport equation.

3.4.4.3. Comments:

- Recent studies [35] [39] have shown that flame surface density and PDF approaches are related. In fact, Σ may be defined as the density of the iso-level $c = c^*$ surface, where c is the progress variable. Then:

$$\Sigma = \overline{|\nabla c| \delta(c - c^*)} = \langle |\nabla c|, c = c^* \rangle p(c^*) \qquad (37)$$

where δ is the Dirac function and $p(c^*)$ is the probability that $c = c^*$. Accordingly, the Σ-equation may be viewed, and closed, as a PDF transport equation [39].

- Probability density function models may be coupled with other approaches. For example, fuel oxidation may be described with a simple algebraic expression or a flame surface density model whereas pollutant formation, having a longer chemical characteristic time, is predicted using a presumed PDF analysis.

3.5. Non Premixed Combustion

Less attention has been focused on non premixed turbulent combustion than on premixed turbulent combustion. In fact, in most practical situations, the reaction may be assumed to be very fast compared to mixing and diffusion times (large Damköhler numbers) corresponding to thin combustion layers ("flamelet"). Non-premixed combustion may also be described using passive scalar variables as previously described (see section 2.3). Accordingly, non-premixed turbulent combustion may be viewed as a mixing problem.

Models for non-premixed combustion are quite similar to those developed in the premixed case and differ only by their ingredients. These models may be summarized as follows:

3.5.1. Algebraic Expressions

The Magnussen model, similar for non-premixed situations to the Eddy-Break-Up model, is the most popular algebraic model. The reaction rate is simply expressed as [44]:

$$\overline{\dot{\omega}}_F = \alpha \overline{\rho} \frac{\varepsilon}{k} \min\left(\tilde{Y}_F, \frac{\tilde{Y}_O}{s}, \beta \frac{\tilde{Y}_P}{1+s} \right) \qquad (38)$$

where Y_F, Y_O and Y_P are respectively the fuel, oxidizer and product mass fractions. The parameter s is the mass stoichiometric coefficient of the chemical reaction Fuel $+s$ Oxidizer $\rightarrow$ Products, and α and β are two model constants.

3.5.2. Flame Surface Density Approach

Flame surface density models, based on a Σ balance equation have been first proposed by Marble and Broadwell [34] to describe non-premixed combustion. This balance equation was initially based on a transport equation for material surfaces and phenomenological descriptions [45]. A recent work [46] has provided a more rigourous framework to construct this equation.

3.5.3. Probability Density Functions

This approach is similar to the one presented for premixed combustion. The probability density function is generally based on the passive scalar Z (see section 2.3) under the assumption of an infinitely fast chemistry (Burke and Schumann limit) because mass fractions and temperature may be expressed as linear functions of Z. In non infinitely fast chemistry, the flow field is described with a PDF based on the passive scalar Z and the dissipation rate χ. Local mass fractions and temperature are then solutions of equations similar to Eq. (9). Assuming that the passive scalar Z and its dissipation are decoupled, the mean value $\overline{Q}$ of any quantity Q is expressed as:

$$\overline{Q} = \int_0^1 \int_\chi Q(Z,\chi)\, p(Z,\chi)\, d\chi dZ = \int_0^1 \int_\chi Q(Z,\chi)\, p(Z)\, p(\chi)\, d\chi dZ \quad (39)$$

In the presumed probability framework, $p(Z)$ is generally assumed to be a β-function whereas $p(\chi)$ is modeled as a gaussian function. The unclosed molecular diffusion term in the PDF balance equation (36) seems to be easier to model in non-premixed situations where the combustion rate is mainly controlled by molecular mixing.

3.6. Comments

This rapid survey of turbulent combustion models needs some further comments:

- Turbulent combustion modeling requires turbulence models for the unknown Reynolds stresses $\overline{\rho u_i'' u_j''}$ in Favre-averaged Navier-Stokes equations. These turbulence models need to take into account turbulence-combustion interactions. While both experimental and theoretical evidence [10] indicate that turbulence is influenced by combustion (flame-generated turbulence, higher viscous dissipation in hot burnt products than in cold reactants flow), little has been done on this subject. In fact, most numerical simulations of practical systems are based on standard k - ε turbulence models where transport equations are cast in terms of Favre averages, without explicit combustion effects. Nevertheless, refinements in turbulence description could be an important step to improve numerical simulations of turbulent reacting flows.

- The combustion models are presented here in their simplest form. They may have to be completed with various submodels to describe ignition, stabilization, pollutant formation, flame-wall interactions...

- A coupling between Σ and PDF balance equations is proposed in [47]. The objective is to take advantages of both approaches in order to describe complex reacting flows. Flame surface density modeling is well suited for thin reaction layers such as fuel oxidation whereas PDF transport equations cannot be easily closed in these regions. On the other hand, a

PDF approach is convenient to described the flame preheat zone and to predict pollutant formation in burnt gas regions.

4.   LARGE EDDY SIMULATIONS FOR REACTING FLOWS

4.1.   Introduction

Large eddy simulation (LES) is a very attractive tool for numerical simulations of fluid flows. Up to now, only few studies have addressed LES of reacting flow fields because of the many time and length scales related to flow characteristics and chemical processes. Most works have been conducted to test the feasibility of large eddy simulations in combustion and very often assumed simplified hypothesis (for example, constant density flows without heat release, two-dimensional flow field,...).

Nevertheless, LES is a promising approach for combusting studies. Most reacting flows exhibits large scale coherent structures. A typical example of LES applications in combustion is combustion instabilities [48]. These instabilities are due to a coupling between heat release, hydrodynamic flow field and acoustic waves. They have to be avoided because they induce noise, variations of the system characteristics, large heat transfers and may lead to the system destruction. LES could be a powerful tool to predict the occurence of such instabilities and to numerically test passive or active control systems. Large eddy simulations could also allow a better description of the turbulence / combustion interactions because, in LES, large structures are explicitly computed and instantaneous fresh and burnt gases zones, where turbulence characteristics are quite different, are clearly identified.

Assuming that G is the LES filter and x the location, any filtered quantity $\overline{Q}$ is defined as:

$$\overline{Q}(x,t) = \int_{-\infty}^{+\infty} Q(x,t)\, G(x-x')\, dx' \tag{40}$$

For reacting flows, a Favre filtering is defined as:

$$\overline{\rho}\widetilde{Q} = \overline{\rho Q} = \int_{-\infty}^{+\infty} \rho Q(x,t)\, G(x-x')\, dx' \tag{41}$$

where $\overline{\rho}$ is the filtered density. The previous definition is similar to the Favre averaging introduced in section 3.

Using this filter for the conservation equations controlling reacting flows introduces unknown quantities to be modeled:

- $\widetilde{u_i u_j} - \widetilde{u}_i \widetilde{u}_j$, the unresolved Reynolds stresses, requiring a subgrid scale turbulence model.

- $\widetilde{u_i Y_k} - \widetilde{u}_i \widetilde{Y}_k$, the unresolved species fluxes. In general, a simple gradient expression is assumed:

$$\widetilde{u_i Y_k} - \widetilde{u}_i \widetilde{Y}_k = -\frac{\nu_T}{S_c} \frac{\partial \widetilde{Y}_k}{\partial x_i} \tag{42}$$

where ν_T is the subgrid kinematic turbulent vicosity and S_c a turbulent Schmidt number

- $\widetilde{u_i T} - \widetilde{u}_i \widetilde{T}$, corresponding to the unresolved heat fluxes. A simple gradient expression, similar to (42) is also assumed.

- $\widetilde{Y_k T} - \widetilde{Y}_k \widetilde{T}$ and $\widetilde{Y_k T^n} - \widetilde{Y}_k \widetilde{T}^n$. These higher species - temperature correlations, which occur when specific heats C_p are expressed in terms of polynomial approximations of T, are usually neglected.

- Filtered reaction rate $\overline{\dot{\omega}}_k$

In the following, our attention will be focussed on the modeling of the filtered reaction rate $\overline{\dot{\omega}}_k$. Unresolved Reynolds stresses are generally described with the well known Smagorinski or Germano dynamic models whereas unresolved turbulent transports are expressed with gradient expressions. No attempt has been yet conducted to take into account counter-gradient transport in large eddy simulations. In the following, the simplest approaches (Arrhenius law, extensions of Favre averaged algebraic expression, simple extension of the Germano dynamic model) will be described first. Then more detailed approaches will be presented for non-premixed (Linear Eddy Model, subgrid scale PDF) and premixed (artificially thickened flame, G-equation) situations. In a last part, some promising ways will be suggested.

4.2. Simplest Approaches

4.2.1. Arrhenius Law Based on Filtered Quantities

A first rough model is to neglect subgrid scale contributions and to write the reaction rate as an Arrhenius law for filtered quantities:

$$\overline{\dot{\omega}}_F = A \overline{\rho}^2 \widetilde{Y}_F \widetilde{Y}_O \widetilde{T}^b \exp \left(-\frac{E}{R\widetilde{T}} \right) \tag{43}$$

Such simplified expression assumes a perfect mixing at subgrid scale level and implicitly suppose that turbulent time scales, τ_t, are shorter than chemical time scales τ_c ($\tau_t \ll \tau_c$). This assumption is generally used for reacting flows in atmospheric boudary layer [21] but is clearly not relevant in most combustion applications.

A more refined approach is to take into account the segregation factor (§ 3.2):

$$\overline{\dot{\omega}}_F = A \overline{\rho}^2 \widetilde{Y}_F \widetilde{Y}_O \widetilde{T}^b \exp \left(-\frac{E}{R\widetilde{T}} \right) \left[1 + \frac{\widetilde{Y_F'' Y_O''}}{\widetilde{Y}_F \widetilde{Y}_O} \right] \tag{44}$$

The segregation factor $I_{sgs} = \widetilde{Y_F''Y_O''}/\tilde{Y}_F\tilde{Y}_O$ is equal to 0 when the fuel and oxidizer are perfectly mixed at the subgrid level and $I_{sgs} = -1$ in a case of an infinitely fast reaction. I_{sgs} may be presumed or found from a balance equation for $\widetilde{Y_F''Y_O''}$ [22]. Once again, such an approximation is quite rough but may be justified when temperature fluctuations are negligible (for example for the dispersion of pollutant in the atmospheric boundary layer).

4.2.2.   Extension of Algebraic Favre Averaged Approaches

A simple idea is to extent Favre algebraic models such as Eddy - Break - Up model (§ 3.4.2) for premixed combustion or Magnussen model (§ 3.5.1) for non premixed combustion to LES [49] [50]. For example, the subgrid scale Eddy-Break-Up model is simply written:

$$\overline{\dot{\omega}}_F = C_{EBU}\bar{\rho}\frac{1}{\tau_t^{SGS}}\frac{\tilde{Y}_F}{Y_F^0}\left(1 - \frac{\tilde{Y}_F}{Y_F^0}\right) \tag{45}$$

where τ_t^{SGS} is a subgrid turbulent time scale, estimated as:

$$\tau_t^{SGS} = \frac{k^{SGS}}{\varepsilon^{SGS}} \approx C_D\frac{\Delta}{\sqrt{k^{SGS}}} \tag{46}$$

where k^{SGS} and ε^{SGS} are respectively the subgrid turbulent kinetic energy and its dissipation, C_D a constant and Δ the filter size. The turbulent kinetic energy k^{SGS} may be estimated from a balance equation.

Such a simple formulation is quite attractive and has not been yet extensively tested. Nevertheless two drawbacks may be suspected. First, the Eddy-Break-up model has known deficiencies in Favre averaged context (reaction rate independent on chemical reaction, overestimation of the reaction rate in zones of strong shears...). In a LES context, the model constant C_{EBU} seems to be strongly dependent on various parameters (flow conditions, mesh size,...).

4.2.3.   Simple Extension of the Germano Dynamic Model

In a recent paper [51], Germano et al. have proposed to extend the Germano dynamic model [52] to reacting flows. In a constant density flow without heat release, the filtered reaction rate $\bar{\dot{\omega}}_F$ is proportional to $\overline{Y_FY_O}$ (Eq. 18). Assuming similarity between two filtering levels (respectively noted $\bar{Q}$ and $\hat{Q}$), the unresolved reaction rate is expressed as:

$$\overline{Y_FY_O} - \overline{Y}_F\overline{Y}_O = k_{fg}\left(\widehat{\overline{Y_F}\overline{Y}_O} - \hat{\overline{Y}}_F\hat{\overline{Y}}_O\right) \tag{47}$$

This analysis is attractive because of its simplicity and its similarity with the Germano dynamic model for unresolved Reynolds stresses $\overline{u_iu_j} - \overline{u}_i\overline{u}_j$. Nevertheless, a-priori tests against DNS data reveal a disappointing result: the similarity constant k_{fg} is found to strongly depend on the mesh size and on the Damköhler number (ratio between turbulent time scale and chemical time scale). In fact, an increase of the Damköhler number corresponds to a thinner reaction zone and the similarity filter $\widehat{Q}$ underestimates the unresolved production term (a large fraction of the chemical production occurs at the subgrid level scale), requiring a higher similarity constant. On the other hand, as the mesh size decreases, the reaction rate estimated from the similarity filter $\widehat{Q}$ increases (an increasing part of the reaction rate occurs between subgrid, $\overline{Q}$, and similarity, $\widehat{Q}$, filter levels) and, accordingly, the similarity constant decreases. Then, such a simple similarity model is not well suited to reacting flows. Length scales effects have to be incorporated (comparison between flame thickness, flame wrinkling and mesh size). The extension of this approach to non-constant density flows with heat release effects is also not obvious.

4.3.   LES Models for non Premixed Combustion

As previously described (§ 2), in non premixed flames, turbulent times τ_t may often be assumed to be quite longer than characteristic chemical times τ_c. Accordingly, such reacting flows are generally controlled by turbulent and molecular mixing processes. Two main ideas have been proposed to describe non premixed combustion flow fields with LES: the Linear Eddy Model derived by Kerstein [53] [54] [55] [56] [57] [58] and the probability density function approach proposed by Gao and O'Brien [59].

4.3.1.   Linear Eddy Model

The Linear Eddy Model (LEM) developed by Kerstein [53] [54] [55] [56] [57] is based on a one dimensional stochastic description of turbulent stirring processes. In a LES framework [58], this analysis is used to describe subgrid scale phenomena.

The **turbulent stirring mechanism** is modeled by a rearrangement process applied to the 1D scalar field. The initial scalar distribution (Fig. 9a) is rearranged on a given segment of size l according to figure 9b ("triplet map"). This process may be viewed as the effect of a single turbulent structure of size l located in x_0. Then, the turbulent mixing is simulated from a stochastic description where vortex locations x_0, vortex sizes l ($l_k \leq l \leq \Delta$, where l_k is the Kolmogorov length scale and Δ the LES filter size) and vortex frequencies λ are specified according to a given turbulence spectra (homogeneous turbulence in general).

Molecular diffusion and chemical processes are described by one dimensional balance equations:

$$\frac{\partial \rho Y_i}{\partial t} = \frac{\partial}{\partial x}\left(\rho D_i \frac{\partial Y_i}{\partial x}\right) + \dot{\omega}_i \tag{48}$$

To summarize, subgrid scale chemical reaction and turbulent mixing are analyzed from a one dimensional DNS where a simple stochastic description of turbulence is achieved. Any complex chemistry features may be easily incorporated in Eq. (48). This approach also provides a direct estimation of filtered

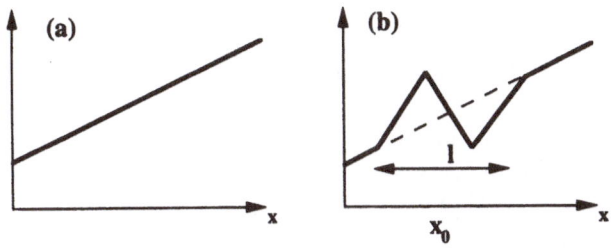

Fig. 9. — "Triplet map" used in the Linear Eddy Model developed by Kerstein to simulate a one-dimensional turbulent stirring process. (a) before mixing; (b) simulated
mixing by a vortex of size l.

mass fractions $\widetilde{Y}_i$ or temperature $\widetilde{T}$ without balance transport equations for
these quantities. Nevertheless, mass fractions and temperature transports be-
tween adjacent mesh cells have to be modeled. The LEM approach may also
be quite time consuming because a one-dimensional DNS is required in each
computational cell.

This approach is probably well suited for large eddy simulations of turbulent
mixing [60] and non premixed combustion, at least when phenomena are con-
troled by mixing [61] [62] [63] [64]. But, despite some attempts [65] [66] [67],
its extension to turbulent premixed combustion is more difficult. As pointed
out by Poinsot et $al.$ [27] and Roberts et $al.$ [28], viscous dissipation and flame
front curvatures play an important role in flame / turbulence interactions and
cannot be accounted for in LEM formulation.

4.3.2. Probability Density Functions

Under the assumption of fast chemistry, the species mass fractions Y_k and the
temperature are functions of the passive scalar Z (see § 2.3). Then, the Favre
filtered fuel mass fraction may be written as [59]:

$$\widetilde{Y}_f(x,t) = \frac{1}{\bar{\rho}} \int_{-\infty}^{+\infty} \rho(x',t) Y_f(Z(x',t)) G(x-x') \, dx' \qquad (49)$$

Introducing the Dirac δ-function, expression (49) may be recast as:

$$\widetilde{Y}_f(x,t) = \frac{1}{\bar{\rho}} \int_0^1 \int_{-\infty}^{+\infty} \rho(\Psi) Y_f(\Psi) \delta(Z(x',t) - \Psi) G(x-x') \, dx' \, d\Psi \qquad (50)$$

leading to:

$$\widetilde{Y}_f(x,t) = \int_0^1 Y_f(\Psi) \underbrace{\frac{1}{\bar{\rho}} \int_{-\infty}^{+\infty} \rho(\Psi) \delta(Z(x',t) - \Psi) G(x-x') \, dx'}_{\text{Filtered PDF } \widetilde{P}(\Psi,x,t)} \, d\Psi \qquad (51)$$

Then:

$$\widetilde{Y}_f(x,t) = \int_0^1 Y_f(\Psi)\, \widetilde{P}(\Psi, x, t)\, d\Psi \qquad (52)$$

where the sub-grid probability density function $\widetilde{P}(\Psi, x, t)$ has been introduced. The problem is now to determine this probability density function. Exactly as for Favre averaged models (§ 3.4.4), this PDF may be either presumed or found from a transport equation.

4.3.2.1. **Presumed sub-grid PDF.** This approach has been firstly proposed, and tested against DNS and experimental data, for reacting flow by Cook and Riley [68]. The authors have found that a β function (§ 3.4.4.1), based on the filtered passive scalar $\widetilde{Z}$ and its unresolved fluctuations $\widetilde{Z''^2}$ is very accurate. They also proposed a similarity model to describe the variance $\widetilde{Z''^2}$ without an additive balance equation:

$$\widetilde{Z''^2} = C_Z \left(\widehat{\widetilde{Z^2}} - \widehat{\widetilde{Z}}^2 \right) \qquad (53)$$

where $\widehat{Q}$ denotes a similarity filter.

In a recent work, Réveillon and Vervisch [69] [70] have proposed a dynamic procedure, similar to the Germano dynamic model [52] to estimate C_Z leading to very promising results without solving a transport equation for the subgrid variance $\widetilde{Z''^2}$.

4.3.2.2. **Transport equation for the sub-grid PDF.** A transport equation for the subgrid scale PDF $\widetilde{P}(\Psi, x, t)$ may be derived [59] and is formally similar to Eq. (36). This formalism has not yet been extensively tested [70].

4.4. LES Models for Premixed Combustion

4.4.1. *Introduction*

A difficult problem is encountered for large eddy simulations of premixed flames: the thickness δ_L of a premixed flame is about 0.1 to 1 mm and is generally smaller than the LES mesh size Δ as plotted on figure 10. The progress variable (i.e. non dimensionalized fuel mass fraction or temperature) is a very stiff variable and the flame front cannot be resolved in the computation, leading to numerical problems. To overcome this difficulty, two approaches have been proposed: simulation of an artificially thickened flame or use of a flame front tracking technique (*G*-equation).

4.4.2. *Artificially Thickened Flames*

The idea is to consider a flame thicker than the actual one, but having the same laminar flame speed S_L [71]. Following simple theories of laminar premixed flame [9] [72], the flame speed S_L and the flame thickness δ_L may be expressed

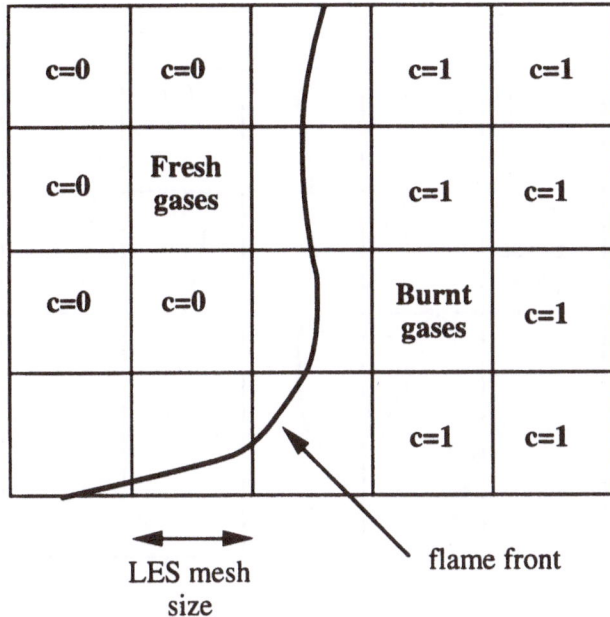

Fig. 10. — Comparison between premixed flame thickness δ_l and LES mesh size Δ. The flame front separates fresh gases (progress variable $c = 0$) from burnt gases ($c = 1$).

as:

$$S_L \propto \sqrt{a\overline{\overline{W}}} \qquad ; \qquad \delta_L \propto \cdot \frac{a}{S_L} \tag{54}$$

where a is the thermal diffusivity and $\overline{W}$ the mean reaction rate. Then, an increase of the flame thickness δ_L by a factor β with a constant flame speed S_L is easily achieved by replacing the thermal diffusivity a by βa and the reaction rate $\overline{W}$ by $\overline{W}/\beta$. If β is sufficiently large, the thickened flame front may then be resolved on the LES computational mesh without subgrid modeling using Eq. (43).

Nevertheless, this technique has a main drawback. The chemical characteristic time $\tau_c = \delta_L/S_L$ is increased by a factor β corresponding to a decrease of the Damköhler number $Da = \tau_t/\tau_c$, ratio of turbulent to chemical time scales, by the same factor β. The turbulence / flame length scale ratio, l_t/δ_L is also decreased by a factor β so that the interaction between turbulent eddies and the flame may be altered in an unpredictable way.

4.4.3. G - Equation

In this approach, the flame surface is described as an infinitely thin propagating surface (i.e. flamelet). The key idea is to track the position of the flame front

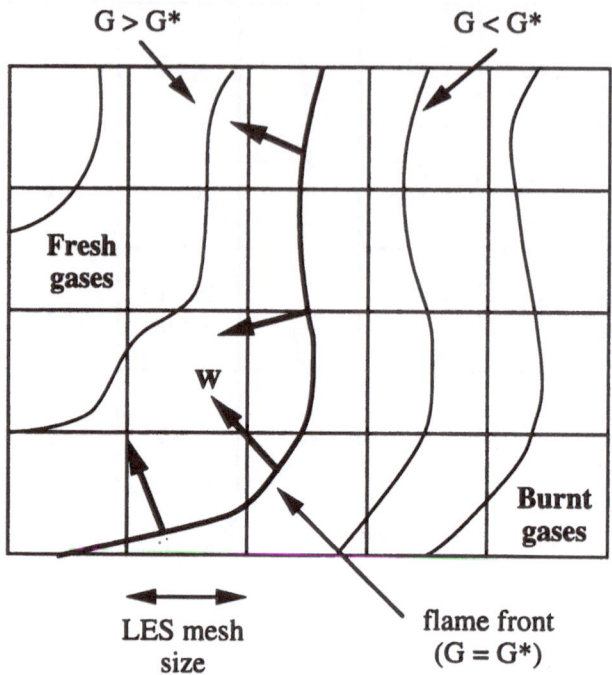

Fig. 11. — Flame front and G-field. w is the displacement speed, relatively to the local flow velocity, of the iso-level G.

using a field variable G. The flame surface is associated to the isolevel $G = G^*$ as plotted on figure 11 [73]. The G-field does not have to follow the gradients of the progress variable c and can be smoothed out to be resolved on the LES mesh. The G-equation is written:

$$\frac{\partial G}{\partial t} + \mathbf{u}.\nabla G = w\,|\nabla G| \tag{55}$$

where w is the local displacement speed of the iso-surface G, relatiye to the flow velocity.

This kinematic description of the combusting flow field leads to various difficulties briefly summarized here and analyzed using direct numerical simulations in [74].

In **constant density flows** (no heat release), the displacement speed w is generally assumed to be constant and equal to the laminar flame speed S_L. As chemistry has no influence on the flow field, each G-iso-surface may be viewed as a flame front. Then, many different flames may be computed in only one simulation ("low cost DNS") [75]. Flame cusps generated by the G-equation are avoided by adding an artificial numerical diffusion in Eq. (55).

In **non constant density flows** when thermal heat release is taken into account, two additive difficulties arise:

- The displacement speed w of the G-isosurface is affected by thermal expansion and should be corrected for density variations even in the case of a steady laminar flame propagating at the constant laminar flame speed S_L, defined relatively to the unburnt gases. This correction is:

$$w = \frac{\rho_u}{\rho} w_u \tag{56}$$

where w_u is the flame displacement speed relative to the unburnt gases of density ρ_u.

- A coupling is required between the G-equation and the species or energy balance equations. In fact, the G-equation provides a kinematic description of the flame front and involves its displacement speed w. The reactant consumption and the heat release rate are controlled by the consumption speed S_c. Of course, w and S_c are related but may be quite different, especially in high flame front curvature zones as pointed out by Poinsot *et al.* [76]. The displacement speed w may also be quite different from the laminar flame speed S_L.

The coupling between the consumption speed S_c and the displacement speed w is a key point in the G-equation formulation. Three approaches have been proposed to overcome this difficulty:

- **flame front tracking technique.** The displacement of the flame front is evaluated from the displacement speed w leading to an estimation of the burnt gases volume and of the thermal heat release [77]. Based on a purely geometric approach, this technique is well suited to two dimensional simulations but its extension to 3D cases is not obvious.

- **Temperature (or energy) reconstruction.** In this approach, the temperature field is directly estimated from the G field as [78]:

$$T = T_u + \frac{Q}{C_p} H (G - G^*) \tag{57}$$

where T_u is the temperature of unburnt gases, Q the reaction heat release and H denotes the Heavyside function (in practical simulations, the Heavyside function is smoothed on the grid mesh). This approach does not require a balance equation for the energy but is not applicable when heat losses or compressibility effects (for example in an internal combustion engines) have to be taken into account.

- **Estimation of the heat release rate from the G-field.** In this approach [74], the G-field is used to estimate the heat release to be incorporated in the balance energy equation from a formulation similar to Eq. (57). Accordingly any other effects (heat losses, compressibility) may be taken into account.

The application of a LES filter to the G-equation (55) leads to:

$$\frac{\partial \overline{G}}{\partial t} + \overline{\mathbf{u}}.\nabla \overline{G} = \underbrace{-\left(\overline{\mathbf{u}.\nabla G} - \overline{\mathbf{u}}.\nabla \overline{G}\right)}_{\text{unresolved transport}} + \underbrace{\overline{w |\nabla G|}}_{\text{front displacement}} \tag{58}$$

where two unclosed terms, corresponding respectively to unresolved transport and to front displacement, have to be modeled. To avoid this description, the G-equation is directly used for the resolved field $\overline{G}$, introducing a subgrid scale turbulent flame speed S_T:

$$\frac{\partial \overline{G}}{\partial t} + \overline{\mathbf{u}}.\nabla \overline{G} = \overline{S}_T |\nabla \overline{G}| \tag{59}$$

The challenge is then to propose a model for $\overline{S}_T$. This closure is generally based on Eq. (1):

$$\frac{\overline{S}_T}{S_L} = 1 + \alpha \left(\frac{\overline{u'}}{S_L}\right)^n \tag{60}$$

where $\overline{u'}$ is the subgrid scale turbulence level that may be estimated as:

$$\overline{u'} \approx \Delta |\overline{S}| = \Delta \sqrt{\left|2\overline{S}_{ij}\overline{S}_{ij}\right|} \tag{61}$$

where S_{ij} are the component of the resolved shear stresses. The constants α and β have to be specified. Im [79] has proposed a dynamic determination for α when $n = 1$.

Equation (59) is a simple formulation corresponding to a simple physical analysis (displacement of the resolved flame front with the displacement speed $\overline{S}_T$). Nevertheless, as already pointed out (§ 2), the turbulent flame speed is not a well defined quantity and no universal model is available. Despite these drawbacks, the G-equation approach is, up to know, the most advanced technique for large eddy simulations of premixed combusting flow fields [77] [78].

4.5. Extensions

We would like, in this section, to propose other possible approaches for large eddy simulations of reactive flows. These approaches are under developments and have not been yet extensively tested but seem very promising.

We have shown (§ 3.4.3) that, under the flamelet assumption, a flame surface density Σ may be introduced to describe the mean reaction rate (Eq. 25). This concept may be extended to LES where the filtered reaction rate $\overline{\dot{\omega}}_k$ may be expressed as:

$$\overline{\dot{\omega}}_k = \langle V_k \rangle_s \overline{\Sigma} \tag{62}$$

where $\overline{\Sigma}$ is the subgrid scale flame surface density. $\langle V_k \rangle_s$ is the reaction rate per unit of flame area averaged along the subgrid scale flame front.

In the G-equation context, the unclosed front displacement term in Eq. (58) may be re-written:

$$\overline{w|\nabla G|} = \langle w \rangle_s \overline{\Sigma}$$
$$= \langle w \rangle_s \Xi |\overline{\nabla G}| \qquad (63)$$

where Ξ is the wrinkling factor of the flame front (i.e. the ratio between the available subgrid flame surface and its projection in the mean propagation direction) introduced by Gosman [80]. The flame surface $\overline{\Sigma}$ or the wrinkling factor Ξ are then to be described. As for Favre averaged modeling (§ 3.4.3), two approaches are available:

- **Algebraic expressions.** These expressions could be based on BML analysis (Eq. 26) or on fractal approach (Eq. 28) but have to compare the filter size Δ with the flame surface cutoff length scales. Similarity or dynamic formulations could be used [81].

- **Balance equation.** There is no theoretical difficulties to derive balance equations for the subgrid flame surface $\overline{\Sigma}$ or flame wrinkling Ξ. In fact, the flame surface density Σ equation (29) may be directly recast in terms of subgrid surface $\overline{\Sigma}$. Quantities $\langle Q \rangle_s$ denote averaging along the flame surface at the subgrid scale level. Of course, the challenge is now to propose relevant models for unclosed terms.

To our mind, this analysis is more promising than the simple use of a subgrid turbulent flame speed $\overline{S_T}$ in a G-equation (Eq. 59). In its simplest form, this approach is similar but may be easily refined.

5. CONCLUSIONS

Modeling and simulations of turbulent combustion is a difficult challenge. In combusting flow fields, various difficulties (strong heat release, large range of time and length scales,...) are added to the complexity of constant-density turbulent flows.

A summary of the most classical Reynolds (or Favre) averaged Navier-Stokes (RANS) description has been proposed. In such analysis, three main elements have to be modeled:

- Reynolds stresses $\widetilde{u_i u_j} - \tilde{u}_i \tilde{u}_j$.

- Turbulent transport of species mass fractions $\widetilde{u_i Y_k} - \tilde{u}_i \tilde{Y}_k$.

- Mean reaction rate $\overline{\dot{\omega}}_k$.

Most works in turbulent combustion modeling have been devoted to propose refined descriptions for the mean reaction rate $\overline{\dot{\omega}}_k$. Few studies have been done on species turbulent transport, generally modeled using a gradient expression whereas counter-gradient turbulent transport is known to occur in some situations. Reynolds stresses are generally described from classical turbulent models, simply re-written in terms of Favre averaged quantities. Combustion effects on the flow (flame induced turbulence generation, higher viscous dissipation...) are not explicitely taken into account.

Large eddy simulations is a very promising tool for combustion flow field description. First, such flows generally exhibit strong large scale coherent structures. LES could also provide a better description of turbulence / combustion interactions because fresh and burnt gases zones, having different turbulence characteristics, are instantaneously identified at the resolved grid level. Nevertheless, LES is only beginning for combustion applications and very few works have already been done on this subject, mainly devoted to feasibility tests (two dimensional simulations, constant density flows,...).

References

[1] Poinsot T., Trouvé A., Candel S., *Progress in Energy and Combustion Science* **21** (1996) 531.

[2] Poinsot T., "Using direct numerical simulation to understand premixed turbulent combustion", Invited lecture, *Twenty sixth Symposium (International) on Combustion* (The Combustion Institute, Pittsburgh, 1996).

[3] Direct numerical simulation for turbulent reacting flows (Baritaud T., Poinsot T. and Baum M. Eds, Editions Technip, Paris, 1996).

[4] Abdel-Gayed R.G., Bradley D., Hamid M.N., Lawes M., Twentieth Symposium (International) on Combustion (The Combustion Institute, Pittsburgh, 1984) p. 505.

[5] Abdel-Gayed R.G., Bradley D., *Combust. Flame* **76** (1989) 213.

[6] Yakhot V., Orszag C.G., Thangam S., Gatski T.B. and Speziale C.G., *Phys. Fluids A* **4** (1992) 1510.

[7] Gouldin F.C., "Combustion intensity and burning rate integral of premixed flames", *Twenty sixth Symposium (International) on Combustion* (The Combustion Institute, Pittsburgh, 1996).

[8] Damköhler G., *Z. Elektrochem.* **46** (1940) 610.

[9] Williams F.A., Combustion Theory (Benjamin Cummings, Menlo Park, 1985).

[10] Bray K.N.C., Libby P.A., Masuya G., Moss J.B., *Combust. Sci. Tech.* **25** (1981) 127.

[11] Shepherd I.G., Moss J.B., Bray K.N.C. Nineteenth Symposium (International) on Combustion (The Combustion Institute, Pittsburgh, 1982) p. 423.

[12] Veynante D., Trouvé A., Bray K.N.C., Mantel T., to be published in *J. Fluid Mech.*, 1996.

[13] Veynante D., Poinsot T., Annual Research Briefs (Center for Turbulence Research, Stanford University - Nasa-Ames, 1995).

[14] Duclos J.M., Veynante D., Poinsot T., *Combust. Flame* **95** (1993) 101.

[15] Bray K.N.C., Champion M., Libby P.A., *Combust. Sci. Tech.* **55** (1987) 139.

[16] Bray K.N.C., Complex Chemical Reactions, (Warnatz J. and Jäger W., Eds., Springer Verlag, 1987).

[17] Bailly P., "Contribution à l'étude de l'interaction turbulence - combustion dans les flammes turbulentes de prémélange à l'aide de modèles du second ordre", Ph'D thesis, Université de Poitiers, 1996.

[18] Bailly P., Garréton D., Simonin O., Bruel P., Champion M., Deshaies B., Duplantier S., Sanquer S., "Experimental and numerical study of a premixed flame stabilized by a rectangular section cylinder", *Twenty-sixth Symposium (International) on Combustion* (The Combustion Institute, Pittsburgh, 1996).

[19] Veynante D., Piana J., Duclos J.M., Martel C., "Experimental analysis of flame surface density models for premixed turbulent combustion", *Twenty-sixth Symposium (International) on Combustion* (The Combustion Institute, Pittsburgh, 1996).

[20] Villasenor R., Pitz R.W., Chen J.Y., "Interaction between chemical reaction and turbulence in supersonic nonpremixed H_2-air combustion", 28th Aerospace Science Meeting, Reno, Nevada 1991, AIAA paper 91-0375.

[21] Nieuwstadt F.T.M., "The large-eddy simulation of the dispersion of pollutants in the atmospheric boundary layer", New Tools in Turbulence Modelling, O. Métais and J. Ferziger org., Ecole de Physique des Houches, May 21-31 1996.

[22] Meeder J.P., Nieuwstadt F.T.M., "Subgrid-scale segregation of chemically reactive species in a neutral boundary layer", Second ERCOFTAC Worshop on direct and large eddy simulation, Grenoble, September 16-19 1996.

[23] Borghi R, Destriau M., La combustion et les flammes (Editions Technip, Paris, 1996), in french.

[24] Borghi R., *Recent Advances in the Aerospace Sciences* (1985) pp. 117-138.

[25] Peters N., Twenty-first Symposium (International) on Combustion (The Combustion Institute, Pittsburgh, 1986) p. 1231-1250.

[26] Peters N., Franke C., *Springer Verlag Series in Synergetics* **48** (1990) 40.

[27] Poinsot T., Veynante D., Candel S., *J. Fluid Mech.* **228** (1991) 561.

[28] Roberts W.L., Driscoll J.F., Drake M.C., Goss L.P., *Comb. Flame* **94** (1993) 58.

[29] Cuenot B., Poinsot T., Twenty-fifth Symposium (International) on Combustion (The Combustion Institute, Pittsburgh, 1994) p. 1383.

[30] Spalding D.B., Combustion and Mass Transfer (Pergamon Press, Oxford, 1978).

[31] Bray K.N.C., Champion M., Libby P.A., Twenty-second Symposium (International) on Combustion (The Combustion Institute, Pittsburgh, 1988) p. 763.

[32] Gouldin F.C., Bray K.N.C., Chen J.Y., *Combust. Flame* **77** (1989) 241.

[33] Gülder O.L., Smallwood G.J., *Combust. Flame* **103** (1995) 107.

[34] Marble F.E., Broadwell J.E., The coherent flame model of non-premixed turbulent combustion. Project Squid Report, TRW-9-PU, 1977.

[35] Pope S.B., *Int. J. Engin. Sci.* **26** (1988) 445.

[36] Candel S.M., Poinsot T.J., *Combust. Sci. Tech.* **70** (1990) 1.

[37] Cant R.S., Pope S.B., Bray K.N.C., Twenty-third Symposium (International) on Combustion (The Combustion Institute, Pittsburgh, 1990) p. 809.

[38] Trouvé A., Poinsot T., *J. Fluid Mech.* **278** (1994) 1.

[39] Vervisch L., Bidaux E., Bray K.N.C., Kollmann W., *Phys. Fluids A* **7** (1995) 2496.

[40] Poinsot T., Echekki T., Mungal M.G., *Combust. Sci. Tech.* **81** (1992) 45.

[41] Veynante D., Duclos J.M., Piana J., Twenty-fifth Symposium (International) on Combustion (The Combustion Institute, Pittsburgh, 1994) p. 1249.

[42] Pope S.B., *Prog. Energ. Combust. Sci.* **11** (1985) 119.

[43] Dopazo C., Turbulent reacting flows (Libby P.A. and Williams F.A. Eds, Academic Publisher, 1994) p. 375.

[44] Magnussen B.F., Mjertager B.H., Sixteenth Symposium (International) on Combustion (The Combustion Institute, Pittsburgh, 1976) p. 719.

[45] Fichot F., Delhaye B., Veynante D., Candel S., Twenty-fifth Symposium (International) on Combustion (The Combustion Institute, Pittsburgh, 1994) p. 1273.

[46] Van Kalmthout E., Veynante D., Candel S., "Direct numerical simulation analysis of flame surface density in non premixed turbulent combustion", *Twenty-sixth Symposium (International) on Combustion*, The Combustion Institute, Pittsburgh, 1996.

[47] Vervisch L., Kollmann W., Bray K.N.C., Mantel T., Proceedings of the Summer Program (Center for turbulence Research, Stanford, 1994) p. 125.

[48] Menon S., Jou W.H., *Comb. Sci. Tech.* **75** (1991) 53.

[49] Fureby C., Löfström C., Twenty fifth Symposium (International) on Combustion (The Combustion Institute, Pittsburg, 1994) p. 1257.

[50] Fureby C., Möller S.I., *AIAA J.* **33 (12)** (1995) 2339.

[51] Germano M., Maffio A., Sello S., Mariotti G., "On the extension of the dynamic modelling procedure to turbulent reacting flows", Second ER-COFTAC workshop on Direct and Large Eddy Simulation, Septembre 16-19, 1996, Grenoble, France.

[52] Germano M., Piomelli U., Moin P., Cabot W.H., *Phys. Fluids* **7(3)** (1991) 1760-1765.

[53] Kerstein A. R., *Comb. Sci. Tech.* **60** (1988) 391.

[54] Kerstein A.R., *Comb. Flame* **75** (1989) 397.

[55] Kerstein A.R., *J. Fluid Mech.* **216** (1990) 411.

[56] Kerstein A.R., *J. Fluid Mech.* **231** (1991) 361.

[57] Kerstein A.R., *Comb. Sci. Tech.* **81** (1992) 75.

[58] Mc Murthy A., Menon S., Kerstein A.R., Twenty-fourth Symposium International on Combustion (The Combustion Institute, Pittsburgh, 1992) p. 271.

[59] Gao F., O'Brien E.E., *Phys. Fluids* **5(6)** (1993) 1282.

[60] McMurthy P.A., Gansauge T.C., Kerstein A.R. and Krueger S.K. *Phys. Fluids A* **5** (1993) 1023.

[61] McMurthy P.A., Menon S., Kerstein A.R., "A subgrid mixing model for nonpremixed turbulent combustion", AIAA 30th Aerospace Science Meeting, Reno, Nevada, 1992, AIAA paper 92-0234.

[62] Menon S., McMurthy P.A., Kerstein A.R., Chen J.Y. *J. Prop. Power* **10** (1994) 161.

[63] Calhoon W.H., Menon S., "Subgrid modeling for large eddy simulations", AIAA 34th Aerospace Science Meeting, Reno, Nevada, 1996, AIAA paper 96-0516.

[64] Mathey F., Chollet J.P. "Sub-grid model of scalar mixing for large eddy simulations of turbulent flows", Second ERCOFTAC workshop on Direct and Large Eddy Simulation, Septembre 16-19, 1996, Grenoble, France.

[65] Menon S., Kerstein A.R., Twenty-fourth Symposium International on Combustion (The Combustion Institute, Pittsburgh, 1992) p. 443.

[66] Menon S., McMurthy P.A., Kerstein A.R., Large eddy simulation of complex engineering and geophysical flows (Galperin B. and Orzag S.A. Eds., Cambridge University Press, Cambridge, 1993) p. 287.

[67] Smith T., Menon S., "Model simulations of freely propagating turbulent premixed flames", *Twenty-sixth Symposium International on Combustion*, The Combustion Institute, Pittsburgh, 1996.

[68] Cook A.W., Riley J.J., *Phys. Fluids* **6(8)** (1994) 2868.

[69] Réveillon J., Vervisch L., *Phys. Fluids* **8(8)** (1996) 2248.

[70] Réveillon J., Vervisch L., "Dynamic subgrid PDF modeling for non premixd turbulent combustion", Second ERCOFTAC workshop on Direct and Large Eddy Simulation, Septembre 16-19, 1996, Grenoble, France.

[71] Butler T.D., O' Rourke P.J., Sixteenth Symposium (International) on Combustion (The Combustion Institute, Pittsburgh, 1977) p. 1503.

[72] Kuo K.K., Principles of Combustion (John Wiley and Sons, 1986).

[73] Kerstein A.R., Ashurst W.T., Williams F.A., *Physical Rev. A* **37(7)** (1988) 2728.

[74] Piana J., Veynante D., Candel S., Poinsot T., "Direct numerical simulation analysis of the G-equation in premixed combustion", Second ERCOFTAC workshop on Direct and Large Eddy Simulation, Septembre 16-19, 1996, Grenoble, France.

[75] Ashurst W.T., Ruetsch G.R., Lund T.S., Proceedings of the Summer Program (Center for Turbulent Research, Stanford University and Nasa-Ames, 1994) p. 151.

[76] Poinsot T., Echekki T., Mungal M.G., *Comb. Sci. Tech.* **81** (1991) 45.

[77] Moser V., "Large eddy simulations of turbulent premixed flames using a capturing/tracking hybrid approach", Sixth International Conference on Numerical Combustion, New Orleans, Louisiana, March 4-6, 1996.

[78] Menon S., "Large eddy simulation of combustion instabilities", Sixth International Conference on Numerical Combustion, New Orleans, Louisiana, March 4-6, 1996.

[79] Im H.T., Annual Research Briefs (Center for Turbulent Research, Stanford University and Nasa-Ames, 1995) p. 347.

[80] Gosman A. D., "Engine aerodynamics", New Tools in Turbulence Modelling, O. Métais and J. Ferziger org., Centre de Physique des Houches, May 21-31 1996.

[81] Piana J., "Etude de l'application de la simulation aux grandes échelles à la combustion turbulente", Ph'D Thesis, Ecole Centrale Paris, 1996.

Modelling Tools for Flow Noise and Sound Propagation through Turbulence

G. Comte-Bellot, C. Bailly
and P. Blanc-Benon

Laboratoire de Mécanique des Fluides et d'Acoustique
CNRS UMR 5509
Ecole Centrale de Lyon
69131 Ecully, France

1. INTRODUCTION

It has been known for a long time that turbulent flows generate noise and that turbulent fields modify the propagation of acoustic waves. However, it is only recently that powerful computers have permitted a large variety of numerical predictions in computational aeroacoustics (CAA). This paper is a survey of the most efficient techniques which are presently available for theoretical analyses and engineering noise predictions.

For flow noise generation, many global characteristics are of interest : levels, spectra, directivities or space-frequency distributions of sound sources. Instantaneous time pressure traces can also illustrate acoustic events associated with special flow patterns. Several approaches will be presented in section 2. Among them, two methods have prominent roles. First, there is the Lighthill analogy for which the equivalent source terms expressing the flow noise can be either directly expressed or approximately modelled as soon as the aerodynamic flow field is obtained. The explicit approach is of interest for low Reynolds number flows, permitting the improvement of knowledge of the physical mechanisms sustaining the noise emission. The modelling approach is required for engineering cases, when Reynolds numbers are high. In both cases, the solutions are often aimed to obtain the radiated acoustic fields, and, as a prerequisite, the Green function has to be known. This is easy in free space, but difficult in a constrained environment, for example when noise from internal flows has to be predicted. The second efficient method makes use of the linearized Euler equations with their associated source terms. The solutions are obtained for the whole domain over which the computation takes place without invoking Green functions. This method is therefore very flexible. In addition, the perturbations contain the vorticity and the entropy modes as well as the acoustic mode. Separation is achieved at the listening point using the sound speed for the acoustic mode, and the flow velocity for the other two modes.

These approaches will be illustrated in section 3 by three examples. The first example deals with the noise generated by isotropic turbulence at modest Reynolds numbers. The straight use of the Lighthill analogy with the source terms given by a direct numerical computation of the turbulent field (DNS) is very rewarding. For the first time, we can obtain the range of turbulent structures mostly responsible of the noise emission. The two other examples, which pertain to engineering situations, deal with the noise emitted by subsonic or supersonic free jets, or jets emerging from a diaphragm in a duct. Here, we will illustrate the complementarity of the Lighthill analogy and the method of the linearized Euler equations.

The second topic of this presentation deals with the propagation over long distances of acoustic waves through turbulence fields, of either velocity or temperature fluctuations. Due to cumulative effects, the transmitted waves present large phase and amplitude fluctuations, as well as noticeable directional changes. Computer simulations permit us to see what is happening everywhere in the field while measurements are only possible at a limited number of points. In section 4, we will learn that long propagation distances mean distances of the order of a hundred times the integral length scale of the turbulent field. In real life applications are numerous. Remote detection arrays in the oceans are subject to phase fluctuations and their correct working conditions require predictions. Annoyance acoustic levels around airports, railroads tracks, or highways, are strongly dependent on weather conditions. For example, above a heated ground, noise can penetrate the shadow or silence zone which would exist in quiet conditions, i.e. without turbulence.

The propagation of acoustic waves over large distances is a multiple scattering problem, single scattering would be for a wave arriving on a small volume of turbulence. In section 4 we will present the most useful prediction tools, the ray equations, the Helmholtz equation with a random index of refraction, and the parabolic equations. Although a linearization is always done relatively to the acoustic variables, turbulent fluctuations appear as a parametric excitation factor. Hence nonlinear turbulence-acoustics coupling terms exist. An ensemble averaging on the equations would lead to higher order nonlinear terms, hence to closure problems similar to those encountered in turbulence, but not so well tractable due to their mixed origin. A new and efficient technique will be therefore described. The acoustic equations are first solved on individual turbulent field realizations, then, and only then, the solutions are ensemble averaged. Two illustrations will be presented : the occurrence of caustics through velocity or temperature turbulent fields, and the scattering of sound into a shadow zone over a heated ground.

2. METHODS FOR AEROACOUSTICS PREDICTIONS

Several approaches permit us to obtain the far field noise generated by turbulent flows. They are not completely equivalent; differences exist in the physics incorporated in the formulation, and also in the required computing time.

2.1. Full-Field Computations

This is a numerical resolution of the compressible Navier Stokes equations in the entire field, i.e. the inner turbulent flow itself and the outer acoustic field. In principle, this would be the perfect approach. However, very large memory capacities are required, because of the intrinsic 3D nature of the field, and the high accuracy schemes and grid strategies needed to reduce the numerical dissipation and dispersion of the traditional CFD codes. Technical details have been made available by Lele [1] or Tam & Webb [2] for instance. Only very large computing facilities will do the job and the best use is for basic cases. Colonius, Lele & Moin studied the noise emitted by two vortices spinning around each other [3] as well as the sound radiated by a two-dimensionnal mixing layer [4]. Mitchell and co-workers [5] investigated the noise radiated by axisymmetric jets.

2.2. Kirchhoff Formulation

Kirchhoff's integral formulation (1883), see Pierce [6], permits us to reduce the computational expenses needed for the full field computations discussed above. Here use is made of a surface S which is assumed to enclose all the nonlinear flow effects and the noise sources. Outside S the acoustic field is governed by a linear wave equation. The radiated pressure is then expressed in terms of quantities prescribed on S. For a stationary surface S, the pressure fluctuation p' in the far field is expressed as :

$$\lim_{x \to \infty} p'(x, t) = \frac{1}{4\pi x} \int_S \left[\frac{1}{c_o} \frac{\partial p'}{\partial t} \frac{\partial r}{\partial n} - \frac{\partial p'}{\partial n} \right] dy. \tag{1}$$

Inside and on S the field is evaluated using a compressible CFD code. The brackets denote evaluation in the retarded (emission) time and n is the normal on S pointing out, so $\partial r / \partial n = \cos \theta$ where θ is the angle between n and the radiation direction $r = x - y$ which reduces to x in the far field. Extensions to uniformly or even arbitrarily moving Kirchhoff surfaces have been made to predict rotor and helicopter noise. An extensive review of these different formulations is given by Lyrintzis [7]. For circular jets the Kirchhoff technique is of more recent use [8, 9, 10]. In practice, a cylindrical surface S surrounds the jet at 10 jet radii away from the jet centerline and the major contributions come from points of S located on lines joining the observer and the source region. Freund, Lele & Moin [11] investigated the errors due to the upstream and downstream ends which have to be left open. There, no linear acoustic data are available because of persistent hydrodynamics effects.

2.3. Lighthill's Analogy

The leading idea to build an acoustic analogy is to combine the conservation equations governing the flow so as to obtain a wave equation, where the left hand side represents a wave operator and the right hand side the associated sources. The pionnering work is due to Lighthill [12], which is here briefly recalled. The continuity and momentum equations are:

$$\frac{\partial \rho}{\partial t} + \frac{\partial (\rho u_i)}{\partial x_i} = 0 \tag{2}$$

$$\frac{\partial (\rho u_i)}{\partial t} + \frac{\partial (\rho u_i u_j)}{\partial x_j} = -\frac{\partial p}{\partial x_i} + \frac{\partial \tau_{ij}}{\partial x_j} \tag{3}$$

where ρ is the density, u the velocity vector, τ the viscous tensor and p the pressure. Let's combine these two equations by differentiating (2) with respect to t and (3) with respect to x_i and substract them, to obtain:

$$\frac{\partial^2 \rho}{\partial t^2} = \frac{\partial^2}{\partial x_i \partial x_j} \left[\rho u_i u_j + p \delta_{ij} - \tau_{ij} \right]. \tag{4}$$

Now, substracting the term $c_o^2 \nabla^2 \rho$ from both sides of (4) where c_o is the reference sound speed in a medium at rest, we have:

$$\frac{\partial^2 \rho}{\partial t^2} - c_o^2 \nabla^2 \rho = \frac{\partial^2}{\partial x_i \partial x_j} \left[\rho u_i u_j + (p - c_o^2 \rho) \, \delta_{ij} - \tau_{ij} \right] = \frac{\partial^2 T_{ij}}{\partial x_i \partial x_j}. \tag{5}$$

Thus in (5) the left hand side is the wave propagation operator with speed c_0 (Dalembertian) and the right hand side represents equivalent sources, i.e. sources which placed in a medium at rest would create the same noise as the turbulent zone. When the Reynolds number is high, of the order of 10^4 to 10^6 as in most practical applications, the viscous stress tensor can be neglected in the source term. Besides, for a perfect gas, fluctuations of pressure, density and entropy are connected by $p' = c_o^2 \rho' + (p_o/c_v) s'$. Thus acoustic generation and propagation without entropy fluctuations are such that $p' = c_o^2 \rho'$ and Lighthill's tensor T_{ij} is then reduced with this assumption to $\rho u_i u_j$.

Solution of (5) gives the pressure (or density) fluctuation arriving in free space at the listening point x at time t :

$$p'(x,t) = \frac{1}{4\pi} \int_V \left[\frac{\partial^2 T_{ij}}{\partial y_i \partial y_j} \right] \frac{dy}{r} \tag{6}$$

where $r = x - y$. The brackets indicate that the source term at y is evaluated at the retarded time $t - r/c_o$. The integral is taken over the turbulent source volume V. Expression (6) can be simplified if x is in the far field, so that :

$$\lim_{x \to \infty} p'(x, t) = \frac{1}{4\pi x} \frac{\partial^2}{\partial x_i \partial x_j} \int_V [T_{ij}] \, dy \qquad (7)$$

and also :

$$\lim_{x \to \infty} p'(x, t) = \frac{1}{4\pi c_o^2 x} \frac{x_i x_j}{x^2} \int_V \left[\frac{\partial^2 T_{ij}}{\partial t^2} \right] dy \qquad (8)$$

with the brackets indicating that the retardation time is used. There are two ways to proceed further:

(i) When the (incompressible) turbulent flow is completely known in time and space (by DNS, LES, or analytical developments) the space-time velocity field can be introduced into expressions (6) to (8). Use of expression (6) seems straightforward as computed velocity fields are often available in space. However (6) introduces truncation effects. The turbulent velocities have therefore to go naturally to zero at the flow boundary, not to be forced to do so by an abrupt cut which creates additional velocity gradients, and hence new sources. All turbulence boxes extended by periodicity to fill in the space present such a problem. Witkowska and Juvé [13] have shown that indeed in that case the use of (6) overestimates the computed noise very significantly. An additional difficulty when using (6) is that the retarded times have to be known very accurately, as pointed out by Crighton [14]. In expression (7) derivatives are taken with respect to the observation point and that is difficult to use in numerical computations. Expression (8) is the most tractable one as soon as a judicious managing of the many time steps, which have to be stored for the retarded times, is developed.

(ii) When the turbulent field is only known by statistical quantities - for engineers it would be most of the time from the standard $K - \epsilon$ codes, with K kinetic energy and ϵ its dissipation rate - a mean square procedure is applied to the solution given by the integral solution (8). One has therefore to deal with second order space-time velocity correlations for the mixed terms involving mean velocity and turbulence - the so-called shear-noise, and fourth order space-time velocity correlations for the terms involving only turbulence - the so-called self-noise [15, 16]. For the space dependence, an isotropy assumption is often acceptable, e.g. for jets [16]. Integral scales and turbulence levels are then built using the data given by the turbulence code. For example, the local integral length scale is approached by $L \sim K^{3/2}/\epsilon$ and the angular frequency by $\omega_t \sim \epsilon/K$. For the time evolution in moving coordinates a Gaussian form eases the computations [17], but more elaborated approaches have been recently developed by Bailly, Candel and Lafon [18] for the mixing noise of supersonic jets. The directivity factor $x_i x_j x_k x_l / x^4$ can be fully expressed using symmetry arguments when they exist. For example, for a round jet, the directivity factor is computed using an azimuthal average, hence giving only terms in θ the angle between the downstream jet axis and the observer x. The final result can then be expressed as an integral over the flow region, with a noise source density per volume unit, involving a Mach number factor and a directivity factor. For example, for a subsonic jet, the power spectral density reads [19] :

$$S(x,\omega) \sim \frac{1}{32\pi^2\sqrt{\pi}c_o^5\rho_o x^2} \int_V \frac{\rho^2\overline{u'^2}L^3}{\omega_t}\omega^4$$

$$\times \left\{ \frac{\overline{u'^2}}{4}\exp\left(-\frac{C^2\omega^2}{8\omega_t^2}\right) + \frac{L^2}{2\pi}\left(\frac{\partial U}{\partial y_2}\right)^2\exp\left(-\frac{C^2\omega^2}{4\omega_t^2}\right)D_\theta \right\}dy \quad (9)$$

where $\partial U/\partial y_2$ is the mean shear, u' the rms value of the velocity fluctuations, L the integral length scale, $D_\theta = \left(\cos^4\theta + \cos^2\theta\right)/2$, the directivity of the shear-noise, and $C = 1 - M\cos\theta$ the Doppler factor with $M = U_c/c_0$ the convection Mach number of the turbulent structures.

2.4. Powell-Hardin and Möhring Analogies

In the Lighthill analogy, the equivalent acoustic sources are expressed in terms of the velocity fields. Powell [20], Howe [21], Hardin [22] and Möhring [23] developed expressions in which the vorticity appears, suggesting that in some processes, vortex pairing for example, vorticity dynamics could have an important role. The Powell - Hardin expression in which velocity and vorticity are still jointly present reads as :

$$\lim_{x\to\infty} p(x,t) = \frac{\rho_o}{4\pi c_o^2}\frac{x_i x_j}{x^3}\int_V \left[\frac{\partial^2}{\partial t^2}y_i(u\times\omega)_j\right]dy \quad (10)$$

and the Möhring expression which concerns only the vorticity is :

$$\lim_{x\to\infty} p(x,t) = \frac{\rho_o}{12\pi c_o^2}\frac{x_i x_j}{x^3}\int_V \left[\frac{\partial^3}{\partial t^3}y_i(y\times\omega)_j\right]dy. \quad (11)$$

These analogies are convenient when the flow is directly expressed in terms of vorticity. For example, Knio, Collorec and Juvé computed the noise emitted by the chaotic motion of three vortices close to a solid wall [24] and Knio and Juvé investigated the noise emission during the head-on collision of deformable ring vortices [25].

2.5. Lilley's Equation

In the previous formulations (6) to (11), mean flow effects on the acoustic waves propagation are not taken into account and it is well known that they modify the aerodynamic noise directivity. However, their integration in a convected wave equation is not easy. Just for a parallel sheared mean flow, Lilley [26] had to establish a third order differential equation which reads :

$$\frac{d}{dt}\left\{\frac{d^2\pi}{dt^2} - \frac{\partial}{\partial x_i}\left(c^2\frac{\partial\pi}{\partial x_i}\right)\right\} + 2\frac{\partial u_i}{\partial x_j}\frac{\partial}{\partial x_i}\left(c^2\frac{\partial\pi}{\partial x_j}\right) \qquad (12)$$

$$= -2\gamma\frac{\partial u_i}{\partial x_j}\frac{\partial u_j}{\partial x_k}\frac{\partial u_k}{\partial x_i} + \text{other terms}$$

where $\pi = \ln p$ is the logarithm of the pressure. The viscous contribution and the entropy fluctuations are here neglected (other terms). Although more exact than expressions (6) to (11) equation (12) cannot be solved explicitely because its free space Green function is unknown. Moreover the pressure fluctuations π in the wave equation are the sum of the acoustic and entropy fluctuations : in general, acoustic and hydrodynamic fluctuations cannot be clearly separated by a wave operator.

2.6.   Linearized Euler Equations : SNGR Model

The Stochastic Noise Generation and Radiation (SNGR) model developed by Bailly, Lafon and Candel [27, 28] provides the time dependent acoustic field from a solution of the linearized Euler equations in which a synthesized turbulent field is used as input in a source term.

The Euler equations for a perfect fluid without entropy fluctuations:

$$\frac{d\rho}{dt} + \rho\frac{\partial u_j}{\partial x_j} = 0 \qquad (13)$$

$$\frac{du_i}{dt} + \frac{1}{\rho}\frac{\partial p}{\partial x_i} = 0 \qquad (14)$$

$$\frac{dp}{dt} + \gamma p\frac{\partial u_j}{\partial x_j} = 0 \qquad (15)$$

are linearized around a stationary mean flow (u_o, p_o) The perturbation fields $u_i' = u_{it} + u_{ia}$ and $p' = p_t + p_a$ contain turbulent and acoustic fluctuations with the additional assumption that $|u_{it}| > |u_{ia}|$ and $|p_t| > |p_a|$. Upon substitution in (14, 15) an order of magnitude analysis yields the following system of two first-order equations [27]:

$$\begin{cases} \dfrac{\partial p'}{\partial t} + u_{jo}\dfrac{\partial p'}{\partial x_j} + u_j'\dfrac{\partial p_o}{\partial x_j} + \gamma p_o\dfrac{\partial u_j'}{\partial x_j} + \gamma p'\dfrac{\partial u_{jo}}{\partial x_j} = 0 \\[3mm] \dfrac{\partial u_i'}{\partial t} + u_{jo}\dfrac{\partial u_i'}{\partial x_j} + u_j'\dfrac{\partial u_{io}}{\partial x_j} + \dfrac{1}{\rho_o}\dfrac{\partial p'}{\partial x_i} - \dfrac{p'}{\rho_o^2 c_o^2}\dfrac{\partial p_o}{\partial x_i} = S_i \end{cases} \qquad (16)$$

where the main source term reads as follows:

$$S_i = -\left\{u_{jt}\frac{\partial u_{it}}{\partial x_j} - \overline{u_{jt}\frac{\partial u_{it}}{\partial x_j}}\right\}. \qquad (17)$$

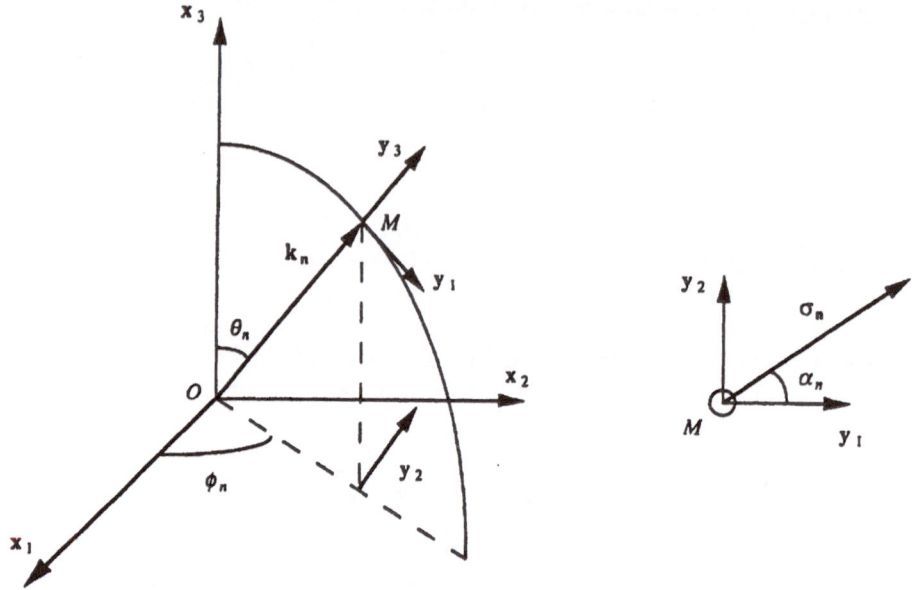

Fig. 1. — Sketch of a 3D Fourier mode for the synthesized turbulent velocity field.

In the left hand side, one uses values of the mean flow field calculated in a previous step as coefficients of the differential system, and on the right hand side, the acoustic source term S is calculated from a synthesized space time turbulent velocity field, given by the sum of N random Fourier modes [28]:

$$u\left(x,t\right) = 2 \sum_{n=1}^{N} \tilde{u}_n \cos\left[k_n \cdot \left(x - tU_c\right) + \psi_n + \omega_n t\right] \sigma_n \qquad (18)$$

Because of the fluid incompressibility k_n is normal to its associated Fourier contribution $\tilde{u}_n$. The field isotropy in 3D requires that the directions of k_n (see Figure 1) and $\tilde{u}_n$ have the following p.d.f. :

$$P(\phi) = \frac{1}{\pi} \quad P(\theta) = \frac{1}{2}\sin\theta \quad P(\alpha) = \frac{1}{2\pi} \qquad (19)$$

with $0 \leq \theta \leq \pi$, $0 \leq \phi \leq \pi$ and $0 \leq \alpha \leq 2\pi$. The field homogeneity is obtained by selecting ψ, with a uniform p.d.f. between 0 and 2π. The modulus of $\tilde{u}_n$ is such that:

$$\tilde{u}_n = \sqrt{E(k_n)\Delta k_n} \qquad \text{with} \qquad k_n = |k_n| . \qquad (20)$$

$E(k_n)$ is the 3D kinetic energy spectrum, approximated by a von Karman expression which describes quite well the entire spectrum :

$$E(k) = \mathcal{A}\frac{\overline{u^2}}{k_e}\frac{(k/k_e)^4}{\left[1 + (k/k_e)^2\right]^{17/6}} \exp\left[-2\left(\frac{k}{k_\eta}\right)^2\right] \tag{21}$$

where k_η is the Kolmogorov length scale. The two parameters $\mathcal{A}$ and k_e are adjusted so as to give K and ϵ from the two relations:

$$K = \int_0^\infty E(k)dk \qquad \text{and} \qquad \epsilon = 2\nu \int_0^\infty k^2 E(k)dk. \tag{22}$$

The time evolution of the turbulence field for every Fourier mode is simply expressed in terms of ω_n with a Gaussian p.d.f. :

$$g(\omega) = \frac{1}{\omega_o\sqrt{2\pi}} \exp\left(-\frac{(\omega-\omega_o)^2}{2\omega_o^2}\right) \qquad \text{with} \qquad \omega_o = 2\pi\frac{\epsilon}{K}. \tag{23}$$

The SNGR model takes into account the mean flow effects on the acoustic waves propagation and the Green function is not required, so noise from internal flows can be treated with this method [29].

3. EXAMPLES OF AEROACOUSTIC PREDICTIONS

3.1. Noise of Isotropic Turbulence

Noise generated by isotropic turbulence is a fundamental case of interest, although not directly accessible to experiments. Analytical approaches are due to Proudman [30] and more recently to Lilley [31]. They provide estimates of the acoustic energy per unit of mass expressed as $\alpha M^5 u'^3/L$. Only computational approaches can offer the value of α as well as time traces and spectra of the radiated acoustic pressure. In the work of Witkowska, Brasseur & Juvé [32] a periodic box of stationary turbulence is considered with the following nondimensionalised characteristics : rms velocity $u' = 0.12$, longitudinal integral length scale $L = 0.42$, longitudinal Taylor microscale $\lambda = 0.23$, Kolmogorov length scale $\eta = 0.026$, turbulent Reynolds number $u'\lambda/\nu = 20$. A 128^3 box of size 2π permits to resolve all the scales satisfactorily. Forcing is achieved at every time step by increasing slightly the amplitude of each Fourier mode in the wave-number range (3-20) around the peak of the 3D kinetic energy spectrum. Expression (8) of the Lighthill analogy is used to obtain the radiated acoustic field, and no spurious effect was due to the forcing procedure which ensures the continuity of the second time derivatives of velocity. Ensemble averages are then done when needed over about ten realizations.

Two results are of interest : (i) the acoustic pressure spectrum peaks at a frequency four times the inverse eddy turn over time (see figure 2 showing that small eddies contribute to the emitted noise and their size, estimated

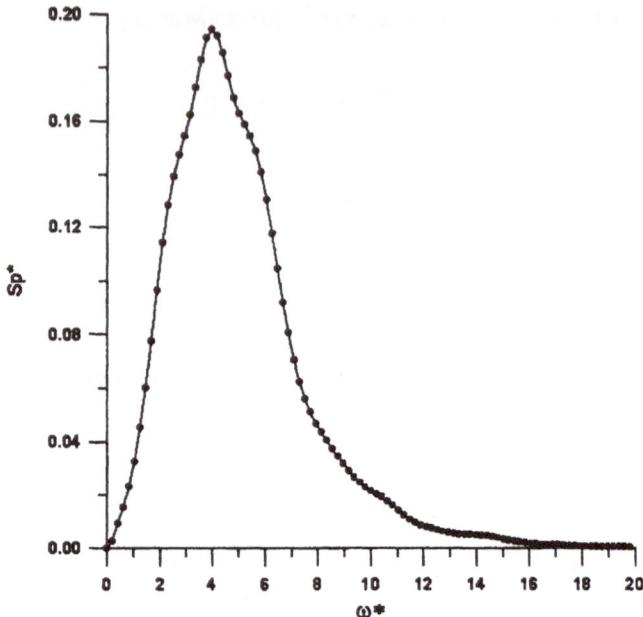

Fig. 2. — Acoustic pressure spectrum radiated by stationary isotropic turbulence, normalized by total acoustic power; $\omega* = \omega L/u'$, from Witkowska [32].

from a Fourier analysis of the velocity signals, is around one-tenth of the most energetic eddies. (ii) the Proudman constant α is found to be 2.1, compared to 2.6 in decaying isotropic turbulence resolved by either a DNS method [32, 33], or a LES approach [34]. The original value established by Proudman is $\alpha = 15$, based on a Heisenberg spectrum and quasi-normality of fourth order correlations. Use of a von Karman spectrum leads to $\alpha = 7$. Release of the quasi-normality constraint gives $\alpha = 3.6$ as recently recomputed by Lilley [31].

3.2. Jet Noise

Jets we have to deal with in engineering applications always have large Reynolds numbers, and noise estimates using modelled acoustic sources from $K-\epsilon$ results are very appropriate. For example, Béchara *et al.* [17] obtained excellent mixing noise estimates for round subsonic jets. It is important to note at once that a global and unique adjustable factor is needed in the model, for both the noise levels and the spectra, and that this factor is adjusted from a single comparison with experiment. This factor is then kept constant for all flow configurations, such as single jets at higher speed or coaxial jets. For example, in the latter case, the ratio between the secondary and primary exit velocities which minimizes the radiated acoustic power is numerically retrieved

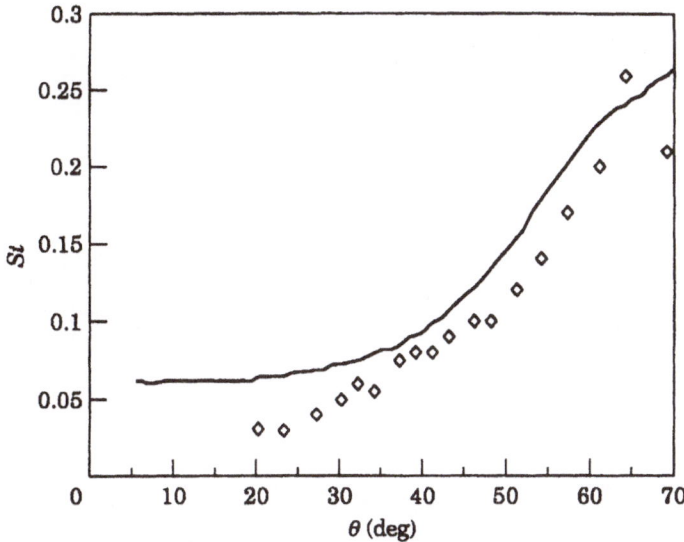

Fig. 3. — Angular dependence of the peak spectral amplitude for a heated supersonic jet, $M = 2.0$ and $T_j/T_o = 2.5$; curve : predictions by Bailly *et al.* [18]; ◊ measurements of Seiner *et al.* [36].

for several nozzle ratios. Bailly, Lafon & Candel extended this approach to supersonic jet mixing noise [19] and also to Mach wave noise [18] using an analytical development of Ffowcs Williams & Maidanik [35]. The model has been applied to a supersonic hot round jet, $M = 2$ and $T_j/T_o = 2.5$, to predict the spectral peak as a function of the observer angle θ, see figure 3. The power spectral density per unit length of the jet shows that the high frequency sources are located near the nozzle, where the convection Mach number M_c is the highest. So, the corresponding Mach wave angle $\theta = \cos^{-1}(1/M_c)$ is larger than the one associated with the low frequency sources located downstream where the convection Mach number is smaller. Figure 3 shows that a very good agreement is obtained with experimental data.

3.3. Diaphragm Noise

Noise generation and propagation by confined turbulent flows is a case where the SNGR model is pertinent. The configuration of a 2D duct obstructed by a diaphragm has already been studied by Van Herpe, Crighton & Lafon [37] in the framework of the Lighthill analogy, but at the cost of a very elaborate Green function. Independently, Bailly, Lafon & Candel [29], resolved the linearized Euler equations (16) using the same aerodynamic characteristics, mean flow, turbulent kinetic energy, see Figure 4, and dissipation. One hundred Fourier modes are used for the synthesized sources described in section 2.6.

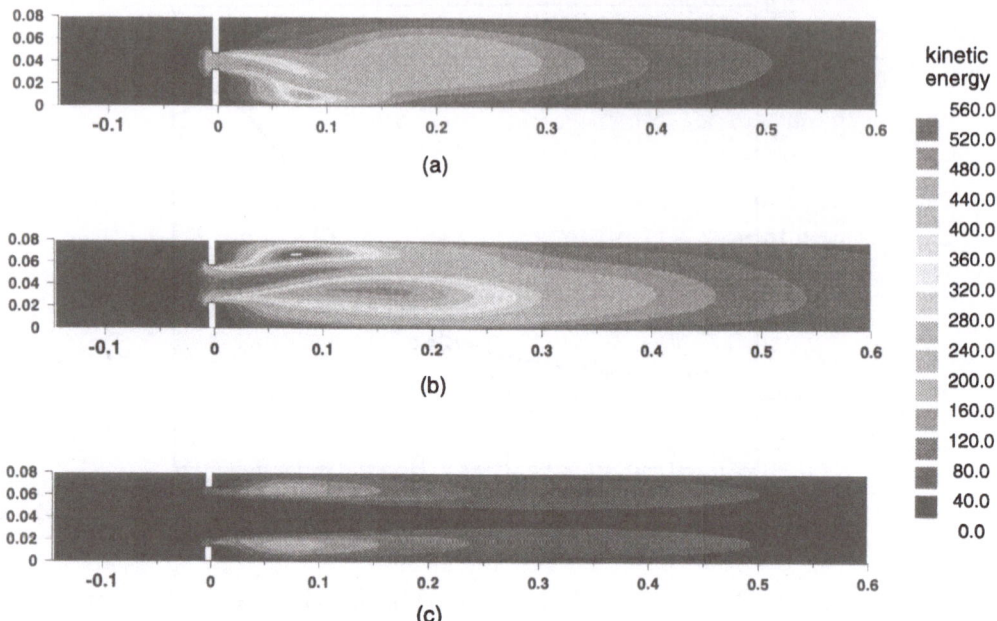

Fig. 4. — Turbulent kinetic energy downstream of a diaphragm in a square duct, section 80 mm x 80 mm, (a) diaphragm width $e = 15$ mm, upstream velocity $U_o = 14$ m/s, (b) $e = 35$ mm, $U_o = 32$ m/s, (c) $e = 55$ mm, $U_o = 55$ m/s, from Bailly et al. [29].

The computation method uses a fractional step scheme and relies on solutions of one-dimensional propagation problems in terms of a weak formulation. Numerical tests indicate that the sound wave propagation is calculated with little dispersion and diffusion. An acoustic grid of 601×32 points with a constant mesh size $(3.0 \times 10^{-3}$m$)$ is used to provide an accurate solution. The radiated acoustic power is obtained by recording the pressure fluctuations at both ends of the computation domain. However acoustic fluctuations as well as vorticity fluctuations are recorded at the outflow boundary. A practical way to achieve the separation is to stretch out the duct length $\mathcal{L}$ in order to increase the difference of travel times $\mathcal{L}/(U_c + c_o) - \mathcal{L}/U_c$ between acoustic and vorticity waves. The acoustic fluctuations are then recorded during a long enough time before the vorticity fluctuations arrive, as illustrated in Figure 5. The expected U^4 law typical of the acoustic radiation in such confined configurations is retrieved and figure 6 shows that the acoustic intensity spectrum is close to the experimental data.

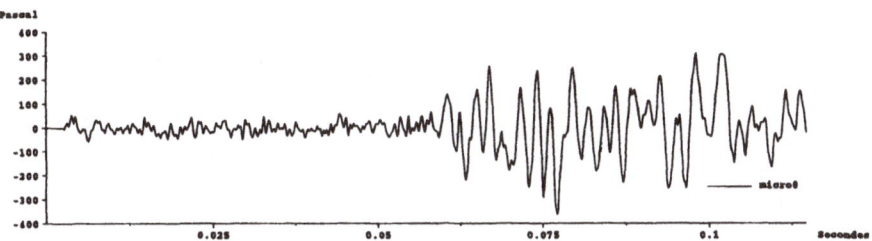

Fig. 5. — Time pressure signal recorded by a microphone far downstream of the diaphragm, $e = 35$ mm, $U_o = 14$ m/s, from Bailly et al. [29].

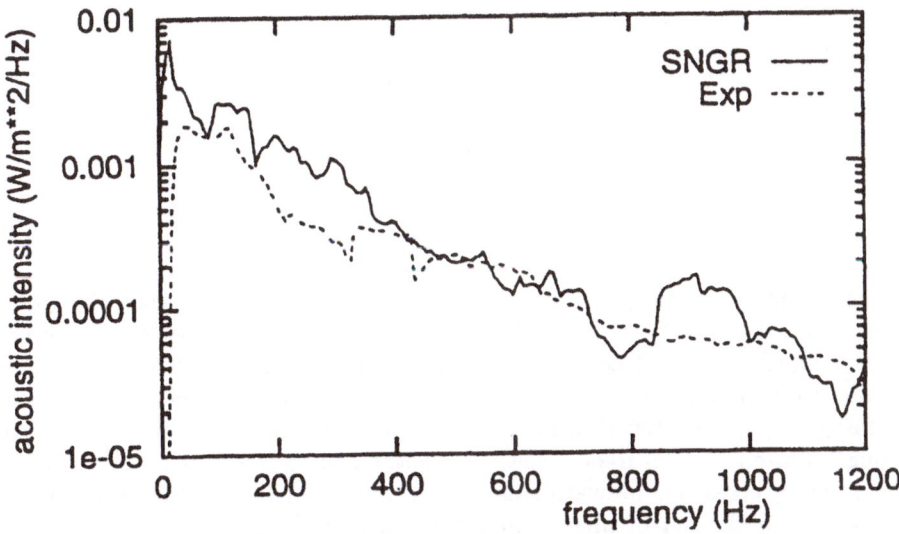

Fig. 6. — Acoustic pressure spectra far downstream of the diaphragm, comparison between SNGR predictions and experimental data, $e = 35$ mm $U_o = 14$ m/s, from Bailly et al. [29].

4. SOUND PROPAGATION THROUGH TURBULENCE

Both velocity and temperature fluctuations introduce inhomogeneities in the medium seen by the acoustic waves. Their effects are similar but not identical, simply because the perturbations have a different nature, i.e. one is a vector, the other a scalar, and they appear differently in the acoustic equations.

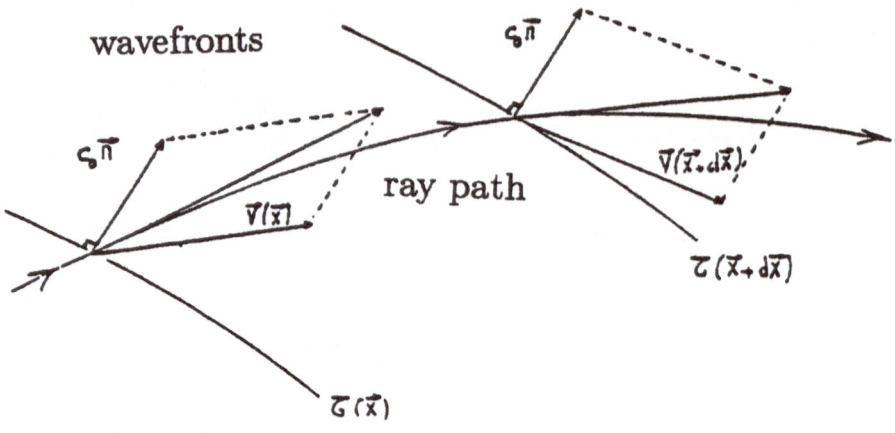

Fig. 7. — Ray path construction in a moving medium.

4.1. Ray Tracing Method

Figure 7 illustrates the way a ray path can be traced out of a point belonging to a wave front when times increases, see [6] for example. The turbulent velocity vector u adds vectorially with the wave velocity cn, where n is the unit vector normal to the wave front. The scalar temperature fluctuation T' affects the amplitude of the sound celerity $c = \sqrt{\gamma r T}$ with which the wave front moves normal to itself. Any deviation of the acoustic ray with respect to the linear uniform propagation existing in a medium at rest is then clearly visible. Although limited to high frequencies, the method has several applications : the obtention of eigenrays between sources and receivers, the prediction of arrival times, the prediction of phase differences as long as caustics have not been reached.

The equations governing the ray trajectory, hence giving the location and inclination of the wave front, have in a moving medium the general form derived by Candel [38]:

$$\frac{dx_i}{dt} = \frac{c_0}{N} \frac{p_i}{N - p_j M_j} + u_i \tag{24}$$

$$\frac{dp_i}{dt} = \frac{c_0}{N} \frac{\partial N}{\partial x_i} - \frac{1}{N} p_j \frac{\partial u_j}{\partial x_i} \tag{25}$$

$$p = \frac{N}{1 + M.n} \tag{26}$$

where x is the position vector, p a nondimensional wave vector pn with $n = k/k_0$, N the refraction index $c_0/c = 1 - T'/2\bar{T}$, u the velocity field and M the flow Mach number u/c_0. The rays are parametrized by the transit time t from the source to a given point. Initial conditions are given by the incident wave front.

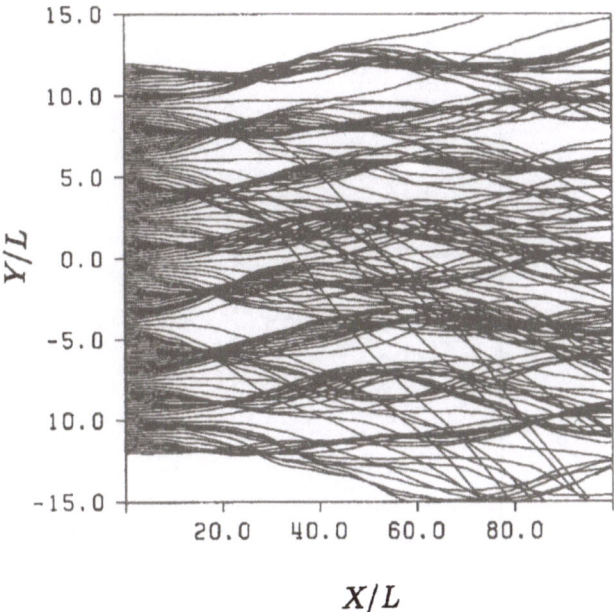

Fig. 8. — Caustics in a 2D turbulent velocity field : ray trajectories, $u'/c_0 = 5.882.10^{-3}$ and $L = 0.10$ m, from Blanc-Benon et al [39].

The resolution of (24) to (26) requires a complete description of the space-time turbulence field. In many cases, DNS data will not be available, so that the synthesized turbulence field (18) developed by Blanc-Benon et al [39] and Karweit et al [40] has two avantages : i) all the spatial derivatives are known analytically; numerical errors are then reduced and computation time is saved in comparison with the usual finite-difference schemes; ii) unlimited fields are offered, thus permitting the cumulative effects to develop as the acoustic waves propagate. In practice the time dependence in (18) can be suppressed because the acoustic transit time is very short compared to the evolution time of turbulence.

Figure 8 illustrates ray paths for an initially plane acoustic wave propagating through a 2D turbulent velocity field whose global characteristics are $u'/c_0 = 5.882.10^{-3}$ and $L = 0.10$ m . The field is synthesized by 50 random Fourier modes uniformly distributed between $0.1L$ and $10L$. The kinetic energy spectrum is such that the longitudinal correlation $f(r)$ is Gaussian, a form which also permits analytical approaches. The occurrence of the first caustics can also be obtained by solving additional equations governing the cross-sectional area of an infinitesimal ray tube (caustics occur when the ray tube section vanishes). Figure 9 clearly shows that caustics occur for a propagation distance of the order of $8L - 15L$. Turbulence boxes whose size is at

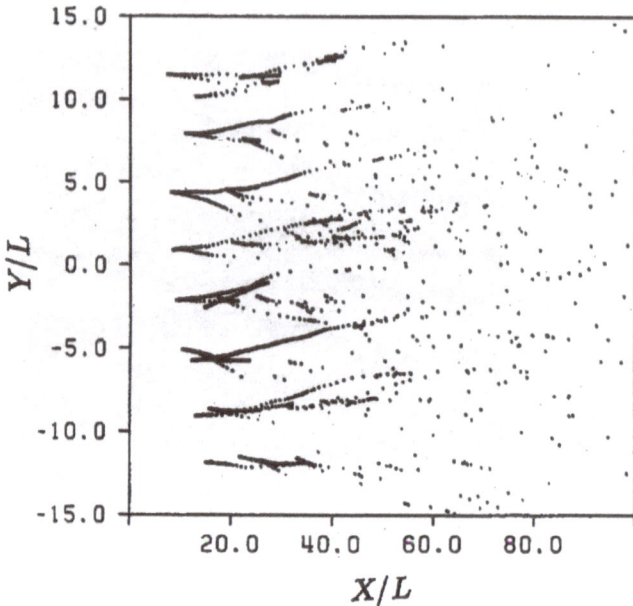

Fig. 9. — Caustics in a 2D turbulent velocity field : spatial localisation, $u'/c_0 = 5.882.10^{-3}$ and $L = 0.10$ m, from Blanc-Benon et al [39].

maximum $10L$ would hardly permit the first caustic to be observed. A slight anisotropic bias would also affect the results as observed by Karweit et al due to the fact that Fourier modes in a box are distributed on a fixed lattice [41].

For temperature fluctuations, the caustics appear a little less rapidly. The comparison takes place with $T'/2\bar{T} = 5.882 \times 10^{-3}$ and $L = 0.10$ m. The rms value T' is such that the two perturbation indexes $T'/2\bar{T}$ and u'/c_0 are equal. For the space correlation we keep a Gaussian form, the spectrum is then accordingly adjusted for a 2D scalar random field. More details can be found in Blanc-Benon, Juvé & Comte-Bellot [39].

4.2. Helmholtz Equation

The ray tracing method we have just seen offers a clear view of the irregular meandering of acoustic trajectories through turbulence fields. Prediction of sound levels is a more difficult task. Sticking to the ray concept, it would require the determination of all the eigenrays joining the source to a given point and additional equations for the evolution of the cross sectional area of ray tubes. This approach remains feasible in 2D and before caustics. In 3D the computations are too long, and when caustics are met, the computations break down. Moreover, shadow zones are beyond reach as diffraction effects are neglected in the ray approximation.

Acoustic pressure levels are hence better predicted using approximate forms of the acoustic wave equation. An exact equation remains beyond possibilities, so that it is generally accepted following the work of Neubert and Lumley [42] and Tatarski [43], that for monochromatic waves:

$$p(x,t) = p'(x) \exp(-i\omega t) \qquad (27)$$

the following Helmholtz equation with a random index is suitable:

$$\left[\nabla^2 + k_0^2 (1 + \epsilon(x,t))\right] p'(x) = 0 \qquad (28)$$

$$\epsilon(x,t) = -2\left(1 - \frac{i}{k_0}\frac{\partial}{\partial x}\right)\left(\frac{u_x(x,t)}{c_0} + \frac{1}{2}\frac{T'(x,t)}{\bar{T}}\right). \qquad (29)$$

Here u_x denotes the velocity component in the direction of propagation x. The imaginary part of ϵ can often be neglected because the scattering of acoustic waves by turbulence is most effective at small angles due to forward scattering, so that :

$$\epsilon(x,t) = -2\frac{u_x(x,t)}{c_0} - \frac{T'(x,t)}{\bar{T}}. \qquad (30)$$

The synthesized turbulence field we used earlier (sections 2.6 and 4.1) is again very useful for solving the Helmholtz equation (28) for either velocity or temperature turbulent fields.

4.3. Parabolic Approximation

The parabolic approximation is deduced from the Helmholtz equation (28) assuming that the space dependence of the acoustic pressure has an envelop $\Psi(x, \rho)$ slowly varying with x:

$$p'(x) = p'(x, \rho) = \Psi(x, \rho) \exp(ik_0 x) \qquad (31)$$

where ρ is the lateral separation. It is then possible to neglect the second derivative $\partial^2\Psi(x, \rho)/\partial x^2$ with respect to the other terms, so that:

$$2ik_0 \frac{\partial}{\partial x}\Psi(x, \rho) + \Delta_\rho \Psi(x, \rho) + k_0^2 \epsilon(x, \rho) \Psi(x, \rho) = 0. \qquad (32)$$

This parabolic equation is easily solved by Fourier techniques or finite difference approximations. In this form its validity is limited to angles within $\pm 15 deg$ relative to the prefered (mean) direction of propagation. Other parabolic equations with wide angle capabilities have been proposed in the literature and used for studying the propagation of sound in a turbulent atmosphere [44].

Let us point out that a crucial nonlinear coupling term $\epsilon(x, \rho)\Psi(x, \rho)$ between acoustics and turbulence exists in equation (32). The synthesized turbulent field already proposed permits to solve (32) on individual realizations,

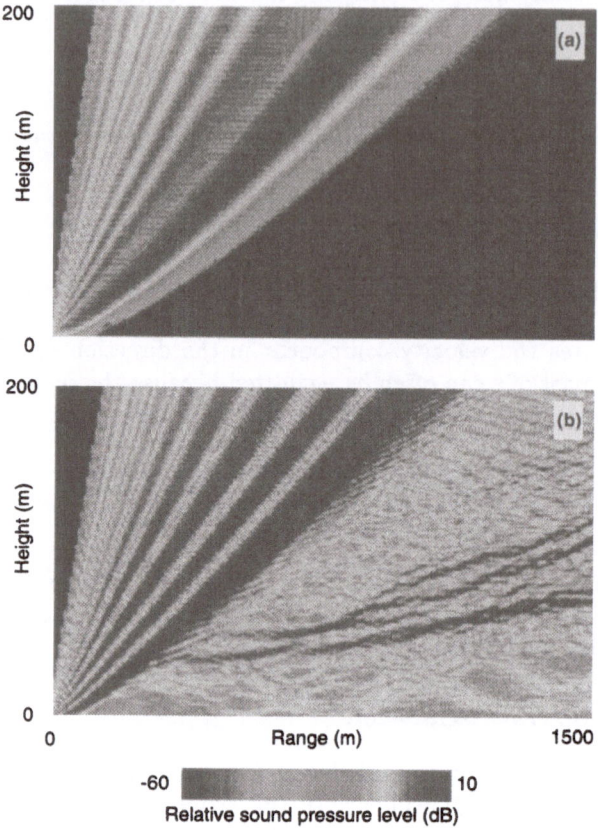

Fig. 10. — Noise scattering in a shadow zone; the upward-refracting index is given by (34), $T'/2\bar{T} = 1.414 \times 10^{-3}$ and $L = 1.1$ m , source height h=3,7 m; frequency f= 424 Hz, from Chevret *et al* [44].

without any additional assumption. By ensemble averaging one is then able to obtain any of the higher order statistical characteristics of the acoustic pressure, such as the usual acoustic intensity $\langle p^2 \rangle / \rho c$ and its normalized variance:

$$\sigma_I = \frac{<I^2> - <I>^2}{<I>^2} \tag{33}$$

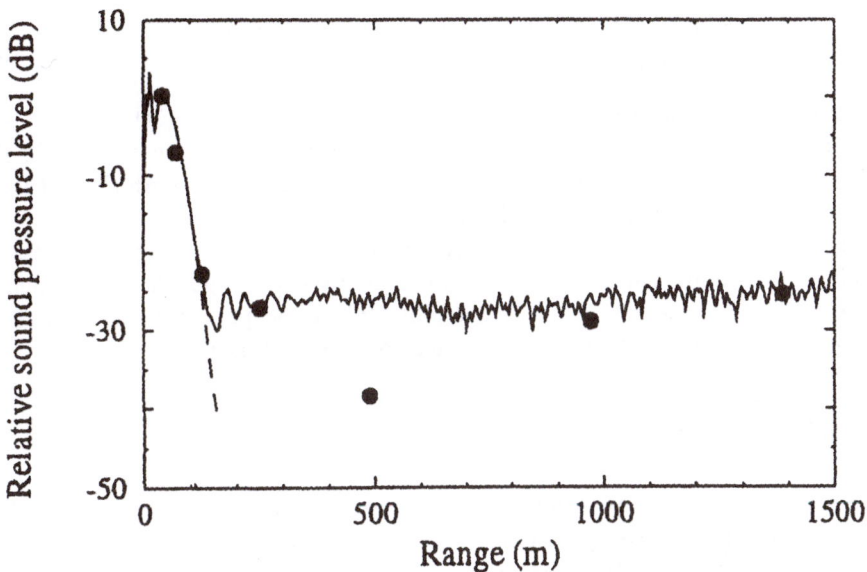

Fig. 11. — Sound levels in the shadow zone, same conditions as in figure 10; curve : predictions by Chevret *et al* [44]; • measurements of Wiener and Keast [45].

which expresses the scintillation index.

This new approach is very advantageous compared to the previous methods which deduced the equations governing the higher moments of Ψ from equation (32) multiplying it by powers of Ψ and averaging, hence generating new nonlinear terms of increasing complexity.

The effect of turbulence on the atmospheric propagation of acoustic waves has been numerically predicted by Chevret, Blanc-Benon and Juvé [44] using the technique we have described. Two situations are of interest: i) the propagation over small distances near the ground where a partial loss of coherence between the direct and the reflected paths tends to smear out the interference pattern, and (ii) the propagation over long distances in a upward-refracting atmosphere. In the absence of turbulence, there would be a deep shadow zone. In the presence of turbulence, sound can penetrate the shadow zone due to diffraction at low frequency and scattering at high frequency. Indeed many in situ experiments report non negligible acoustic levels in that zone (see for example [45]).

Figure 10 illustrates the shadow zone in the absence of turbulence (a) and the filling-in due to turbulence scattering (b). These results have been recently obtained by Chevret, Blanc-Benon and Juvé [44]. The mean sound profile is given by:

$$\begin{cases} c = 340 - 2\ln(z/0.006) & z \geq 0.01\text{m} \\ c = 340 - 2\ln(0.01/0.006) & z < 0.01\text{m} \end{cases} \qquad (34)$$

Turbulence is due to temperature fluctuations, $T'/2\bar{T} = 1.414 \times 10^{-3}$ and $L = 1.1$ m. The source is 3.7 m above the ground and its frequency is 424 Hz. These specifications have been chosen to permit comparisons with the outdoors experiments of Wiener and Keast [45].

Figure 11 shows a good agreement betweeen the predicted mean pressure levels obtained by Chevret, Blanc-Benon et Juvé [44] and the experimental data collected by Wiener and Keast [45]. Thus the procedure developed to quantify the acoustic-turbulence interaction is very satisfactory.

5. CONCLUSION

The acoustic community needs full space-time velocity and temperature fields to make progress in the prediction of both the noise generated by turbulent flows and the perturbations brought by turbulence to sound levels. DNS predictions give useful results, however at still too low Reynolds numbers. At the high Reynolds numbers occurring in almost all engineering applications, the synthesized turbulence fields, made of random Fourier modes adjustable to any turbulence spectra, have shown a great utility. In our group, we continue to enlarge the technique, improving different aspects, such as the time evolution of the synthesized field and the codes resolving the linearized Euler equations in 3D.

References

[1] Lele S.K., *J. Comput. Physics* **103** (1992) 16.

[2] Tam C.K.W., Webb J.C., *J. Comput. Phys.* **107** (1993) 262.

[3] Colonius T., Moin P.,Lele S.K., *Dept. Mech. Eng., Stanford University*, TF-65, 1995.

[4] Colonius T., Lele S.K., Moin P., *15th Aeroacoustics Conference*, Long Beach, CA, AIAA-93-4328, 1993.

[5] Mitchell B.E., Lele S.K., Moin P. *Dept. Mech. Eng., Stanford University*, TF-66, 1995.

[6] Pierce A.D., Acoustics : An Introduction to Its Physical Principles and Applications, McGraw Hill, 1996.

[7] Lyrintzis A.S., *J. Fluids Engineering* **116** (1994) 665.

[8] Soh W.Y., *32nd Aerospace Science Meeting*, Reno, N.V., AIAA-94-0138, 1994.

[9] Lyrintzis A.S., Mankbadi R.R., *AIAA Journal* **34** (1996) 413.

[10] Gamet L., Ph.D. Thesis, *Ecole Nationale Supérieure de l'Aéronautique et de l'Espace*, 1996-183.

[11] Freund J.B., Lele S.K., Moin P., *AIAA Journal* **34** (1996) 909.

[12] Lighthill M.J., *Proc. Roy. Soc. London* **A 211** (1952) 564.

[13] Witkowska A., Juvé D., *C.R. Acad. Sci. Paris* **318 (II)** (1994) 597.

[14] Crighton D., *Prog. Aerospace Sci.* **16** (1975) 31.

[15] Ffowcs Williams J.E., *Phil. Trans. Roy. Soc.* **A 225** (1963) 469.

[16] Ribner H.S., *J. Fluid Mech.* **38** (1969) 1.

[17] Béchara W., Lafon P., Bailly C., Candel S., *J. Acoust. Soc. Am.* **97** (1995) 3518.

[18] Bailly C., Candel S. & Lafon P., *J. Sound Vib.* **194** (1996) 219.

[19] Bailly C., Lafon P., Candel S., *Acta Acustica* **2** (1994) 101.

[20] Powell A., *J. Acoust. Soc. Am.* **36** (1964) 177.

[21] Howe M.S. *J. Fluid Mech.* **71** (1975) 625.

[22] Hardin J., *ASME Symposium on Noise anf Fluid Engineering*, Atlanta, G.A., 1977.

[23] Möhring W., *J. Fluid Mech.* **85** (1978) 685.

[24] Knio O.M., Collorec L., Juvé D., *J. Comput. Phys.* **116** (1995) 226.

[25] Knio O.M., Juvé D., *C.R. Acad. Sci. Paris* **322 (IIb)** (1996) 591.

[26] Lilley G.M., *Air Force Aero Propulsion Laboratory*, **IV**, AFAPL-TR-72-53, 1972.

[27] Bailly C., Ph.D. Thesis, *Ecole Centrale Paris*, 1994-19.

[28] Bailly C., Lafon P., Candel S., *16th AIAA Aeroacoustics Conference*, Münich, Germany, AIAA-95-092, 1995.

[29] Bailly C., Lafon P., Candel S., *17th Aeroacoustics Conference*, State College, PA, AIAA-96-1732, 1996.

[30] Proudman I., *Proc. Roy. Soc. London* **A 214** (1952) 119.

[31] Lilley G.M., *J. Sound Vib.* **190** (1996) 463.

[32] Witkowska A., Brasseur J.G., Juvé D., *16th AIAA Aeroacoustics Conference*, Münich, Germany, AIAA-95-037, 1995.

[33] Sarkar S., Hussaini M.Y., *ICASE* Report 93-74, 1993.

[34] Witkowska A., Ph.D. Thesis, *Ecole Centrale de Lyon*, 1994-58.

[35] Ffowcs Williams J.E., Maidanik G., *J. Fluid Mech.* **21** (1965) 641.

[36] Seiner J.M., Ponton M.K., Jansen B.J., Lagen N.T., AIAA Paper 92-02-046, 1992.

[37] Van Herpe F., Crighton D., Lafon P., *16th Aeroacoustics Conference*, Münich, Germany, AIAA-95-035, 1995.

[38] Candel S.M., *J. Fluid Mech.* **83** (1977) 465.

[39] Blanc-Benon Ph., Juvé D., Comte-Bellot G., *Theoret. Comput. Fluid Dynamics* **2** (1991) 271.

[40] Karweit M.J., Blanc-Benon Ph., Juvé D., Comte-Bellot G., *J. Acoust. Soc. Am.* **89** (1991) 52.

[41] Karweit M., Blanc-Benon Ph., *J. Computational Acoustics* **3** (1995) 203.

[42] Neubert J.A., Lumley J.L., *J. Acoust. Soc. Amer.* **48** (1970) 1212.

[43] Tatarskii V.I., The Effect of a Turbulent Atmosphere on Wave Propagation I.P.S.T. Keter, Jerusalem, 1971.

[44] Chevret P., Blanc-Benon Ph., Juvé D., *J. Acoust. Soc. Amer.* (1996), in press.

[45] Wiener F.M., Keast D.N., *J. Acoust. Soc. Am.* **31** (1959) 724.

Vortices in Compressible LES and Non-Trivial Geometries

P. Comte *

Turbulence Simulation & Modelling Team
L.E.G.I./Institut de Mécanique de Grenoble
B.P. 53
F 38041 GRENOBLE Cedex 9
FRANCE

1. INTRODUCTION

The distinction between Direct Numerical Simulation and Large-Eddy Simulations seems to be fading away with time: it is usually accepted that resolution of all scales of motion down to the very Kolmogorov scale is a waste of computational resources, cut-off at about 5η being quite sufficient in practice (see e.g. [1]). Some go further and rely upon numerical dissipation to supersede a subgrid-scale-turbulence model ([2], [3]). On the other hand, LES are often performed in industrial configurations involving domains and Reynolds numbers so large that near-wall turbulence cannot be sufficiently resolved ([4], [5], [6]...). Law-of-the-wall models then have to be used, as in one-point closures. Finally, some SGS models (*e.g.* [7]) make use of a transport equation for the subgrid kinetic energy, which is also reminiscent of one-point closures.

Our team strategy is, for the moment, among the most traditional ones: we try to work out SGS models which are as code-independent as possible, checking

---

(*) Institut National Polytechnique de Grenoble (INPG), Université Joseph Fourier (UJF) and Centre National de la Recherche Scientifique (CNRS).

them systematically in conjunction with non-dissipative pseudo-spectral methods. We make use of EDQNM closures in incompressible isotropic turbulence
to derive spectral eddy-viscosity and eddy-diffusivity coefficients in the spirit
of Kraichnan [8]. Analogous coefficients local in the physical space are then
proposed, striving to be consistent with their spectral counterparts as much as
is permitted by Heisenberg's principle: in addition to locality in the physical
space, a certain locality is required in the spectral space in order to prevent
the model from acting as long as there is no energy at the cut-off scale. Otherwise, transition could be hampered, especially in the case of low-Mach-number
wall-bounded flows for which the most unstable modes are of viscous origin.

This short presentation aims at showing examples of what can be obtained
with a supercomputer and simple eddy-viscosity models. We are of course
aware of the shortcomings of such models, such as the assumption that the
subgrid stress tensor is aligned with the resolved strain rate tensor. This has
been experimentally found to be wrong, even at large Reynolds numbers [9].
Using eddy-viscosity models thus implies that we previlege the energetical aspect (extract the "right" amount of energy from the "dynamical system" we
solve) with respect to the phase information (among which the principal axes of
the subgrid stress tensor). For the design of our models, it is therefore consistent to import information from two-point stochastic closures such as EDQNM,
which correctly reproduces the phenomenology of incompressible isotropic turbulence, at least at the energetic level.

One controversial question among is the role that numerical dissipation can
play in the turbulence-modelling process, either through the nature of the
scheme or the mesh size. In our LES, the solutions to the equations solved
do contain a certain level of kinetic energy in the smallest resolved scales. This
is often criticized on the ground that all numerical methods behave badly in the
small scales (even the spectral methods blur the phase information at the highest wavenumber). Validation then has to be performed on physical grounds,
through comparison with experimental data, predictions of stability theories or
numerical results obtained with different methods. Note that Leonard[10], who
coined the expression *Large-Eddy Simulation*, proposed a formalism thanks to
which no energy would be left in the smallest resolved scales([1]). To the other
extreme, some claim that numerical dissipation can play the role of a subgrid-
scale turbulence model, and sometimes that of the molecular viscous terms too
(approaches refered to as *Monotonically Integrated LES*, *Built-In LES*, and so
on).

This paper is organized as follows: in sections 2 and 3 is presented the most
straightworward extension of the LES formalism to compressible flows in curvi-

---

([1]) The Navier-Stokes equations are first convolved through a continuous low-pass
filter which commutes with the time and space derivatives. The resulting equations
are then closed thanks to a subgrid-scale turbulence model. This closed system of
equations is eventually discretized onto a grid which is finer than the cut-off scale of
the filter, so that the result can be checked to be independent of the mesh size (but
of course not of the filter's cut-off scale).

linear co-ordinates. A fully-explicit (2,4) solver is then briefly described, with emphasis put on some details of implementation that are usually overlooked (such as the way to discretize the metrics). Finally, two applications are presented: the transition to turbulence of a high-Mach number boundary layer (section 5) and the acoustically-driven vortex shedding in a model solid rocket engine (section 6). The models used in these 2 cases are not described here, the reader is refered to Lesieur and Métais [11] or Lesieur [12].

2. COMPRESSIBLE L.E.S. FORMALISM

In Cartesian co-ordinates, the compressible Navier-Stokes equations can be cast in the so-called fast-conservation form

$$\frac{\partial U}{\partial t} + \frac{\partial F_1}{\partial x_1} + \frac{\partial F_2}{\partial x_2} + \frac{\partial F_3}{\partial x_3} = 0 \quad , \tag{1}$$

with

$$U = {}^T(\rho, \rho u_1, \rho u_2, \rho u_3, \rho e) \quad , \tag{2}$$

ρe being the total energy defined by, for an *ideal* gas,

$$\rho e = \rho \, C_v \, T + \tfrac{1}{2}\rho(u_1^2 + u_2^2 + u_3^2) \tag{3}$$

The fluxes F_i read, $\forall i \in \{1, 2, 3\}$,

$$F_i = \begin{pmatrix} \rho u_i \\ \rho u_i u_1 - \sigma_{i1} \\ \rho u_i u_2 - \sigma_{i2} \\ \rho u_i u_3 - \sigma_{i3} \\ \rho e u_i - u_j \sigma_{ij} - k\dfrac{\partial T}{\partial x_i} \end{pmatrix} \quad , \tag{4}$$

$k = \rho C_p \kappa$ being the thermal conductivity (and κ the thermal diffusivity). The components σ_{ij} of the stress tensor are given by the Newton law

$$\sigma_{ij} = -p \, \delta_{ij} + \mu S_{ij} \quad , \tag{5}$$

in which

$$S_{ij} = \left[\frac{\partial u_j}{\partial x_i} + \frac{\partial u_i}{\partial x_j} - \frac{2}{3}(\nabla.u)\delta_{ij} \right] \tag{6}$$

denotes the deviatoric part of the strain-rate tensor. Bulk viscosity is neglected (Stokes hypothesis), as commonly accepted except in extreme thermodynamical situations. This yields

$$F_i = \begin{pmatrix} \rho u_i \\ \rho u_i u_1 + p \, \delta_{i1} - \mu S_{i1} \\ \rho u_i u_2 + p \, \delta_{i2} - \mu S_{i2} \\ \rho u_i u_3 + p \, \delta_{i3} - \mu S_{i3} \\ (\rho e + p)u_i - \mu u_j S_{ij} - k\dfrac{\partial T}{\partial x_i} \end{pmatrix} \tag{7}$$

The Sutherland empirical law

$$\mu(T) = \mu(273.15) \left(\frac{T}{273.15}\right)^{1/2} \frac{1 + S/273.15}{1 + S/T} \tag{8}$$
$$\text{with } \mu(273.15) = 1.711 \ 10^{-5} Pl \text{ and } S = 110.4K$$

and its extension to temperatures lower than 120 K :

$$\mu(T) = \mu(120) \ T/120 \quad \forall \ T < 120 \quad , \tag{9}$$

are prescribed for molecular viscosity. Conductivity $k(T)$ is obtained assuming the Prandtl number $Pr = C_p \mu(T)/k(T)$ constant and equal to 0.7, as in air at ambiant temperature. The equation of state

$$p = R \rho T \tag{10}$$

closes the system, with $R = C_p - C_v = \frac{\mathcal{R}}{\mathcal{M}} = 287.06 \ Jkg^{-1}K^{-1}$ for air.

2.1.   LES Filtering Procedure

As in the incompressible regime, whatever the numerical method used, the discretization of the above equations introduces a cut-off scale Δ which is by hypothesis larger than the Kolmogorov scale. The simplest way of accounting for this is to picture the numerical treatment as a low-pass filter of width Δ, defined, in the case of uniform cubic meshes of side Δ, by [10]

$$\overline{\phi}(x) = \left(\frac{1}{\Delta}\right)^3 \int \prod_{i=1}^{3} G\left(\frac{x_i - x_i'}{\Delta}\right) \phi(x') dx_1' dx_2' dx_3' \tag{11}$$

for a given variable ϕ, with

$$G(\xi) = \begin{cases} 1 \ , \ \text{if } |\xi| \leq \frac{1}{2} \\ 0 \ , \ \text{otherwise} \ . \end{cases} \tag{12}$$

The operator $\overline{}$ commutes with the space and time derivatives. Convolution of the above equations therefore yields

$$\frac{\partial \overline{U}}{\partial t} + \frac{\partial \overline{F}_1}{\partial x_1} + \frac{\partial \overline{F}_2}{\partial x_2} + \frac{\partial \overline{F}_3}{\partial x_3} = 0 \quad , \tag{13}$$

with

$$\overline{\rho e} = \overline{\rho \ c_v \ T} + \frac{1}{2}\overline{\rho(u_1^2 + u_2^2 + u_3^2)} \tag{14}$$

and

$$\overline{p} = \overline{\rho \ R \ T} \tag{15}$$

At this level, it is convenient to introduce the density-weighted (or Favre [13]) filter $\tilde{}$ defined, for a given variable ϕ, by

$$\tilde{\phi} = \frac{\overline{\rho\phi}}{\overline{\rho}} \quad . \tag{16}$$

We then have

$$\overline{U} = {}^T(\overline{\rho}, \overline{\rho u_1}, \overline{\rho u_2}, \overline{\rho u_3}, \overline{\rho e}) \quad , \tag{17}$$

and the *resolved total energy*

$$\overline{\rho e} = \overline{\rho \tilde{e}} = \overline{\rho} \, C_v \, \tilde{T} + \tfrac{1}{2}\overline{\rho(u_1^2 + u_2^2 + u_3^2)} \quad . \tag{18}$$

The resolved fluxes $\overline{F_i}$ read

$$\overline{F_i} = \begin{pmatrix} \overline{\rho \tilde{u}_i} \\ \overline{\rho u_i u_1} + \overline{p}\,\delta_{i1} - \overline{\mu S_{i1}} \\ \overline{\rho u_i u_2} + \overline{p}\,\delta_{i2} - \overline{\mu S_{i2}} \\ \overline{\rho u_i u_3} + \overline{p}\,\delta_{i3} - \overline{\mu S_{i3}} \\ \overline{(\rho e + p)u_i} \qquad - \overline{\mu S_{ij}u_j} - k\dfrac{\overline{\partial T}}{\partial x} \end{pmatrix} \quad , \tag{19}$$

with the filtered equation of state

$$\overline{p} = \overline{\rho} R \tilde{T} \quad . \tag{20}$$

2.2.   The Simplest Possible Closure

The usual subgrid-stress tensor $\overline{\overline{T}}$ of components

$$T_{ij} = -\overline{\rho u_i u_j} + \overline{\rho}\tilde{u}_i\tilde{u}_j \tag{21}$$

is introduced and split into its isotropic and deviatoric parts, the latter being noted $\overline{\overline{\tau}}$:

$$T_{ij} = \underbrace{T_{ij} - \frac{1}{3}T_{ll}\delta_{ij}}_{\tau_{ij}} + \frac{1}{3}T_{ll}\delta_{ij} \quad . \tag{22}$$

Equations (19) and (18) then read

$$\overline{F_i} = \begin{pmatrix} \overline{\rho}\tilde{u}_i \\ \overline{\rho}\widetilde{u_i u_1} + (\overline{p} - \frac{1}{3}T_{ll})\,\delta_{i1} - \tau_{i1} - \overline{\mu S_{i1}} \\ \overline{\rho}\widetilde{u_i u_2} + (\overline{p} - \frac{1}{3}T_{ll})\,\delta_{i2} - \tau_{i2} - \overline{\mu S_{i2}} \\ \overline{\rho}\widetilde{u_i u_3} + (\overline{p} - \frac{1}{3}T_{ll})\,\delta_{i3} - \tau_{i3} - \overline{\mu S_{i3}} \\ \overline{(\rho e + p)u_i} \qquad\qquad - \overline{\mu S_{ij}u_j} - k\dfrac{\overline{\partial T}}{\partial x} \end{pmatrix} \quad . \tag{23}$$

and

$$\overline{\rho \tilde{e}} = \overline{\rho}\, C_v \tilde{T} + \frac{1}{2}\overline{\rho}\,(\widetilde{u_1}^2 + \widetilde{u_2}^2 + \widetilde{u_3}^2) - \frac{1}{2}T_{ll} \tag{24}$$

There are two options for the treatment of the uncomputable term T_{ll}:

- simply neglect it, arguing as in [14] that it can be re-written as $T_{ll} = \gamma M_{\text{sgs}}^2\overline{p}$, in which the *subgrid Mach number* M_{sgs} can be expected to be small when M_∞ is small.

- model it, as proposed by Yoshizawa [15], in a way which is consistent with the model chosen for $\overline{\overline{\tau}}$ (see e.g. [16]). Note that this was the initial choice of Erlebacher et al. [17].

We will here choose the first option, as in [18], bringing another argument: the incompressible LES formalism (see e.g. [19]), often introduces the *macro-pressure*

$$\varpi = \bar{p} - \frac{1}{3}T_{ll} \quad . \tag{25}$$

It thus seems a good idea to re-write equation (24) as

$$\widetilde{\rho e} = \bar{\rho}\, C_v \left(\tilde{T} - \frac{1}{2C_v\bar{\rho}} T_{ll} \right) + \frac{1}{2}\bar{\rho}\, (\widetilde{u_1}^2 + \widetilde{u_2}^2 + \widetilde{u_3}^2) \tag{26}$$

and introduce a *macro-temperature*

$$\vartheta = \tilde{T} - \frac{1}{2C_v\bar{\rho}} T_{ll} \quad , \tag{27}$$

computable out of $\overline{U}$ thanks to equation (26). The filtered equation of state (20) then reads

$$
\begin{aligned}
\varpi &=& \bar{\rho}R\vartheta + \left(\frac{R}{2C_v} - \frac{1}{3} \right) T_{ll} \\
&=& \bar{\rho}R\vartheta + \frac{3\gamma - 5}{6}\, T_{ll} \quad .
\end{aligned}
\tag{28}
$$

Thus, for gases like argon or helium (for which $\gamma \approx 5/3$ at $T = 298K$), the contribution of T_{ll} to equation (28) is quite negligible whatever the Mach number, which makes ϖ computable in all cases. It is extremely tempting to generalize this to air (for which $\gamma \approx 1.4$) by assuming

$$\varpi \simeq \bar{\rho}R\vartheta \quad . \tag{29}$$

In other words, the first option amounts to assume $\frac{3\gamma - 5}{6}\, \gamma M_{sgs}^2 \ll 1$ in the equation of state only, which sounds slightly less stringent than assuming $\gamma M_{sgs}^2 \ll 1$ everywhere.

Considering from now on ϖ computable, it is sensible to involve it in the definition of a subgrid heat-flux vector, noted $\boldsymbol{Q}$, of components

$$Q_i = -\overline{(\rho e + p)u_i} + (\widetilde{\rho e} + \varpi)\tilde{u}_i \quad . \tag{30}$$

Provided acceptable models are proposed for $\overline{\overline{T}}$ and $\boldsymbol{Q}$, the *resolved fluxes* already look more tractable:

$$
\overline{F_i} = \begin{pmatrix}
\bar{\rho}\tilde{u}_i \\
\bar{\rho}\tilde{u}_i\widetilde{u_1} + \varpi\,\delta_{i1} - \tau_{i1} - \overline{\mu S_{i1}} \\
\bar{\rho}\tilde{u}_i\widetilde{u_2} + \varpi\,\delta_{i2} - \tau_{i2} - \overline{\mu S_{i2}} \\
\bar{\rho}\tilde{u}_i\widetilde{u_3} + \varpi\,\delta_{i3} - \tau_{i3} - \overline{\mu S_{i3}} \\
(\widetilde{\rho e} + \varpi)\tilde{u}_i \quad - Q_i - \overline{\mu S_{ij}u_j} - \overline{k\frac{\partial T}{\partial x}}
\end{pmatrix} \quad . \tag{31}
$$

The remaining non-computable terms are viscous terms, which can be considered of less importance when the Reynolds number is sufficiently large. We therefore simply replace (31) by

$$
\overline{F_i} \simeq \begin{pmatrix} \overline{\rho \widetilde{u}_i} \\ \overline{\rho \widetilde{u}_i \widetilde{u}_1} + \varpi\, \delta_{i1} - \tau_{i1} - \mu \widetilde{S}_{i1} \\ \overline{\rho \widetilde{u}_i \widetilde{u}_2} + \varpi\, \delta_{i2} - \tau_{i2} - \mu \widetilde{S}_{i2} \\ \overline{\rho \widetilde{u}_i \widetilde{u}_3} + \varpi\, \delta_{i3} - \tau_{i3} - \mu \widetilde{S}_{i3} \\ (\overline{\rho \widetilde{e}} + \varpi)\widetilde{u}_i \quad - Q_i - \mu \widetilde{S}_{ij}\widetilde{u}_j - k\dfrac{\partial \vartheta}{\partial x} \end{pmatrix} , \tag{32}
$$

in which μ and k are linked to ϑ through the Sutherland relation (8) and the constant Prandtl number assumption $Pr = C_p \mu(\vartheta)/k(\vartheta) = 0.7$.

The system is finally closed with the aid of variable-density eddy-viscosity and diffusivity models, in the form

$$
\tau_{ij} \simeq \overline{\rho}\nu_t \widetilde{S}_{ij} \tag{33}
$$

$$
Q_i \simeq \overline{\rho}\,\frac{\nu_t}{Pr_t}\,\frac{\partial \vartheta}{\partial x_i} , \tag{34}
$$

expressions for $\nu_t(\widetilde{u})$ are provided in [12] (see also [11]). In practice, the models used here are the *structure-function model* and its *filtered* and *selective* extensions, with constant the turbulent Prandtl number 0.6.

3. EXTENSION TO CURVILINEAR CO-ORDINATES

When the domain is no longer cubic or parallelepipedic, it is still convenient to use body-fitted co-ordinates, that is, co-ordinates (ξ_1, ξ_2, ξ_3) such that each boundary of the domain corresponds either to constant ξ_1, ξ_2 or ξ_3. An appropriate grid generator can provide a set of vectors $\boldsymbol{\xi}$ which are the co-ordinates of the cell vertices or centres. Assume that the domain (hereafter refered to as "physical domain") can be remapped onto a cubic domain (called "computational") meshed with a uniform grid of spacing Δ as in the above sections. Let $\boldsymbol{x}$ be the co-ordinates of the cell vertices or centres of these cubic meshes. There exists a mapping function $\boldsymbol{h}$ such that

$$
\boldsymbol{\xi} = \boldsymbol{h}(\boldsymbol{x}) \quad ; \quad \boldsymbol{x} = \boldsymbol{h}^{-1}(\boldsymbol{\xi}) , \tag{35}
$$

and characterized by its Jacobian

$$
J = \det \begin{pmatrix} \dfrac{\partial \xi_1}{\partial x_1} & \dfrac{\partial \xi_1}{\partial x_2} & \dfrac{\partial \xi_1}{\partial x_3} \\[2mm] \dfrac{\partial \xi_2}{\partial x_1} & \dfrac{\partial \xi_2}{\partial x_2} & \dfrac{\partial \xi_2}{\partial x_3} \\[2mm] \dfrac{\partial \xi_3}{\partial x_1} & \dfrac{\partial \xi_3}{\partial x_2} & \dfrac{\partial \xi_3}{\partial x_3} \end{pmatrix} , \tag{36}
$$

which satisfies

$$d\xi_1 d\xi_2 d\xi_3 = J dx_1 dx_2 dx_3 \quad . \tag{37}$$

To each nodal variable $\phi(\boldsymbol{x})$ of the "computational" (i.e. cubic) domain corresponds a nodal variable $\psi(\boldsymbol{\xi})$ of "physical" domain, such that $\psi(\boldsymbol{\xi}) = \phi(\boldsymbol{x})$. Application of filter $\overline{}$ onto ϕ yields, thanks to (11) and (37)

$$\overline{\psi}(\boldsymbol{\xi}) = \left(\frac{1}{\Delta}\right)^3 \int \prod_{i=1}^{3} G\left(\frac{[\boldsymbol{h}^{-1}(\xi)]_i - [\boldsymbol{h}^{-1}(\xi')]_i}{\Delta}\right) \psi(\boldsymbol{\xi}') J^{-1} d\xi_1' d\xi_2' d\xi_3' \quad , \tag{38}$$

with the filter kernel G still defined by (12). It can then be proved that this new operator $\overline{}$, defined in the "physical" domain, commutes with the partial derivatives with respect to ξ_i up to second order (see e.g. Ghosal and Moin [20], who coinded the expression *Second-Order Commuting Filter*).

Straightforward application of the chain rule

$$\frac{\partial}{\partial x_i} = \frac{\partial}{\partial \xi_1}\frac{\partial \xi_1}{\partial x_i} + \frac{\partial}{\partial \xi_2}\frac{\partial \xi_2}{\partial x_i} + \frac{\partial}{\partial \xi_3}\frac{\partial \xi_3}{\partial x_i} \tag{39}$$

to (13) yields, after some manipulations ([22], see also [23] or [24] for details),

$$\frac{\partial \hat{U}}{\partial t} + \frac{\partial \hat{F}}{\partial \xi_1} + \frac{\partial \hat{G}}{\partial \xi_2} + \frac{\partial \hat{H}}{\partial \xi_3} = 0 \quad , \tag{40}$$

with

$$\hat{U} = \overline{U}/J \tag{41a}$$

$$\hat{F} = \frac{1}{J}\left[\left(\frac{\partial \xi_1}{\partial x_1}\right)\overline{F_1} + \left(\frac{\partial \xi_1}{\partial x_2}\right)\overline{F_2} + \left(\frac{\partial \xi_1}{\partial x_3}\right)\overline{F_3}\right] \tag{41b}$$

$$\hat{G} = \frac{1}{J}\left[\left(\frac{\partial \xi_2}{\partial x_1}\right)\overline{F_1} + \left(\frac{\partial \xi_2}{\partial x_2}\right)\overline{F_2} + \left(\frac{\partial \xi_2}{\partial x_3}\right)\overline{F_3}\right] \tag{41c}$$

$$\hat{H} = \frac{1}{J}\left[\left(\frac{\partial \xi_3}{\partial x_1}\right)\overline{F_1} + \left(\frac{\partial \xi_3}{\partial x_2}\right)\overline{F_2} + \left(\frac{\partial \xi_3}{\partial x_3}\right)\overline{F_3}\right] \quad , \tag{41d}$$

The chain rule has to be used again to express all the derivatives which arise in the fluxes $\hat{F}$, $\hat{G}$ and $\hat{H}$ (see section below). Note also that vector $\hat{U}$ is still a function of the cartesian co-ordinates x_i and time t.

4.   NUMERICS

The system (40) is solved on this grid by means of a (2,4) extension of the fully-explicit McCormack scheme devised by Gottlieb and Turkel [21], in the form

$$
\overline{U}^{(1)}_{i,j,k} = \overline{U}^n_{i,j,k} - J^{(p)}_{i,j,k}
\left\{
\begin{array}{l}
\frac{\Delta t}{\Delta \xi_1}\left[\ \frac{7}{6}\left(\hat{F}^n_{i+1,j,k} - \hat{F}^n_{i,j,k}\right)\right.\\[4pt]
\left.- \frac{1}{6}\left(\hat{F}^n_{i+2,j,k} - \hat{F}^n_{i+1,j,k}\right)\right]\\[8pt]
+\frac{\Delta t}{\Delta \xi_2}\left[\ \frac{7}{6}\left(\hat{G}^n_{i,j+1,k} - \hat{G}^n_{i,j,k}\right)\right.\\[4pt]
\left.- \frac{1}{6}\left(\hat{G}^n_{i,j+2,k} - \hat{G}^n_{i,j+1,k}\right)\right]\\[8pt]
+\frac{\Delta t}{\Delta x_3}\left[\ \frac{7}{6}\left(\hat{H}^n_{i,j,k+1} - \hat{H}^n_{i,j,k}\right)\right.\\[4pt]
\left.- \frac{1}{6}\left(\hat{H}^n_{i,j,k+2} - \hat{H}^n_{i,j,k+1}\right)\right]
\end{array}
\right\}
\tag{42a}
$$

$$
\overline{U}^{n+1}_{i,j,k} = \ \frac{1}{2}\left[\overline{U}^{(1)}_{i,j,k} + \overline{U}^n_{i,j,k}\right]
\tag{42b}
$$

$$
-\frac{1}{2}J^{(c)}_{i,j,k}
\left[
\begin{array}{l}
\frac{\Delta t}{\Delta \xi_1}\left[\ \frac{7}{6}\left(\hat{F}^{(1)}_{i,j,k} - \hat{F}^{(1)}_{i-1,j,k}\right)\right.\\[4pt]
\left.- \frac{1}{6}\left(\hat{F}^{(1)}_{i-1,j,k} - \hat{F}^{(1)}_{i-2,j,k}\right)\right]\\[8pt]
+\frac{\Delta t}{\Delta \xi_2}\left[\ \frac{7}{6}\left(\hat{G}^{(1)}_{i,j,k} - \hat{G}^{(1)}_{i,j-1,k}\right)\right.\\[4pt]
\left.- \frac{1}{6}\left(\hat{G}^{(1)}_{i,j-1,k} - \hat{G}^{(1)}_{i,j-2,k}\right)\right]\\[8pt]
+\frac{\Delta t}{\Delta x_3}\left[\ \frac{7}{6}\left(\hat{H}^{(1)}_{i,j,k} - \hat{H}^{(1)}_{i,j,k-1}\right)\right.\\[4pt]
\left.- \frac{1}{6}\left(\hat{H}^{(1)}_{i,j,k-1} - \hat{H}^{(1)}_{i,j,k-2}\right)\right]
\end{array}
\right]
$$

As mentioned in [25] and recalled in [23], the metrics $\partial \xi_i / \partial x_j$ which arise in the fluxes and Jacobians above have to be discretized in such a way that unwanted cross-terms cancel out, otherwise the scheme is not consistent.

Firstly, they have to be expressed as analytic functions of the metrics $\partial x_\ell / \partial \xi_m$ of the inverse transform h^{-1}, in order to eliminate all derivatives with respect to the x_i's. Secondly, the inverse metrics are discretized, the only 3-point stencil which works in the present case is

$$
\left(\frac{\partial x_\ell}{\partial \xi_1}\right) =
\left\{
\begin{array}{l}
\dfrac{-\,1/6\ x_{\ell,i+2,j,k} + 8/6\ x_{\ell,i+1,j,k} - 7/6\ x_{\ell,i,j,k}}{\Delta \xi_1}\\[4pt]
\text{in the predictor step (42a), and}\\[10pt]
\dfrac{7/6\ x_{\ell,i,j,k} - 8/6\ x_{\ell,i-1,j,k} + 1/6\ x_{\ell,i-2,j,k}}{\Delta \xi_1}\\[4pt]
\text{in the corrector step (42b)}
\end{array}
\right.
\tag{43}
$$

This is only first-order accurate, and acceptable only when the grid is quasi-orthogonal (i.e. $\partial x_\ell / \partial \xi_m \approx \delta_{\ell m}$ almost everywhere). Otherwise, 5-point stencils at least have to be used.

In the same way, the chain rule (39) has to be applied to eliminate all derivatives with respect to x_1, x_2 and x_3 from the fluxes F_i. This introduces metrics to be evaluated as said above, together with derivatives of velocity and temperature with respect to ξ_1, ξ_2 and ξ_3. Consistency then determines the way these derivatives should be discretized.

4.1.   Boundary Conditions

The boundary conditions are based on a decomposition into characteristics, in the spirit of Thompson ([26], [27]) and Poinsot and Lele ([28]). The Riemann invariants of outgoing characteritics are extrapolated, whereas the incoming ones are either prescribed (*e.g.* at the inflow boundary) or set to zero (*non-reflective* or *open* boundary condition). For example, going back to cartesian co-ordinates for the sake of simplicity, in the case of a boundary perpendicular to the direction x_1, the Euler equations are recast in their quasi-linear form

$$\frac{\partial V}{\partial t} + A \frac{\partial V}{\partial x_1} = 0, \quad \text{with} \quad V = {}^T(\rho, \rho u_1, \rho u_2, \rho u_3, p) \quad . \tag{44}$$

The matrix A is, as per usual, diagonalized in the form $\Lambda = L^{-1}AL$. Assuming L to be locally constant and introducing the vector $W = LV$, system (44) decouples into 5 equations of the form

$$\frac{\partial w}{\partial t} + \lambda \frac{\partial w}{\partial x_1} = 0 \quad , \tag{45}$$

to be solved at the boundary point N through the semi-implicit scheme

$$\frac{w_N^{n+1} - w_N^n}{\Delta t} + \frac{\lambda_N^n + |\lambda_N^n|}{2} \left[\frac{w_N^{n+1} - w_{N-1}^{n+1}}{\Delta x_1} \right] + \tag{46}$$

$$\frac{\lambda_N^n - |\lambda_N^n|}{2} \left[\frac{w_{N+1}^{n+1} - w_N^{n+1}}{\Delta x_1} \right] = 0 \quad .$$

For the outgoing characteristics ($\lambda_N^n > 0$), the values of w_N^{n+1} are obtained from that of λ_N^n, w_N^n and w_{N-1}^{n+1}, which are supposed to be known. For the incoming characteristics ($\lambda_N^n < 0$), it is necessary to prescribe w_{N+1}^{n+1} in order to pull out w_N^{n+1}. This is done by considering the nature of the boundary condition (adherence, free slip, periodicity, prescribed flow rate, non-reflectivity, inter-block matching...). V_N^{n+1} is finally deduced from W_N^{n+1} assuming simply $L_N^{n+1} = L_N^n$.

5.   AN ACADEMIC CASE: THE FORCED TRANSITION OF A HIGH-MACH NUMBER BOUNDARY LAYER

In odrer to advocate the straightforward closure of the compressible LES equations presented in section 2, we herefter present a temporal simulation of the transition to turbulence of a high-Mach number boundary layer over an adiabatic flat plate. The Mach and Reynolds numbers are $M_\infty = 4.5$ and $Re_{\delta_i} = 10000$ (δ_i denotes the initial displacement thickness), which matches a case which has been extensively investigated at ICASE (see e.g. Ng and Erlebacher [29]). For such a Mach number, the dominant instability is inviscid and two-dimensional (Mack's second mode [30]), because the solution to the laminar similarity equations exhibit a generalized inflection point, i.e. a distance y_S to the wall where the local angular momentum $\rho\omega = \rho du/dy$ is maximum (in magnitude). Consequently, Kelvin-Helmholtz-like vortices of period $\lambda_a = 2.8\ \delta_i$ form around $y_S = 1.05\ \delta_i$, as shown below as the result of a 2D simulation initialized by the laminar similarity solution for a wall temperature $T_w = 180K$, perturbed by small-amplitude white noise. The domain's size is $L_x = 4\ \lambda_a = 11\ \delta_i$ and $L_y = 20\ \delta_i$ for a resolution $40 \times 70 \times 36$, most of the points are concentrated between the wall and y_S, the first mesh line away from the wall is at $y = 0.024\ \delta_i$.

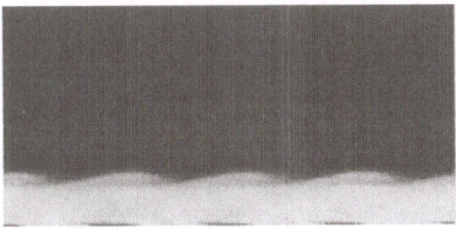

Fig. 1. — vorticity map after the growth of the second mode (the top part of the domain is not shown). The production of vorticity of both signs at the wall results from baroclinic effects induced by the reflexion of acoustic waves.

A 3D DNS and a LES with the *structure-function model* in its four-neighbour formulation (see again [11] or [12]) are now performed, in a smaller domain $L_x = 2\ \lambda_a = 5.5\ \delta_i$, $L_y = 7.15\ \delta_i$ and $L_z = 6.28\ \delta_i$ (the preferential wavelength of the subharmonic mode of secondary instability found in [29]), with the resolution $40 \times 70 \times 36$. Both runs start from the same initial conditions: the fluctuations at the same timestep as for figure 1 are rescaled to an amplitude $A = 8\%$ of U_∞ and sprinkled with 3D white noise of amplitude $10^{-4} U_\infty$. The DNS blows up at $t = 390\ \delta_i/U_\infty$, but the LES continues further. We stopped it at $t = 450\ \delta_i/U_\infty$, after transition is completed. Figure 2 shows the time evolution of the prominent modes. Mack's second mode then appears as $(2,0)$ and the oblique subharmonic of [29] as $(1,1)$. Notice in particular that both the DNS and the LES give about the same growth rate for this mode, viz., $\approx 1.7\ 10^{-2} U_\infty/\delta_i$, which is in acceptable agreement with Ng and Erlebacher [29] who find $2.5\ 10^{-2} U/\delta_i$ for $A = 6\%$. However, the most interesting fact is the resonance of the x-independent (i.e. purely spanwise) mode $(0,2)$, which

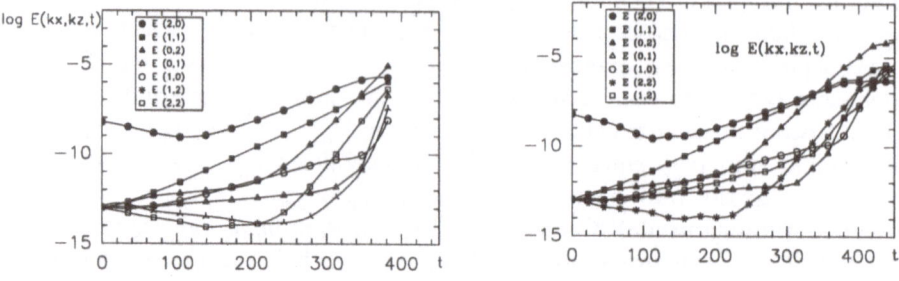

Fig. 2. — Energy of the prominent modes for the DNS (left) and the LES (right).

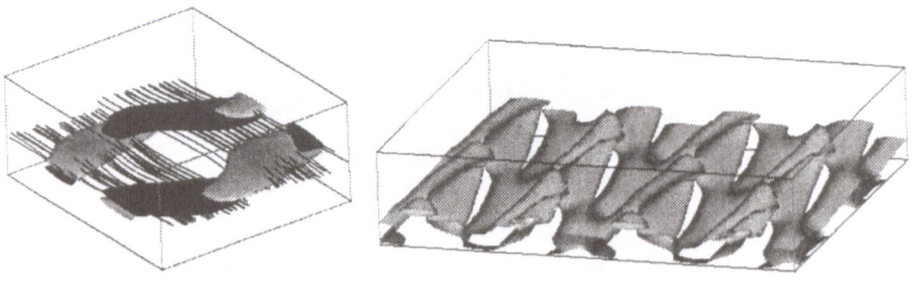

Fig. 3. — **left:** Vortex lines at $t = 350\ \delta_i/U_\infty$, with the isosurfaces $\omega_x = 0.1U_\infty/\delta_i$ in clear and $\omega_x = -0.1U_\infty/\delta_i$ in dark.
right: Isosurface of spanwise vorticity $\omega_z = -U_\infty/\delta_i$, still at $t = 350\ \delta_i/U_\infty$. In order to increase the resemblance with Fig. 11 of [35], the domain is instanced twice thanks to spanwise periodicity.

shoots up as from $200\delta_i/U_\infty$.

Until $t \approx 350\ \delta_i/U_\infty$ (the time origin is the beginning of the 3D calculations), this resonance appears essentially in the form of streks of weak vorticity normal to the wall ($\omega_y \approx \pm 0.03U_\infty/\delta_i$, see Figure 6 of [31]), which supports its interpretation in terms of Squire modes. The vortical structure of the flow remains dominated by Kelvin-Helmholtz-like vortices at y_S, slowly going three-dimensional as mixing layers at $M_c \approx 1$ ([32], [33], [34]), in the form of Λ-vortices facing each other; The left plot of figure 3 shows the most intense vortex lines together with isosurfaces of streamwise vorticity ω_x. Another representation of the flow at the same instant is given in the right plot: from isosurfaces of spanwise vorticity (or vorticity norm), one can make out staggered Λ-vortices stretched into

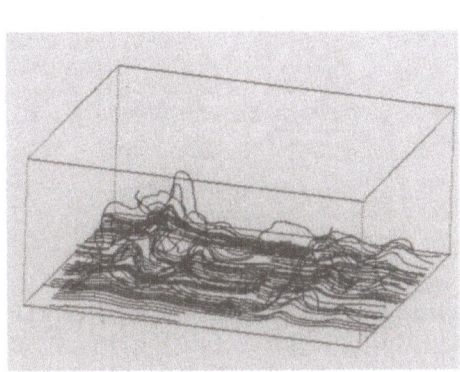

Fig. 4. — **left:** Vortex lines at $t = 390 \; \delta_i/U_\infty$. **right:** Corresponding isosurfaces of vorticity normal to the wall $\omega_y = 0.7 \; U_\infty/\delta_i$ in clear and $\omega_y = -0.7 U_\infty/\delta_i$ in dark.

Y-layers, as proposed by Adams and Kleiser [35] (see also [36]). Both views show the DNS results, but the LES ones are almost identical.

This is only the beginning of the transition process: at $t = 390 \; \delta_i/U_\infty$, when the DNS blows up because of the onset of small-scale turbulence, the LES shows the skin-friction coefficient lifting off (from $0.5 \; 10^{-3}$ up to $3.8 \; 10^{-3}$) while the shape factor H_{12} decreases from 14.5 down to 9.5, as expected from the empirical formula proposed in [37]

$$H_{12} = H_{inc} + 0.4 \; M_\infty^2 + 1.222 \; \frac{T_w - T_{ad}}{T_\infty} \quad , \tag{47}$$

in which $H_{inc} = 1.4$ denotes the incompressible counterpart of H_{12} and T_{ad} the adiabatic recovery temperature at the wall, equal to the wall temperature $T_w = 180K$ in our case.

After transition, the flow pattern is too messy to render properly with vortex lines in black and white. However, just at the beginning of it, *i.e.* $t \approx 400 \; \delta_i/U_\infty$, a fairly well organized streaky pattern is observed at the wall (figure 4), which can be interpreted as the result of the temporary emergence of mode (0,2). Later on, some of this organization persists, with a striking resemblence with incompressible boundary layers (see *e.g.* [38]). In particular, the velocity profile after transition exhibits a logarithmic zone which is not very different from its incompressible counterpart (not shown here).

Finally, figure 5 shows the instantaneous Reynolds stress profiles $\langle \rho \rangle \langle u'u' \rangle (y, t)$ and $\langle \rho \rangle \langle u'v' \rangle (y, t)$ normalized by $\rho_\infty U_\infty^2$, in which $\langle \rangle$ denotes the streamwise and spanwise average over the box. $\langle \rho \rangle \langle u'u' \rangle$ remains about 4 times as large as $\langle \rho \rangle \langle u'v' \rangle$. During the laminar stage (*i.e.* up to $t \approx 370 \; \delta_i/U_\infty$), the curves grow almost self-similarly with a peak around y_S. Between 360 and 420 δ_i/U_∞, the

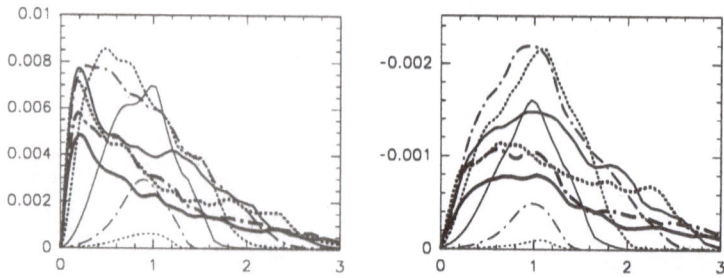

Fig. 5. — Time evolution of the Reynolds stress between 350 and 450 δ_i/U_∞. The heavier the line the later. **left:** $\langle\rho\rangle\langle u'u'\rangle(y,t)/(\rho_\infty U_\infty^2)$ **right:** $\langle\rho\rangle\langle u'v'\rangle(y,t)(\rho_\infty U_\infty^2)$.

peak of $\langle\rho\rangle\langle u'u'\rangle$ gets more and more acute. Meanwhile, it shifts towards the wall and settles down at about $y \approx 0.2\ \delta_i \approx 10$ wall units. In contrast, the profile of $\langle\rho\rangle\langle u'v'\rangle$ flattens in time with no visible shift towards the wall.

6.   TOWARD INDUSTRIAL APPLICATIONS : THE VORTEX SHEDDING INSIDE A SIMPLIFIED SOLID ROCKET ENGINE

We are participating in an operation set up by CNES and ONERA concerning the control of the vibrations induced by vortex shedding within the solid-propellant boosters of the European launcher ARIANE V. We show below preliminary simulations performed with the code described above, in a simplified planar test case, with the grid shown below (Fig. 6).

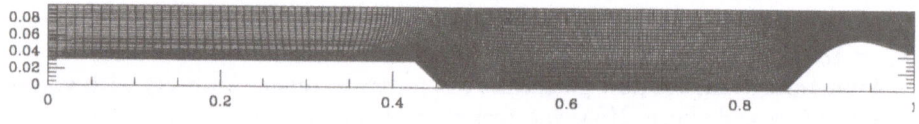

Fig. 6. — Grid of the C1 test case (length $L = 0.47m$, radius $H = 0.045m$, resolution 318×31 points

The step is made of burning propellant, at a flame temperature of $3387\ K$ and a mass flow rate, normal to the walls, of $21.2\ kg/m^2/s$. Pressure $p = 4.66\ bar$ is prescribed at the upstream end. The outlet is a nozzle and the outflow boundary conditions are supersonic. The burnt gases are characterized by the following parameters: $\gamma = 1.14$, $R = 299.53\ J/kg/K$, $\mu_{mol} = 9.\ 10^{-5}\ Pl$ et $Pr = 1$.

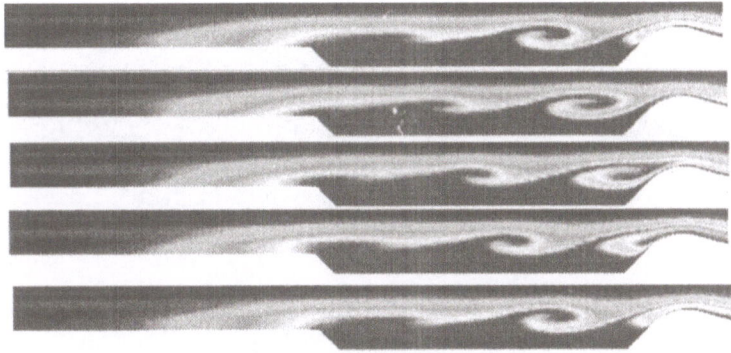

Fig. 7. — Contour maps of entropy at 5 equally spaced instants, in a low-Reynolds number 2D DNS.

With such values, 2D simulations are not possible without flux limiters or artificial viscosities. With a viscosity 8 times as large, they become possible without such limiters, and Figure 7 shows the resulting vortices, in time evolution. In such a case, the code gives approximately the same results as the second-order Mc Cormack code SIERRA of ONERA [39].

In 3D at the true viscosity and with the filtered structure function model described above, the advantages of the (2,4) scheme become evident. The following figures correspond to a LES at a spanwise resolution of 90 points equally spaced over the span $L_z = \pi\ H \approx 0.141\ m$, with periodic boundary conditions. The initial condition consists of the 2D flow shown above, taken at a given instant of the steady regime, with low-amplitude white noise (of amplitude 10^{-4} the speed of sound at the surface of the propellant) on all the components of $\hat{U}$. Without this perturbation, the flow would have remained 2D, which proves that the code is not "noisy". After having reached the steady regime, which took 50 hours of Cray 90 at 450 Mflops (corresponding to $8ms$ of real time), time series are recorded for $5ms$. Figure 8 shows an animation of an isosurface of the magnitude of the vorticity vector. Streamwise vortices are not only visible inbetween the large Kelvin-Helmholz billows, but also at the wall of the nozzle. These are likely to result from a Dean-Görtler instability of the detached boundary layer, which re-attaches in the convergent part of the nozzle (Fig. 9).

The statistics are in global agreement with the experimental data. In particular, we found kinetic energy and pressure spectra which exhibit a fundamental peak around 2500Hz, and its successive harmonics. More precisely, Figure 10 shows a comparison between the present LES and the 2D calculation just above. In the 3D case, the spectra are more developed, in particular in the low frequency, and the fundamental frequency is lower (2300Hz versus 2670). This proves that the streamwise vortices which periodically impinge the nozzle (as seen in Figure 8 affect and lower the shedding frequency of the quasi-2D

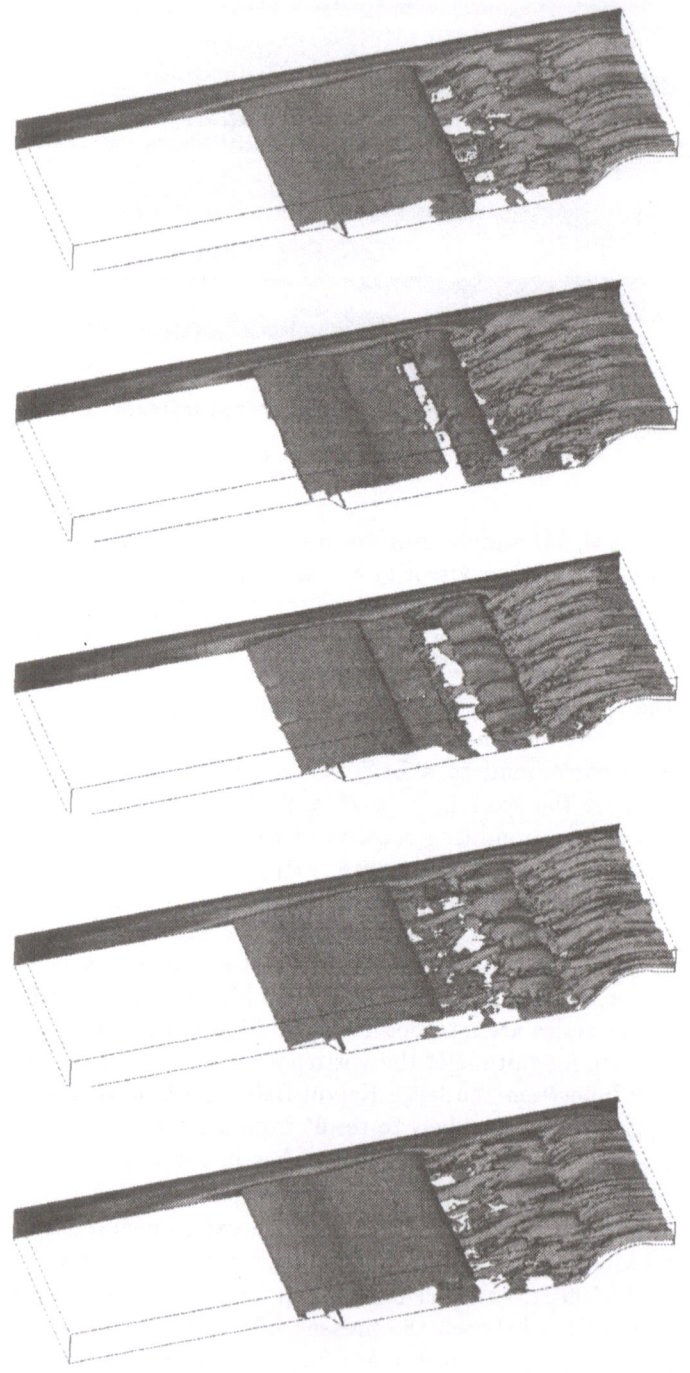

Fig. 8. — One period of the vortex shedding sequence in LES, in an "almost indus-trial" configuration.

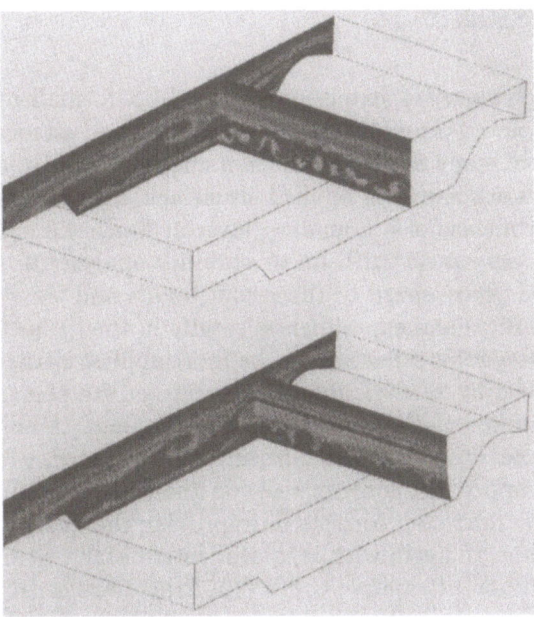

Fig. 9. — Maps of the entropy field. The top view shows a cross section of the Görtler vortices, the bottom one the streamwise vortices which connect the KH billows.

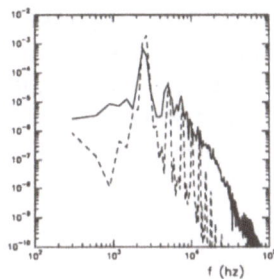

Fig. 10. — Temporal kinetic energy spectra recorded in the middle of the booster. The solid line corresponds to the LES and the dashed line to the 2D DNS.

Kelvin-Helmholtz billows. Although the reasons for this are not yet clear to us, this is of crucial importance for the design of the anti-vibration protections of the rocket's control systems, and illustrates the importance of taking three-dimensionality into account, even when the largest vortices are expected to be two-dimensional.

7. CONCLUSION

The most straightworward extension of the LES formalism to compressible
flows in curvilinear co-ordinates has been recalled and advocated. Despite the
evident crudity of some approximations made, informative results have been
obtained both in academic an applied situations. In particular, the complete
transition to turbulence of a boundary layer at Mach 4.5 has been simulated,
with acceptable agreement with linear stability analyses in the linear regime
(growth rate and phase speed of the second mode and the dominant mode of
secondary instability) and experimental results in the turbulent regime (shape
factor, log law, Reynolds stresses). In the more applied situation of a transonic
mixing layer forced by acoustic modes (the case of the planar model booster),
LES have contributed to the improvement of the understanding of the mechanism of vortex shedding: in particular, intense streamwise vortices have been
made out, with important consequences on the pressure (and thrust) fluctuation spectra. The next step consists of improving the quality of the results, in
academic situations to begin with. Experimental databases and incompressible
simulations with non-dissipative codes (spectral) are needed more than ever,
in order to improve upon the subrid-scale turbulence models and the perturbations injected at the upstream boundary.

Acknowledgments

The results presented here have been obtained by F. Ducros [40] and J. Silvestrini [41]) during their PhD's in Grenoble. Most of the computational time
used for the 3D calculations has been freely allocated by IDRIS, the CNRS
computing center. The study of the vortex shedding in the boosters of Ariane
V is under the CNES/ ONERA contract n° 22.492/DA/A1.CC1.

References

[1] Moin P., Kim J., *J. Fluid Mech.* **155** (1985) 441.

[2] Porter D.H., Pouquet A., Woodward P.R., *Phys Fluids* **6** (1994) 2133.

[3] Kawamura T., Kuwahara K., *AIAA Paper* **85-0376** (1985) .

[4] Silveira-Neto A., Grand D., Métais O., Lesieur M., *J. Fluid Mech.* **256**
     (1993) 1.

[5] Pereira J.C.F., Sousa J.M.M., *74th. AGARD Fluid Dynamics Symposium
     on "Application of Direct and Large-Eddy Simulation to transition and
     turbulence"*, Chania, 1994.

[6] Hoffmann G., Benocci C., *74th. AGARD Fluid Dynamics Symposium on
     "Application of Direct and Large-Eddy Simulation to transition and tur-
     bulence"*, Chania, 1994.

[7] Ghosal S., Lund T.S., Moin, P., Annual Research Briefs (Center for Turbulence Research, Ames Research Center and Stanford University, 1992) p. 3.

[8] Kraichnan R.H., *J. Atmos. Sci.* **33** (1976) 1521.

[9] Liu S., Meneveau C., Katz J., *J. Fluid Mech.* **275** (1994) 83.

[10] Leonard A., *Adv. Geophys.* **18A** (1974) 237.

[11] Lesieur M., Métais O., *Ann. Rev. Fluid Mech.* **28** (1996) 45.

[12] Lesieur M., this volume.

[13] Favre A., *J. de Mécanique* **4** (1965) 361.

[14] Erlebacher G., Hussaini M.Y., Speziale C.G., Zang T.A., *J. Fluid Mech.* **238** (1992) 155.

[15] Yoshizawa Y., *Phys. fluids* **29** (1986) 2152.

[16] Moin P., Squires K., Cabot W., Lee S., *Phys. Fluids A 3* **11** (1991) 2746.

[17] Erlebacher G., Hussaini M.Y., Speziale C.G., Zang T.A., *ICASE Report* **87-20** (1987) .

[18] Normand X., Lesieur M., *Theor. and Comp. Fluid Dyn.* **3** (1992) 231.

[19] Lilly D.K., in Computational Fluids Dynamics, Les Houches; session *LIX*, 1993 Lesieur, Comte, Zin-Justin eds., Elsevier Sciences Publishers, B.V., 1996, p. 353.

[20] Ghosal S., Moin, P. *J. Comp. Phys.* **118** (1995) 24.

[21] Gottlieb D., Turkel E., *Math. Comp.* **30** (1976) 703.

[22] Viviand H. *Rech. Aeros.* **1** (1974) 65.

[23] Fletcher C.A.J. Computational techniques for fluid dynamics 2, (Springer series in Computational Physics, 1988), p. 484.

[24] Vinokur, M. *J. Comp. Phys.* **14** (1974) 105.

[25] Thompson J.F., Warsi Z.U.A, Mastin, Numerical grid generation, foundations and applications, North-Holland, Amsterdam, 1995.

[26] Thompson K.W., *J. Comp. Phys.* **68** (1987) 1.

[27] Thompson K.W., *J. Comp. Phys.* **89** (1990) 439.

[28] Poinsot T.J., Lele S.K., *J. Comp. Phys.* **101** (1992) 104.

[29] Ng L.L., Erlebacher G., *Phys. Fluids A* **4** (1992) 710.

[30] Mack L.M., Boundary-layer stability theory, Jet Propulsion Lab., Pasadena, Calif., rep. **900-277**, 1969.

[31] Ducros F., Comte P., Lesieur M., in Proceedings of Turbulent Shear Flows 9, Springer-Verlag, 1995, 283.

[32] Sandham N.D., Reynolds W.C., *J. Fluid Mech.* **224** (1991) 133.

[33] Comte P., Fouillet Y., Lesieur M., *Revue Scientifique et Technique de la Défense* **3** (1992) 43.

[34] Fouillet, Y. *Contribution à l'étude par expérimentation numérique des écoulements cisaillés libres : effets de compressibilité*, Thèse de l'Institut National Polytechnique de Grenoble, 1991.

[35] Adams N.A., Kleiser L., *Transitional and Turbulent Compressible flows;* *ASME* **151** (1993) 101.

[36] Adams N.A., Kleiser L., *J. Fluid Mech* **317** (1996) 301.

[37] Michel R., Couches limites - frottement et transfert de chaleur, Cours E.N.S.A.E., 1967.

[38] Ducros F., Comte P., Lesieur, M., *J. Fluid Mech.* **326** (1996) 1.
[39] Lupoglazoff N., Vuillot F., AIAA Paper **92-0776**, 30th AIAA Aerospace Sciences Meeting Reno, USA
[40] Ducros, F., *Simulations numériques directes et des grandes échelles de couches limites compressibles*, Thèse de l'Institut National Polytechnique de Grenoble, 1995.
[41] Silvestrini, J.H., *Simulation des grandes échelles des zones de mélange; application à la propulsion solide des lanceurs spatiaux*, Thèse de l'Institut National Polytechnique de Grenoble, 1996.

LES-DNS: The Aeronautical and Defense Point of View

P. Sagaut[1], T. H. Lê[1],
P. Moschetti[2]

[1] ONERA,
29, Av. de la Division Leclerc 92320 Châtillon,
France
[2] DGA/DRET,
26, Bd Victor 00460 ARMÉES,
France

1.   INTRODUCTION

Computational Fluid Dynamics (CFD) plays a crucial role in the industrial context for improving the capacity and the competitivity of the Research and Developement (R & D) teams, in order to reduce costs and delay during the design procedure of sophisticated aeronautical equipments.

To reach these objectives, CFD is used for the two following inquiries:

- *Prediction* of the flow, in order to have access to some parameters of interest for engineering studies (e.g. drag and lift coefficients, wind load on structures, characteristic frequencies). This can be done by looking at mean values of these parameters (this is entirely satisfactory for the study of design issues corresponding to established flows), or by looking at peak values of these parameters to analyse off-design situations for safety studies.

- *Analysis* of the flow, to get a deeper insight into the underlying physics, to be able to identify the predominant dynamical mechanisms which must be controlled and/or manipulated to enhance the performance of the equipments. Such an analysis requires a fine description of both the locally spatial and temporal organization of the flow, to identify basic events responsible for momentum, heat and mass transfer such as coherent waves or structures.

To capture the complete physical phenomena of the flow, advanced CFD is based on the Navier-Stokes equations, which take into account both convective and viscous mechanisms and also thermal aspects. Three approximations are used to solve numerically the Navier-Stokes equations (see others contributions in the present volume):

- The use of the *Reynolds Averaged Navier-Stokes* equations (RANS), which is a fully statistical approach: only the statistical average of the solution is computed (often reduced to a time-average by invoking ergodicity), and all the higher-order statistical moments of the variables are not directly computed but must be modelled. This approach is well adapted for predictive studies of the first inquiry (means values), but its use for studies of the second inquiry is still problematic (see Gatski in the present volume). The reliability of the results for analysis studies is far more uncertain, because the averaging process may create some artefacts (see Ferziger in this volume) which have no physical relevance at all.

- The *Direct Numerical Simulation* technique (DNS) [9], which is a fully deterministic approach: all the spatial and temporal scales are computed by solving the full unsteady Navier-Stokes equations without any model. So it provides the total information on the flow, and then DNS is of great benefit, when it is applicable, for studies of interest. But it should be noticed that DNS requires that the mesh size and the time step are small enough to correctly represent the smallest dynamical scales of the flows, and then involves so huge number of grid points and time steps that it can not be applied to most of the practical studies, because of the restrained capacity of available computers.

- The *Large Eddy Simulation* concept (LES) [9] [8], which is an hybrid deterministic/statistical approach based on the scale separation abstract idea: the large scales of the flow, which are anisotropic and are sensitive to the boundary conditions are directly computed in a way similar to DNS, while the small scales of the flow, which are assumed to be much more isotropic and universal than the previous ones thanks to the local isotropy hypothesis, are parametrized using a statistical model called subgrid scale model. The potentialities of LES for prediction and analysis are similar to those of DNS, but restricted to the large scales (which in practice are the scales represented on the computational mesh), if it is assumed that the subgrid scale modeling is good enough to maintain a good representation of the dynamics of the large scales.

Each of these approaches has its own advantages and drawbacks, that will be discussed in the paper. The text is organized as follows: first, we will list the identified needs for improved knowledge and tools, and the relative advantages and limitations of the various data sources will be discussed. Then, the ONERA strategy concerning the development of LES and DNS techniques will be presented. After that, we will briefly take stock of the current status of these two techniques in our own institute, and a special care will be devoted to show that LES and DNS are already able to furnish some valuable informations for studies of practical interest. Before concluding, we will try to check some specific points linked to the extension of LES to engineering problems, and to propose a list of requirements that should be fulfilled by an accurate industrial CFD tool dealing with fully unsteady problems.

2. THE IDENTIFIED NEEDS

It has already been noticed in the previous section that the role of CFD in the industry is to reduce the costs and the delay during the design stage, and to facilitate technological improvements. Precisely, that reduction of the R & D stage global cost corresponds to the fact that CFD must be able to minimize the number of testing prototypes in wind tunnel. The challenge is obviously to be able to design numerically the right prototype, without any complementary wind tunnel test. So the preliminary CFD stage must provide solutions to most technological problems which could be faced. That implies that the CFD tool has to take into account as many physical mechanisms as possible: convective and viscous effects to get a relevant description of the vorticity generation and transport, heat and mass transfer, but also fluid-structure interaction for buffet, flutter and vibro-acoustics phenomena, noise production for aero-acoustics, combustion ...

At present time, no existing CFD code is able to provide satisfactory answers to all these problems in their integrity, and the principal reasons for that are:

- Most of industrial CFD tools belong to the RANS family, and so are of poor interest for dealing with intrisically unsteady problems (acoustics, fluid-structure interaction ...).

- Turbulence models are based on restricted assumptions (equilibrium, isotropy) that are not in good accordance with "real-life" turbulence, and then fail to predict the flow behaviour.

- Numerical methods adapted to unsteady calculations in very complex geometries are still to be specified (numerical scheme, mesh generation and adaptation ...).

The improvement of the efficiency of the CFD tools for R & D applications can be thought in the following way:

- To reduce the delay of response of the already existing tools: this can be
  done by improving the numerical algorithms and increasing the computing capabilities.

- To take into account more physical phenomena: this involves a change
  in the physical model (basic equations, turbulence model, coupling with
  structure and/or acoustics solvers)

The most promizing way is the second one, because it is the one which offers
the possibility to find solutions to some unanswered technological problems, by
extending the range of investigation of CFD. It involves new developments concerning the modeling, to improve classical models, i.e. to extend their range of
application to non-equilibrium situations and to take into account extra source
terms (rotation, stratification, thermodynamics, ...). To make these efforts
possible, *4-D data bases* (time and space dimensions) are necessary, to define
and to assess the models on generic flows. These data bases must be accurate
and complete enough to avoid any problem concerning the interpretation of the
results.

Another point, which is a necessary condition for technological improvement,
is the carrying out of fundamental studies in order to define reliable theoretical
models, providing a better understanding of basic dynamical mechanisms to
be able to manipulate and control them. For this task, well documented data
bases are needed too.

*So one of the main objective for the next years must be to define, to generate
and to manage data bases, which are necessary to improve our physical knowledge and modeling.*

To generate such data, three investigation sources are commonly used:

- *Wind tunnel experiments.* The main advantages of that way of generating
  data is that realistic configurations can be investigated (very complex geometries, correct values of relevant parameters like Reynolds and Mach
  numbers ...), and that it allows fast data acquisition, a very important
  factor for the realization of parametrical studies. The major drawbacks
  come from the metrology problems. None of the known measuring techniques is defect-free: some are intrusive techniques, which modify the
  flow, while the others rely on the probability density function of velocity
  of some particles advected by the flow, inducing some errors due to the
  forces acting on these particles (drag forces, Faxen and Basset terms ...).
  Generally, it is difficult to get fully satisfactory data in heavily separated
  regions and near solid boundaries. Another point is that no experimental
  techniques can give access to all the quantities of interest: most of the
  measurement techniques deal with velocity moments, but it remains to
  measure with the same degree of accuracy other quantities like temperature or density, or complex quantities like the spatial gradients of the
  previous one. Despite these limitations, wind tunnel experiments are the

only way to investigate very complex flows for realistic regimes, which are still far out of range of DNS and LES, today.

- *DNS.* The main advantages of DNS are data quality and data quantity. It give access to all the quantities at every location of the computational domain. But there are some severe drawbacks. The first one is the limitation of the range of application of DNS, because of the available computing capabilities. To our knowledge only academic flows have been computed, at moderate Reynolds number (typically $\approx 10^3$). The second one is similar to a metrological problem: because of the cost of the simulation, statistically converged high-order moments of the solution are very difficult to obtain. Almost all DNS users limit themselves to second-order moments. The last major drawback deals with the validation of the computation: one has to check the independency of the results with respect to the numerical aspects, to guaranty their physical reliability. This is a very expensive task, which can become problematic when dealing with large scale computations which can not be repeated as often as necessary. Nevertheless, DNS seems to be a privileged tool to perform fundamental studies.

- *LES.* The main advantage of LES versus DNS is that it allows to compute flows at larger Reynolds numbers (currently $O(10^4)$, with some trials up to 10^6 [5]), but still for simplified geometries. But it introduces new error sources: now some scales are modelled and not computed. This induces a loss of accuracy of the results, resulting in an increase of the uncertainity on interpretation of results.

It should be noticed that what is generally called a "complex geometry" by LES and DNS practioners is very simple compared to the "simplified geometries" as defined by engineers in the industry. *An important point is that none of these three techniques is self-sufficient: to get a safe data base, both numerical and experimental approaches must be used, in order to allow cross-validations.*

3. THE STRATEGY

The development of new tools for modern CFD is then a large-scale task, involving theoreticians, numericists and experimentalists. That effort is a large scale one, which must be supported at least at a national level to be able to remain competitive and competent when compared to non-european research programs.

Focusing on the numerical part of the work, the generation of the necessary data sets implies that the five following requirements can be fulfilled: experimental data availability, numerical methods adequacy, physical models consistency, human resources availability and high performance computing development.

To enhance its own potentiality, ONERA has developed a partnership politics. Three levels of collaboration can be distinguished:

- An *organic collaboration* with external research groups or institutions on identified topics, concretized by hiring associate collaborators (e.g., C.Basdevant from ENS/LMD for pollutant transport, C.Cambon from ECL/LMFA for rotation effects, P.Comte from IMG/LEGI for LES, M.Deville from EPFL/DGM for numerical methods, P.L.Ta from CNRS/LIMSI for LES, Y.Maday from UPMC/LAN for coupling methods ...)

- A *practical collaboration* with external research groups or institutions via collaborative studies, mostly with a financial support from DRET (e.g., CEAT-Poitiers for advanced signal analysis and coherent structures eduction, ECL/LMFA for compressible subgrid scale modelling, LEGI for general studies on subgrid scale modelling, CNRS/LIMSI for subgrid scale modelling and shock-capturing schemes for DNS and LES, ENSCPB for local mesh refinement techniques, ...)

- A *european collaboration* with others research institutions through the GARTEUR-EG 30 on the DNS and LES of the flow around a wing.

4. DNS AND LES AT ONERA: STATUS

DNS and LES represent two different classes of studies: they are research topics, principally on numerical methods and subgrid-scale models, but they are already CFD tools to perform some studies of industrial interest. These two aspects are detailed in the sequel.

4.1. Research on DNS and LES

The first part of the research deals with the numerical aspect of these techniques. Main topics are:

- *Realistic unsteady inflow conditions*: one of the major problem that has to be solved to extend the range of application of LES and DNS is the generation of inflow boundary conditions able to take into account both turbulent stresses and spatio-temporal coherence. Two techniques are commonly used: to compute the complete transition process or to approximate the turbulent flow by surperimposing a random noise and a mean velocity profile. The first technique produces very good results but is not easy useful to complex geometries, while the second one can be employed everywhere but may induce large errors. So a new method, which will take into account coherent structures has to be defined.

- *"Non-reflecting" outflow boundary conditions* : this is a common research topic in CFD, but the massive development of aero-acoustic studies amplifies the need for clean outflow conditions.

- *Solid wall representation for high-Reynolds computations*: for these flows, it becomes very expensive to solve the dynamics of the boundary layer and to use no-slip conditions. A first possibility is to keep no-slip conditions by refining locally the mesh: such algorithms are now increasing in popularity, but are still rare in LES or DNS computations. The second one is to use wall models, which are generally based on canonical boundary layers flows, and their use becomes problematic if we are interested in interfacial phenomena like heat transfer.

- *Convection and subgrid-scale terms formulation and discretization*: recent studies (see Moin in this volume) have demonstrated that the sensitivity of the results of DNS and LES is much more pronounced than for RANS calculations. The greater insentivity of RANS calculations comes from the fact that the modelled terms are several orders of magnitude higher than those in LES. Fully satisfactory solutions are still to be found.

- *Shock capturing scheme for LES and DNS*: most of the recorded simulations deal with shock-free flows. So there is only a poor knowledge on the ability of shock capturing schemes to produce good DNS and LES results.

The second part deals with the fundamental aspects of the LES approaches. These efforts have already led to original results, like the Mixed Scale Model (MSM) [10] or the Time Filtered Large Eddy Simulation (TFLES) [1]. Presently, efforts are concentrated on the following points:

- *Anisotropy effects*: subgrid scale models are based on an isotropy assumption for both subgrid scales and filter (i.e. comptutational mesh). These criteria are not fulfilled in practical cases (anisotropic flows on distorted meshes), and some improvements are needed to get better results.

- *Subgrid scale models for passive scalar and temperature for low-Mach number flows*: subgrid scale modeling efforts have been devoted to simple aerodynamic flows, and more efforts are needed to take into account for temperature and stratification effects in LES. Two problems arise: the definition of a reliable mathematical model to describe the coupling between momentum and temperature (Boussinesq model or more complex ones), and the account of volumic forces, like stratification effects, on subgrid scales.

- *Compressibility effects on subgrid scales*: this is the extension of the previous point to steep density variations. For example, this should be the case near shock waves.

- *Subgrid scale corrections for special topics*: concerning aero-optics, aeroacoustics or lagrangian particle tracking fields: up to now, the modeling effort has been focused on the parametrization of the kinetic energy transfer between resolved scales and subgrid scales. But some application fields deal with another fundamental physical process (density variation for aero-optics, pressure variation for aero-acoustics, subgrid transport for dispersed particles), and the subgrid scale contributions remain to be investigated and modelled.

4.2.   DNS and LES as CFD tools

DNS and LES are already used as CFD tools to perform studies of practical interest. Most of these studies are multidisciplinary ones, in which coherent structures play an important role.

To illustrate the use of these two techniques, we will present in the following some applications, grouped with respect to the field of application.

4.2.1.   *Transition and control*

The analysis of transient phenomena, like transition, requires some accurate unsteady simulations. Two examples are described below: the study of the boundary layer receptivity to external forcing in the presence of vorticity generator [2], and the dynamics of countrail vortices [12]. In that two cases, the goal is to understand the transition mechanism, in order to control it.

4.2.1.1.   Leading edge receptivity. Leading edge receptivity of laminar boundary layer is investigated with DNS of a 2D incompressible flow. The external forcing induces the existence of a forced Stokes layer solution. The simulation was able to predict the development of a Tollmien-Schlichting wave when a vorticity generator (roughness, rib, suction, blowing) was added. The computed wave behind a rib is shown for example on figure 1.

4.2.1.2.   Aircraft wake dynamics. An aircraft wake is made up of two counter-rotating vortices. These are dissipated in the far field at a rate depending on atmospheric conditions and vortex properties. They can either last a very long time or suffer violent instabilities that can lead to their rapid disruption. Both DNS and LES are used to analyse the latter phenomenon. A view of the two vortices subjected to the Crow instability is reported on figure 2.

4.2.2.   *Aero-optics*

The efficiency of the optical devices of the airborn weapon systems may happen to be severely restricted by optical effects induced by the aerodynamical flow. For high speed conditions, coherent structures which are created near solid boundaries generate large density variations at some characteristic frequencies, which scramble the target's image. To be able to control that phenomenon, DNS is used to study the basic physical phenomena over a simplified configuration [3]. This is illustrated by figure 3, where the instantaneous density field is shown.

4.2.3.   *Aerothermics*

The improvement of the performances of turbomachinery requires an exact description of heat exchanges between metallic parts and the fluid, especially for

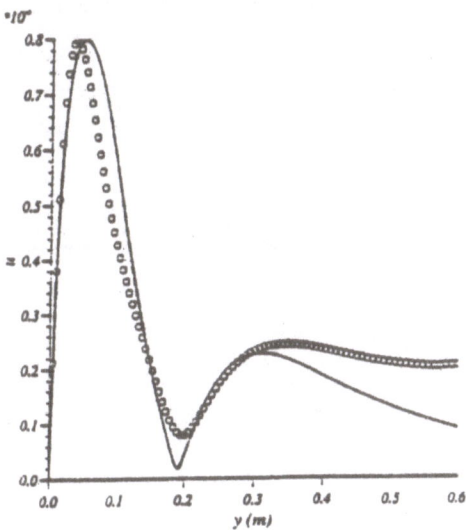

Fig. 1. — DNS of transitioning boundary layer with a rib (incompressible case) - comparison of the computed Tollmien-Schlichting wave (symbol) with the Orr-Sommerfeld theory (line)

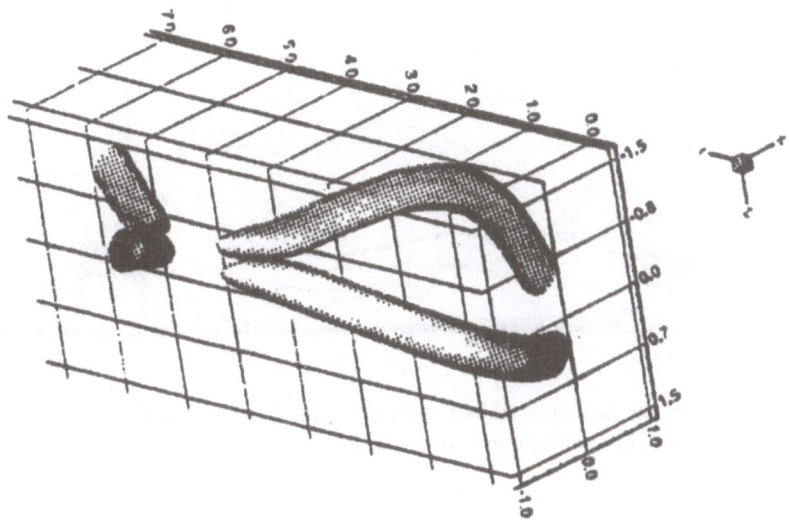

Fig. 2. — DNS of contrail vortices subjected to the Crow instability (incompressible case) - topology of the flow after the reconnexion

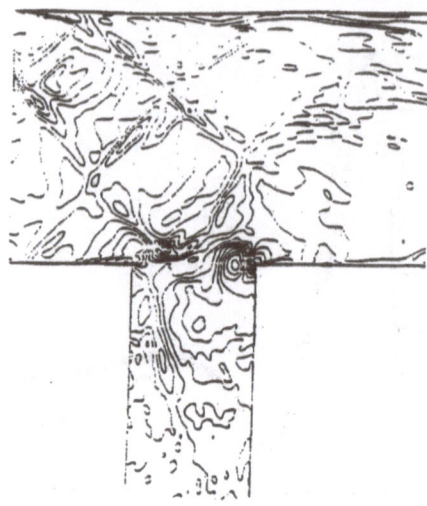

Fig. 3. — DNS of the flow over a cavity (compressible case) - instantaneous iso-values of density

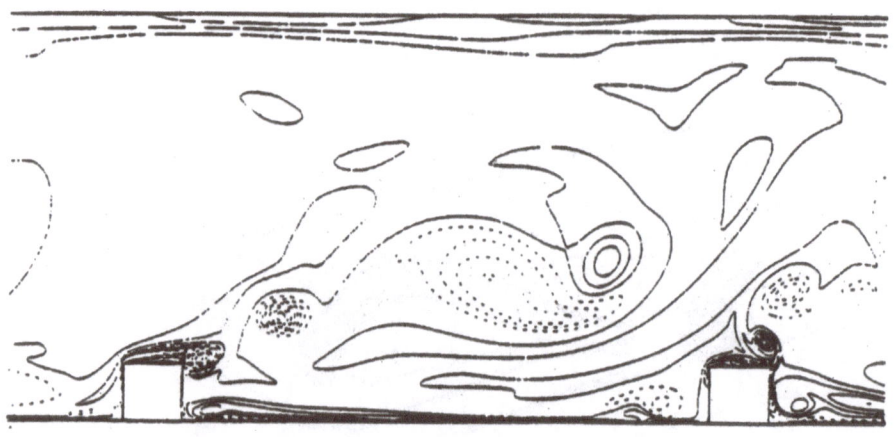

Fig. 4. — DNS of the flow in a ribbed cooling channel (compressible case) - instantaneous iso-values of the y component of vorticity

the design of the cooling devices. The governing phenomena are very complex: separation and recirculation, diffusion and turbulent mixing, secondary flows induced by rotation or curvature effects. To optimize the cooling devices, DNS and LES are used to get an accurate description of the physical phenomena,

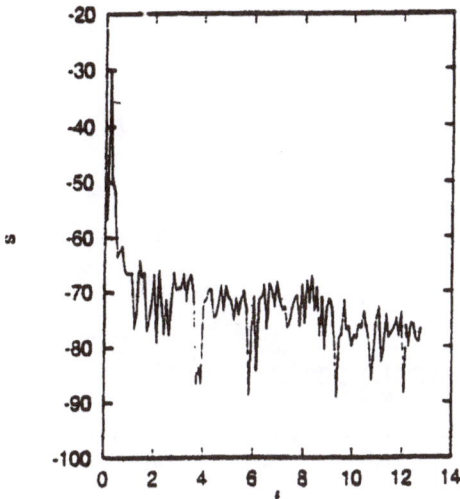

Fig. 5. — LES of the flow around a blunt trailing edge (incompressible case) - temporal spectra of static pressure at the wall

especially the contribution of coherent structures to the heat transfer process [7] [6]. The flow in a cooling channel with ribs is shown on figure 4.

4.2.4. Acoustics

The determination of the acoustic properties of engines and aircraft is now a very important task for the R & D teams. There are various noise generation mechanisms: noise induced by structure vibration ("structural noise"), noise produced by non-linear vortex dynamics ("aerodynamical noise"), and the noise generated by the shock-vortex interaction ("shock noise"). Each of these phenomena is presently investigated via DNS or LES.

4.2.4.1. Structural noise The unsteady vorticity dynamics near solid boundaries results in an unsteady forcing of solid structures, inducing a vibration of the latter. The vibrating structure then radiates some noise, which is computed by coupling the LES code to a structure code, which computes the structure displacement, and to an acoustic code, which predicts the far-field radiated noise [13]. An example of temporal pressure spectra at the surface of a blunt body is shown on figure 5.

4.2.4.2. Aerodynamical noise. The far field noise generated by the non-linear interactions between the different dynamical scales of the flow is computed, in the low-Mach number case, by the Kirchhoff or the Lighthill acoustic analogies. To use these two techniques, the flow field is first computed using LES or DNS, and the acoustic sources are evaluated, and then the far field noise is

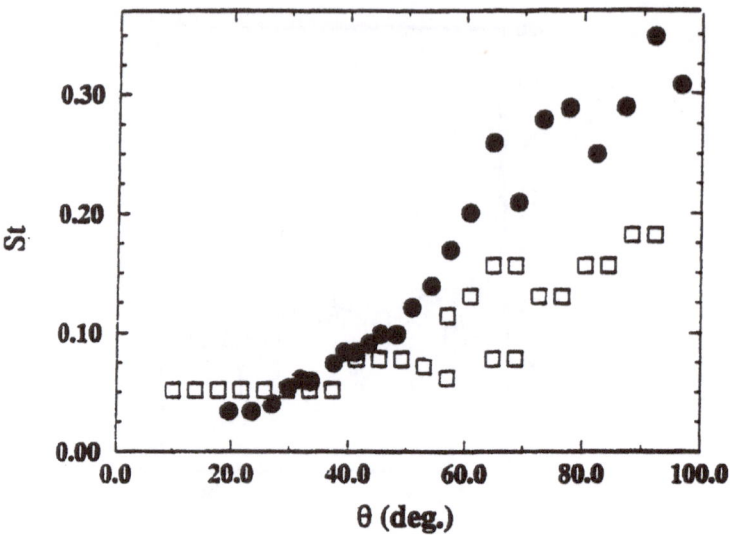

Fig. 6. — LES of supersonic round jet (compressible case) - angular distribution of the dominant frequency of the far field radiated noise computed with the Kirchhoff analogy - circle: experiments, square: computation

computed by solving an integral problem. The far field noise level computed by coupling LES and the Kirchhoff analogy for the round jet flow is compared to experimental data on figure 6 [4].

4.2.4.3. Shock noise. The pressure field is subjected to complex variations during the shock-vortex interaction, resulting in an acoustic wave generation. This phenomenon is predicted by DNS of the full compressible Navier-Stokes equations [11]. An example of such a calculation is shown on figure 7, where an instantaneous pressure field is plotted.

4.2.5. Combustion

Another very important application field is combustion. The improvement of the engine performances relies upon a clear understanding on the interaction between chemical and aerodynamical phenomena. LES is used to analyse the role of the large-scale structures in the mixing process for the supersonic round jet flow [1]. This simulation is illustrated by figure 8, where computed time-averaged mass fractions are compared to experimental data.

5.   THE INDUSTRIAL CHALLENGE

The last part of the present discussion will deal with the particular features of what should be a CFD tool for unsteady industrial computations. Industrial

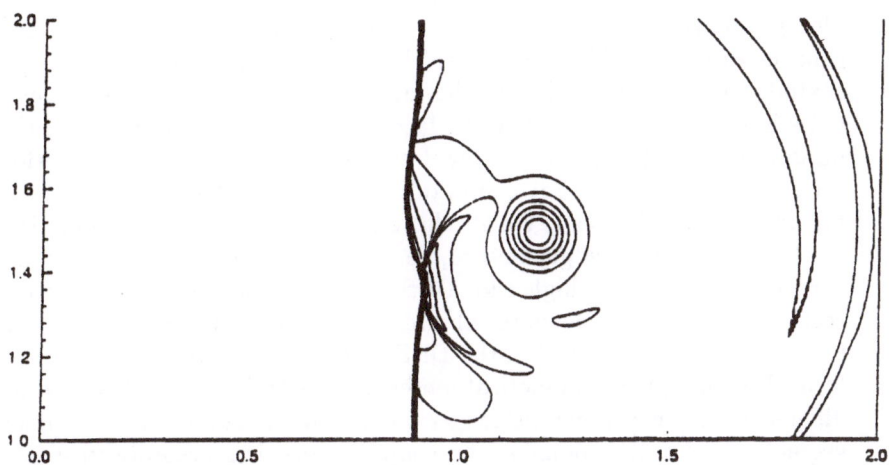

Fig. 7. — DNS of the shock-vortex interaction (compressible case) - instantaneous pressure field

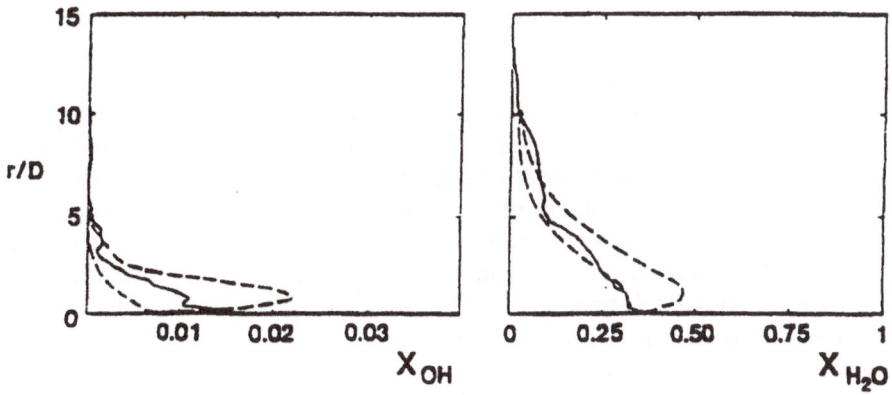

Fig. 8. — LES of a reactive supersonic round jet flow (compressible case) - time-averaged radial distribution of two species - full line: computation, dashed line: experiments

computations mean here computations of flows of direct practical interest for R & D teams in the industry. These flows differ a lot from the academic flows computed by DNS and LES approaches, and we will see that lots of *new problems* come from the differences between these two categories.

Industrial flows main characteristics are:

- *Complex configurations* These flows generally correspond to very com-

plex geometries. They are fully three-dimensional, without any homogeneous direction. They are hybrid laminar/transitional/turbulent and
spatially developing flows, and the information on the boundary conditions is very poor. So we have to face here towards the problem of the
mesh type (structured, unstructured or hybrid) and the mesh generation
(grids for accurate unsteady computations are more difficult to define that
for steady computations). We also have to define an acceptable numerical
method to handle such geometries, i.e. to find a compromise between accuracy of the method (higher for DNS and LES than for RANS) and the
cost, and to develop adequate subgrid scale models. It is very important
to keep in mind that most of the DNS and LES computations were performed using optimal numerical methods (spectral methods, Hermitian
differences ...), which can no longer be used when dealing with complex
geometries. Another point is the coupling with less accurate unsteady
methods, like unsteady-RANS calculations, in order to handle very large
scale computation with acceptable boundary conditions.

- *High Reynolds number.* Realistic flows are most of the time high Reynolds
  number flows ranging from 10^3 to 10^8, corresponding to much higher values than those currently handled by LES. So appears here the problem
  that for these flows, the percentage of the total kinetic energy contained
  in the subgrid scale will be higher than for classical LES, and the ac-
  cumulated experience has shown that the classical subgrid scale models
  produce large error in this case. So the models must be modified to improve the results. Industrial configurations involve solid boundaries, and
  here appears the problem of the representation of the boundary layer,
  which can be very thin. Various ways can be defined to take into account
  for solid boundaries: wall models, wall laws or mesh refinement. None
  of these three solutions gives fully satisfactory results, and no obvious
  solution is known at present time.

The last comment is about the CFD tool itself. The industrial context implies
that the same CFD tool will be used by a large number of users for various
application fields, a much more higher number than in a research team, and
that these users will not be as aware as specialists of the tricks of modeling and
numerics. The standard policy "one application for each code and one code
for each application", which is applied by most of the research teams, must
be changed. There is a need for an homogeneization of the numerical tools,
looking for a compromise between the quality of the results and the generality
of the method.

6.   RECOMMENDATIONS

As we have just seen the new challenge, for LES in near future and for DNS
in far future, is the industrial applications of interest in companies. In all the

areas we have mentioned, the industrial stakes are very high and should stimulate both numerical research and more fundamental research. Such research effectively governs progress in the different fields.

We repeat that it seems to be of prime importance to work in multidisciplinary teams, combining knowledge and talents of experts in fluid mechanics with those of numericists, while experimentalists must not forget to borrow the best suited measuring techniques from physics.

References

[1] Billet G., Technical Memorandum 1996-3, ONERA, 1996

[2] Casalis G., Private Communication, 1996

[3] Deron R., Labbe O., Delorme Ph., Report RTS 6/4525 PY, ONERA, July 1995

[4] Gamet L., PhD Thesis, ENSAE, April 1996

[5] Jansen K., *CTR Annual Research Briefs* **1** (1994) 161.

[6] Labbe O., Technical Report RSF 29/3419 AY, ONERA, February 1995

[7] Labbe O., Technical Report RTS 17/4368 AY, ONERA, May 1996

[8] Lesieur M., Metais O., *Ann. Rev. Fluid Mech.* **28** (1996) 45.

[9] Rogallo R., Moin P., *Ann. Rev. Fluid Mech.* **16** (1984) 99.

[10] Sagaut P., PhD Thesis, University Paris 6, June 1995

[11] Sagaut P., Lenormand E., Klahr D., Technical Report RTS 18/4368 AY, ONERA, September 1996

[12] Sipp D. , DEA Report, ONERA, July 1996

[13] Troff B., Technical Report RT 34/7257 PYA, ONERA, November 1995

Turbulence Modelling in Aeronautical Flows

B. Aupoix[1], J. Cousteix[1], G. Chevalier, S. Viala[1] and P. Malecki[1]

[1] *ONERA–CERT* *Aerothermodynamics Department*
B.P. 4025
31055 Toulouse Cedex 4   France

1.   INTRODUCTION

Flows of interest in the aerospace industry present a wide variety of challenges for the turbulence models. High Reynolds number flows are encountered on wings or fuselage while flows over turbine blades can be affected by low Reynolds effects. Subsonic as well as super- and hypersonic flows are to be predicted. Adverse pressure gradients and separation are one of the key problems for flows over wings while strongly accelerated flows are encountered on turbine blades or in rocket nozzles.

The present paper is devoted to wall flows. Although the goal is mainly the prediction of the viscous drag and the wall heat flux, a good prediction of the mean and turbulent flow properties are sought to get confidence in turbulence models. Let us just remind that a 1% accuracy on the total airplane drag is required by airplane manufactors, which is presently beyond the capabilities of any turbulence model.

2.   TURBULENCE MODELS

2.1.   Introduction

For the sake of simplicity, we will discuss only the dynamical problem. As usual, any quantity f is split into a mean or ensemble average value $F = < f >$ and a fluctuation f' such as $f = F + f'$. When the momentum equation is ensemble

averaged, new terms appear due to the above Reynolds decomposition; this is
the Reynolds stress tensor $-\rho < \underline{u}' \otimes \underline{u}' >$.

Models are thus required to express the Reynolds stress tensor. A first model
hierarchy can be introduced according to the way the Reynolds stress are expressed. The first level is the Boussinesq approximation, or eddy viscosity
model which relates the deviatoric part of the Reynolds stress tensor to the
deviatoric part of the mean deformation tensor with the help of an eddy viscosity μ_t as

$$-\rho < \underline{u}' \otimes \underline{u}' > +\frac{2}{3}\rho k \underline{\underline{1}} = \mu_t \left(\operatorname{grad} \underline{U} + {}^t\operatorname{grad} \underline{U} - \frac{2}{3}\operatorname{div} \underline{U}\,\underline{\underline{1}} \right) \qquad (1)$$

where $k = \frac{1}{2} < \underline{u}' \cdot \underline{u}' >$ is the turbulent kinetic energy. As this approximation
yields poor predictions of the diagonal stresses, the second level is made with
more elaborate anisotropic eddy viscosity models in which the Reynolds stress
tensor is related to the turbulence scales k and L, the mean velocity gradient and sometimes an extra quantity (Struct) linked to the Reynolds stress
anisotropy as

$$-\rho < \underline{u}' \otimes \underline{u}' >= \mathcal{F} \left(k, L, \operatorname{grad} \underline{U}, \operatorname{Struct} \right) \qquad (2)$$

This approach is presently blooming and seems to be very promising. However,
no example of application will be given in the present paper. At last, the third
level consists of Reynolds stress models in which transport equations are used
to compute each component of the Reynolds stress tensor.

A second hierarchy deals with the way the velocity and length scales are
prescribed. Let us first remind that the eddy viscosity can be expressed as the
product of a density, a turbulence velocity scale and a turbulence length scale
as

$$\mu_t = \rho \mathcal{U} \mathcal{L}$$

The models are usually sorted according to the number of transport equations
they use to evaluate these turbulent scales. The first class corresponds to
zero-equation or algebraic models. They assume that the turbulent field is at
equilibrium with the mean flow and usually relate the turbulence length scale
to some mean flow length scale such as the wall distance or the boundary layer
thickness and the turbulence time scale to the mean flow time scale $|\operatorname{grad} \underline{U}|^{-1}$.
Mixing length models are a typical example. The second class is one-equation
models. In most of these models, one transport equation is used to evaluate the
turbulent kinetic energy as its transport equation requires little modelling. The
turbulence length or time scale is still given by the equilibrium assumption.
Another recent and popular kind of one-equation turbulence model directly
solves a transport equation for the eddy viscosity. At last, two-equation models
solve two transport equations to evaluate the two turbulence scales and so get
rid of ad-hoc equilibrium assumptions and are more general. The most popular
two-equation models are the $k - \varepsilon$ models in which ε is the turbulent kinetic

energy dissipation rate and the $k-\omega$ models in which ω is the specific dissipation rate ε/k. Many other two-equation models exist, which may solve an equation for $q = \sqrt{k}$ to compute the velocity scale and for $\tau \sim k/\varepsilon$, kL or L to compute the length scale.

2.2. Two-Equation Models

Let us consider, as an example, the high Reynolds number form of the $k - \varepsilon$ model. Similar derivations can be performed with any two-equation model. The transport equations for the turbulent kinetic energy and its dissipation rate, together with the eddy viscosity relation read

$$\rho \frac{Dk}{Dt} = \underbrace{-\rho < \underline{u}' \otimes \underline{u}' >: \operatorname{grad} \underline{\underline{U}}}_{P_k} - \varepsilon + \operatorname{div} \left[\left(\mu + \frac{\mu_t}{\sigma_k} \right) \operatorname{grad} k \right] \quad (3)$$

$$\rho \frac{D\varepsilon}{Dt} = (C_{\varepsilon_1} P_k - C_{\varepsilon_2} \rho \varepsilon) \frac{\varepsilon}{k} + \operatorname{div} \left[\left(\mu + \frac{\mu_t}{\sigma_\varepsilon} \right) \operatorname{grad} \varepsilon \right] \quad (4)$$

$$\mu_t = C_\mu \frac{\rho k^2}{\varepsilon} \quad (5)$$

where $\frac{D}{Dt}$ indicates a substantial derivative. The transport equation for the turbulent kinetic energy (3) is derived from the Navier–Stokes equations. Only the turbulent diffusion term is modelled. It is possible to derive a transport equation for the dissipation rate from the Navier–Stokes equations but no term-by-term modelling is possible. Therefore, the modelled transport equation is based upon an analogy with the turbulent kinetic energy transport equation.

Among the five constants which appear in the above model, C_{ε_2} is the only one which appears in the decay of isotropic turbulence. However, experiments as well as theory can only provide a range for this constant, not a definite value.

The idealized description of the logarithmic region, in an incompressible two-dimensional boundary layer without pressure gradients yields another important relation between the coefficients. This description first assumes that the advection is negligible near the wall. Moreover, in the logarithmic region, viscous transport is negligible compared to turbulent transport. The momentum equation can then be simplified to

$$- < u'v' >= \frac{\tau_w}{\rho} = u_\tau^2 \quad (6)$$

Moreover, experiments show that the turbulent kinetic energy is proportional to the main Reynolds stress

$$- < u'v' >= 2a_1 k \qquad a_1 \sim 0.15 \quad (7)$$

These two assumptions, together with the kinetic energy transport equation, yield

$$P_k = - < u'v' > \frac{\partial u}{\partial y} = \varepsilon \quad (8)$$

i.e. the production balances the dissipation. This also leads to the relation (5) with

$$C_\mu = (2a_1)^2 \tag{9}$$

If the mean velocity profile is assumed to follow a logarithmic law, the mean velocity gradient reads

$$\frac{\partial u}{\partial y} = \frac{u_\tau}{\kappa y} \tag{10}$$

which, together with the above assumptions and the dissipation transport equation, yields the following relation

$$C_{\epsilon_2} - C_{\epsilon_1} = \frac{\kappa^2}{2a_1 \sigma_\epsilon} \tag{11}$$

At last, free shear flows or the wake region of the boundary layer can be used to optimize the values of the turbulent diffusion coefficients σ_k and σ_ϵ.

When low Reynolds number flows or flows near walls are considered, extra damping terms are introduced in the model. The near-wall expansions of the velocity components [18] give information about the behaviour of all the turbulent quantities very close to the wall.

The final emerging image is that the models are strongly tuned on an idealized description of the logarithmic region. The outer part of the boundary layer, which is somehow similar to a wake flow, is well described by tuning the diffusion part of the models with respect to free shear flows. Recent models also account for the near-wall behaviours as given by Taylor's expansions [18]. On the contrary, the buffer layer, where most of the turbulent kinetic energy production takes place, is not at all described. The physics of this region is presently well documented but no physical information has been introduced in the models.

2.3.   Reynolds Stress Models

The Reynolds stress transport equation can be derived from the Navier–Stokes equations. It reads

$$\rho \frac{D < \underline{u}' \otimes \underline{u}' >}{Dt} = -\rho < \underline{u}' \otimes \underline{u}' > \cdot \underline{\underline{\text{grad}}} \, \underline{U} - \rho \underline{\underline{\varepsilon}} + \underline{\underline{\Pi}} + \text{div} \, [\cdots] \tag{12}$$

where the last term includes both the viscous and turbulent diffusion. $\underline{\underline{\varepsilon}}$ stands for the dissipation tensor while $\underline{\underline{\Pi}}$ is the pressure/strain term

$$\underline{\underline{\Pi}} = < p' \left(\underline{\underline{\text{grad}}} \, \underline{u}' + {}^t \underline{\underline{\text{grad}}} \, \underline{u}' \right) > \tag{13}$$

they are usually modelled together in terms of the turbulent kinetic energy dissipation rate $\varepsilon = \frac{1}{2} \text{trace} \left(\underline{\underline{\varepsilon}} \right)$, the Reynolds stress tensor and the mean velocity gradient tensor.

3. AIRFOIL PERFORMANCES AT HIGH ANGLES OF ATTACK

3.1. Introduction

A large database of the flow over a two-dimensional wing profile for high angles of attack has been acquired in the ONERA F1 and F2 wind tunnels to document the flow behaviour near stall [6, 7]. Wall pressure and skin friction distributions have been measured. The flow velocity has been measured by LDV so that mean and fluctuating data are available. The lift coefficient is deduced from pressure integration on the profile, the drag coefficient from a far wake momentum balance.

This flow has been computed using a Navier–Stokes solver [2, 3]. Boundary layer computations, using the Navier–Stokes pressure distribution, have been used to check the grid accuracy in the viscous region. The efforts on the wing are deduced from a global momentum balance. Three turbulence models have been tested:
- Michel's mixing length model [17],
- A two-layer model using Norris and Reynolds one-equation model near the wall and the $k - \varepsilon$ model in the outer region of the boundary layer [2],
- Wilcox $k - \omega$ model [20] together with the BSL and SST blendings of the $k - \omega$ and $k - \varepsilon$ models proposed by Menter [16].

3.2. Results

The predicted lift coefficient evolution versus angle of attack is plotted in figure 1. No model gives a good prediction of the stall angle but a model ranking is obvious. The worse prediction is achieved with the mixing length model and the two-layer model which do not predict stall. The $k - \omega$ model predicts stall but for an angle of attack three degrees too large. The Menter's models give intermediate predictions between the $k - \omega$ and the $k - \varepsilon$ models.

This analysis can be confirmed by looking at the velocity profile evolution along the suction side as plotted in figures 2 and 3. The angle of attack of $13°$ .has been selected as it is close to stall and is the largest angle of attack for which the flow remains two-dimensional in the experiment. The two-layer model better predicts the velocity profiles up to $x/c \sim 40\%$ but is unable to reproduce the shape of the velocity profile in adverse pressure gradient. The $k - \omega$ model overestimates the initial boundary layer thickness but reproduces the pressure gradient effect. The BSL and SST models, which blend $k - \omega$ and $k - \varepsilon$ models, give results between these two models. Similar conclusions can be drawn on the Reynolds stresses profiles where the $k - \omega$ overestimates the boundary layer thickness as well as the maximum turbulent stress at the beginning of the wing and yields the best predictions in the forward region.

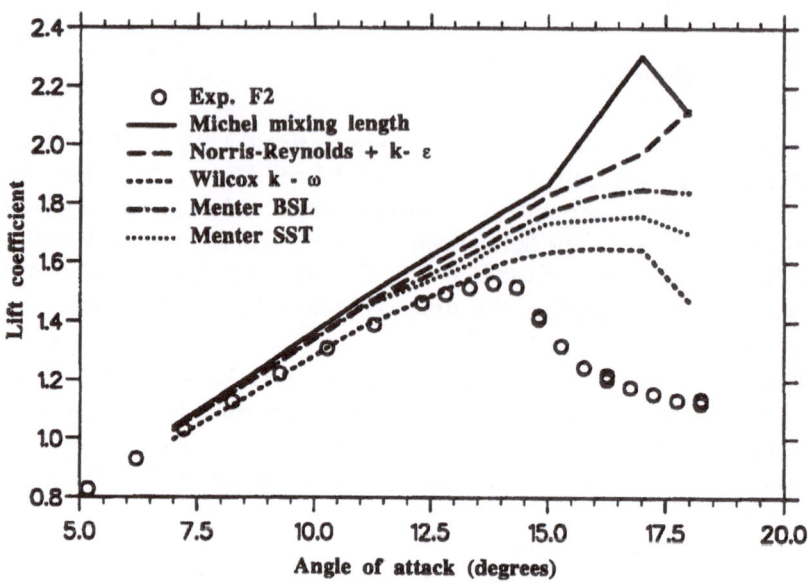

Fig. 1. — A profile – Lift versus angle of attack

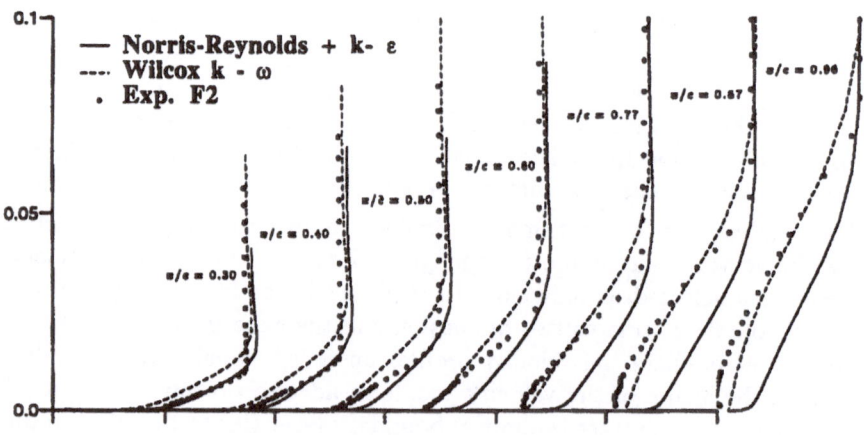

Fig. 2. — A profile – Mean velocity profiles predictions – $\alpha = 13°$ – $k - \varepsilon$ and $k - \omega$ models

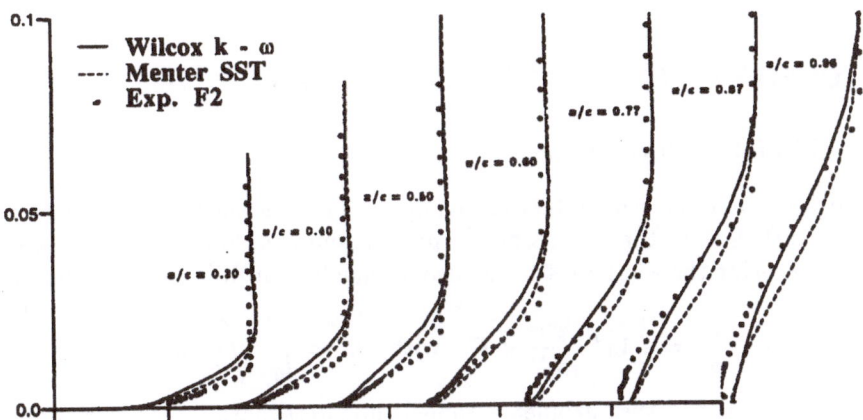

Fig. 3. — A profile – Mean velocity profiles predictions – $\alpha = 13°$ – $k - \omega$ and SST models

3.3.   Analysis

Two arguments can be used to explain why the $k-\omega$ model yields better results than the $k - \varepsilon$ models in adverse pressure gradient flows.

First, the $k - \varepsilon$ model can be written as a $k - \omega$ model. It then differs from the original $k - \omega$ model mainly by a $\underline{\mathrm{grad}\,k} \cdot \underline{\mathrm{grad}\,\varepsilon}$ term in the ω equation. The rôle of this term is to decrease the turbulent kinetic energy and hence the turbulent viscosity in the $k-\omega$ model compared to the $k-\varepsilon$ model. Therefore, the boundary layer is less turbulent and more prone to separate. This argument is exact only if the velocity profile is unchanged, which is not true. Therefore, the missing term both acts to reduce the turbulence and to change the velocity profile [2, 3].

The analysis of the model consistency with the logarithmic region presented in section 2.2 only holds for zero pressure gradient boundary layers. For adverse pressure gradient flows, Huang and Bradshaw [8] have shown that the $k - \omega$ model correctly predicts the logarithmic region while the $k-\varepsilon$ model does not. This also explains the better predictions achieved with the $k-\omega$ model. It must however be pointed out that Menter's model, which behaves like a $k-\omega$ model near the wall and like a $k - \varepsilon$ model in the outer part of the boundary layer, should also predict the correct logarithmic region but yields poor predictions compared to the original $k - \omega$ model.

However, it must be reminded that the $k - \omega$ model predictions are very sensitive to the external value of the specific dissipation ω, as pointed out e.g. by Menter [15]. In the present study, the value of the specific dissipation at the boundary layer edge was fixed as $\omega = \sqrt{k}/(0.0036)$ where δ is the boundary layer thickness.

4. COMPRESSIBLE BOUNDARY LAYERS

4.1. Model Performances

For high Mach number boundary layers, up to at least Mach 5 or 6, experiments as well as DNS show that the turbulent motion remains close to incompressible. Moreover, using the van Driest transformation, a logarithmic region is retrieved as

$$\frac{u^*}{u_\tau} = \frac{1}{\kappa} \ln \frac{\rho_w \, y u_\tau}{\mu_w} + C \quad \text{with} \quad u^* = \int_0^u \sqrt{\frac{\rho}{\rho_w}} \, du \tag{14}$$

However, most of the turbulence models underestimate the skin friction as the Mach number increases. Wilcox [21] showed that the solution for the Couette flow near the wall can be obtained as

$$\frac{u^*}{u_\tau} = \frac{1}{\kappa_\varphi} \ln \frac{\rho_w \, y u_\tau}{\mu_w} + \frac{1}{\kappa_\varphi} \left(\frac{\rho}{\rho_w} \right)^{n_\varphi} + C$$

where κ_φ depends upon the friction Mach number $M_\tau = u_\tau / \sqrt{\gamma R T_w}$ while the exponent n_φ of the extra term is equal to $\frac{5}{4}$ for the $k - \varepsilon$ model and $\frac{1}{4}$ for the $k - \omega$ model.

This analysis was improved by Huang et al. [9] who investigated the equilibrium in the logarithmic region for a general $k - \varphi$ model where φ is any length scale variable of the form

$$\varphi = \rho^n k^m \varepsilon^l$$

which satisfies an equation of the production/destruction/diffusion form. Compared to the analysis presented in section 2.2, two major changes occur. First the turbulent kinetic energy is related to the wall shear stress as

$$\rho k = \frac{1}{2a_1} \rho_w u_\tau^2$$

while the velocity gradient reads

$$\frac{\partial u}{\partial y} = \sqrt{\frac{\rho}{\rho_w}} \frac{u_\tau}{\kappa y}$$

Therefore, the density profile plays a rôle in the balance which becomes

$$\frac{2a_1 \sigma_\varphi}{l \kappa^2} (C_{\varepsilon_2} - C_{\varepsilon_1}) = 1 + \frac{1}{l^2} \left[d_1 \frac{y}{\rho} \frac{\partial \rho}{\partial y} + d_2 \frac{y^2}{\rho} \frac{\partial^2 \rho}{\partial y^2} + d_3 \left(\frac{y}{\rho} \frac{\partial \rho}{\partial y} \right)^2 \right]$$

$$\text{with} \quad d_1, d_2, d_3 = \text{fn}(n, m, l, C_{\varepsilon_1}, \frac{\sigma_\varphi}{\sigma_k})$$

Model	Type	κ	d_1	d_2	d_3	$d_1 + d_2 + d_3$
Jones (1972)	$k - \varepsilon$	0.41	3.041	0.54	-0.06	2.44
Launder (1974)	$k - \varepsilon$	0.43	2.872	0.37	0.19	2.69
Chien (1982)	$k - \varepsilon$	0.42	2.755	0.26	0.37	2.87
Nagano (1990)	$k - \varepsilon$	0.42	2.35	-0.15	0.98	3.48
Myong (1990)	$k - \varepsilon$	0.40	2.3	-0.20	1.05	3.55
So (1991)	$k - \varepsilon$	0.39	3.9	1.40	-1.35	1.15
Wilcox (1988)	$k - \omega$	0.41	0.56	0.06	-0.33	0.17
Smith (1995)	$k - L$	0.41	0.75	0.	0.	0.75

Table I. — Logarithmic law coefficients for some popular turbulence models

The standard balance for incompressible flows is retrieved but no longer holds for compressible flows as the d coefficients are not null, whatever the model. It has been observed that $\frac{y}{\rho} \frac{\partial \rho}{\partial y}$ is of order unity as well as $\frac{y^2}{\rho} \frac{\partial^2 \rho}{\partial y^2}$ which is negative. Therefore, the lower the term $d_1 - d_2 + d_3$, the better the model predicts the logarithmic region. The values of these constants for some popular turbulence models are given in table I. All $k - \varepsilon$ models, except the recent So et al. recent which uses unusual coefficients, behave poorly. The $k - \omega$ model gives low values of all the coefficients while the $k - L$ model, which has been derived for high Mach number flows, nullifies two coefficients.

This is evidenced in figure 4 on which the profiles predicted by various models are plotted in wall variables for the same Reynolds number and compared to the logarithmic law. The $k - \varepsilon$ model correctly predicts the logarithmic law in incompressible flow (Mach = 0.1) but not at Mach 5. The $k - L$ model is superior but the best agreement is achieved with the $k - \omega$ model which is however unable to exactly predict a logarithmic law.

4.2. Model Correction

In order to force the turbulence model to predict the correct logarithmic region, an extra term has been added in the dissipation equation [1, 19]. For a $k - \varepsilon$ model, it reads

$$\frac{\mu_t \varepsilon}{\sigma_\varepsilon} \left[d_1 \frac{y}{\rho} \frac{\partial \rho}{\partial y} + d_2 \frac{y^2}{\rho} \frac{\partial^2 \rho}{\partial y^2} + d_3 \left(\frac{y}{\rho} \frac{\partial \rho}{\partial y} \right)^2 \right] \tag{15}$$

This term decreases in the outer region of the boundary layer but may require some damping near the wall, according to the $k - \varepsilon$ model. Tests have been conducted with the Launder and Sharma [11] and the Chien [4] models and gave similar results. The velocity profile now reproduces the correct logarithmic law, as shown in figure 5. This also improves the skin friction prediction whatever the Mach number and the wall temperature [19].

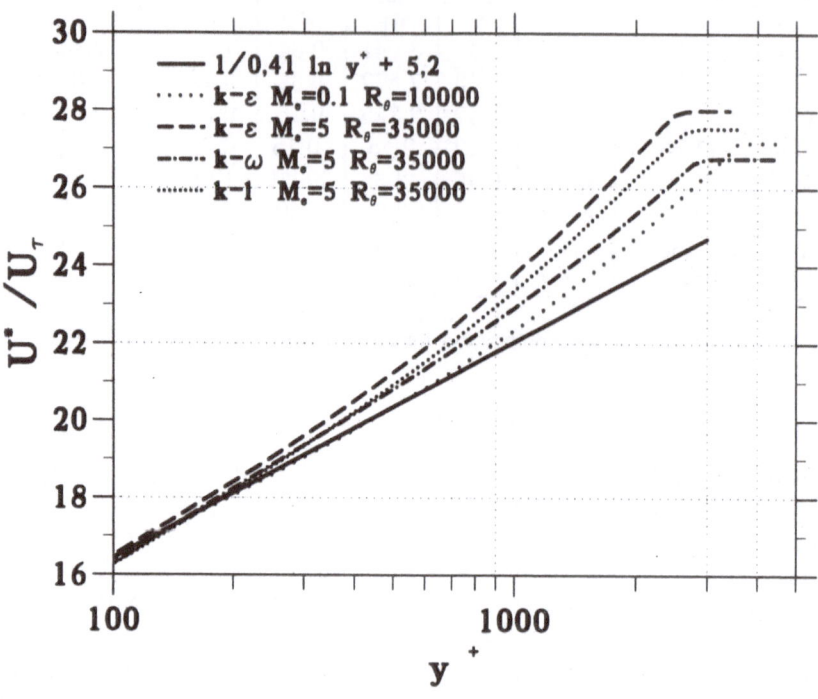

Fig. 4. — Comparison of transformed velocity profiles predictions for a zero pressure gradient boundary layer on an adiabatic wall

4.3. Analysis

A more physical form for the correction has to be found. In the logarithmic region, both ρk and $\rho^{3/2} \varepsilon y$ are constants. Therefore, the density gradients in (15) can be rewritten as products of turbulent kinetic energy and dissipation gradients. However, the physical meaning of these terms is not clear. As the production and destruction terms in the dissipation equation correctly reproduce homogeneous flows, are they a model for an inhomogeneous contribution to the production or the destruction of dissipation or a missing diffusion term?

5. THREE-DIMENSIONAL BOUNDARY LAYERS

5.1. Introduction

Three-dimensional boundary layer are a big challenge for turbulence models. As shown above, the turbulence models are tuned with respect to the logarithmic

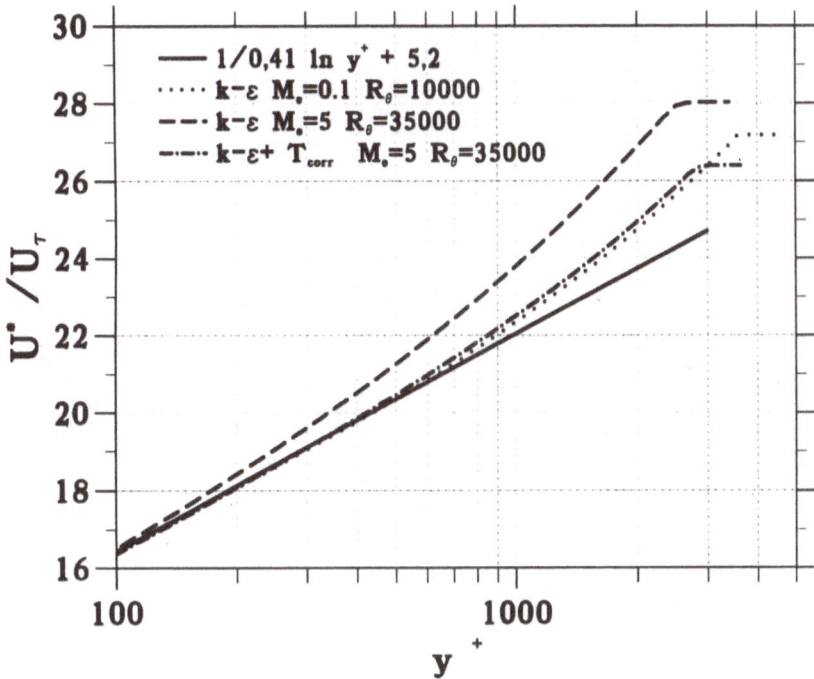

Fig. 5. — Transformed velocity profiles predictions with and without corrections

region. In three-dimensional flows, a logarithmic region seems to exist but its correct definition is still in debate. While in two-dimensional flows the logarithmic region always starts at the same distance $y^+ = \rho y u_\tau / \mu$ from the wall, its location is more variable in three-dimensional flows.

In two-dimensional boundary. layers, the eddy viscosity concept only gives a good description of the main turbulent shear stress $< u'v' >$. In three-dimensional boundary layers, the eddy viscosity assumption forces the main Reynolds stresses vector $(< u'v' >, < v'w' >)$ to be parallel to the the corresponding velocity gradient vector $\left(\frac{\partial U}{\partial y}, \frac{\partial W}{\partial y} \right)$, which is at variance with experiments.

5.2. Model/Experiment Comparisons

A large variety of turbulence models has been tested with respect to the experiment of the boundary layer on a flat plate below an infinite swept wing conducted by van den Berg et al. at NLR [13, 14]. Both direct mode, i.e. imposing the edge velocity, and inverse mode, i.e. imposing the displacement

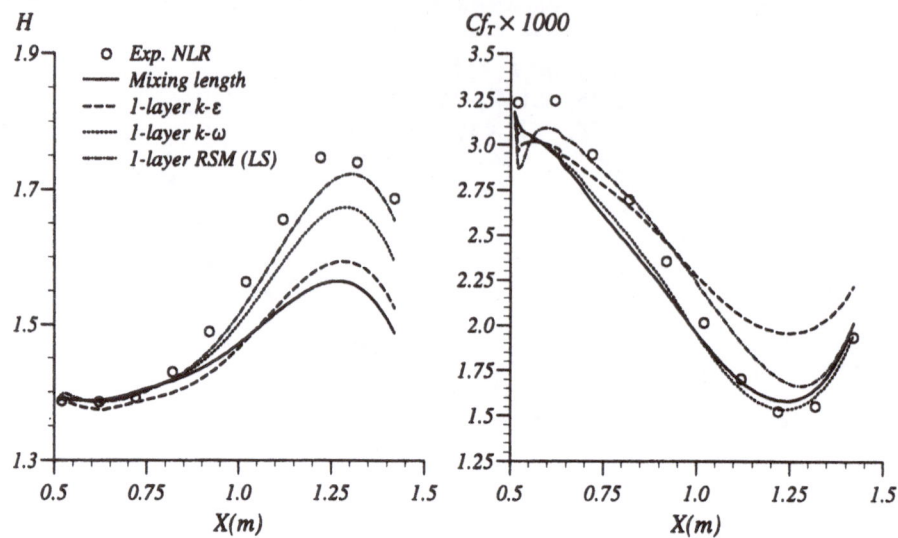

Fig. 6. — van den Berg et al. experiment – Shape factor and skin friction predictions – One layer models

thicknesses, were used to solve the boundary layer equations. Only the inverse mode results will be discussed here as they are not affected by the singularity which exists in the boundary layer equations near the separation point, when they are used in direct mode [5].

Mixing length models, the $k - \varepsilon$ or $k - \omega$ models and the Launder and Shima [12] Reynolds stress model were solved down to the wall. Figure 6 shows the longitudinal shape factor and total skin friction coefficient predictions. The mixing length model gives good predictions for the skin friction but not for the shape factor. As expected for such an adverse pressure gradient flow, better results are achieved with the $k - \omega$ model than with the $k - \varepsilon$ model, but these results are very sensitive to the edge value for the specific dissipation. The Reynolds stress model also gives good results.

More elaborate Reynolds stress models have been compared using a two-layer approach. A variant of the Norris and Reynolds model is used in the wall region while a Reynolds stress model, still accounting for the wall damping in a way similar to the one proposed by Gibson and Launder, is used in the outer part of the boundary layer. Figure 7 compares the predictions of the two-layer version of the $k - \varepsilon$ model to the Launder, Reece and Rodi model [10] and its isotropic version (IP). The Reynolds stress modelling definitely improves the prediction. However, better agreement is achieved with the simplest IP Model which is moreover more stable.

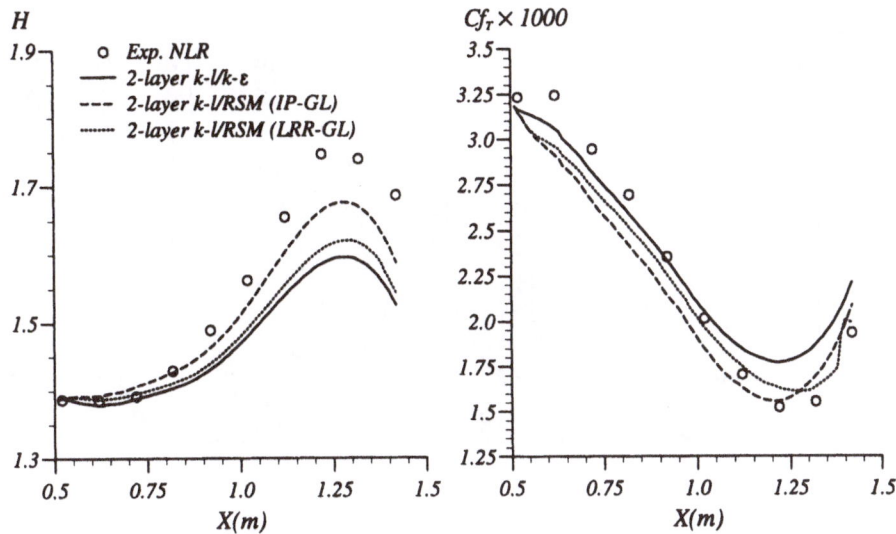

Fig. 7. — van den Berg et al. experiment – Shape factor and skin friction predictions – Linear Reynolds stress models

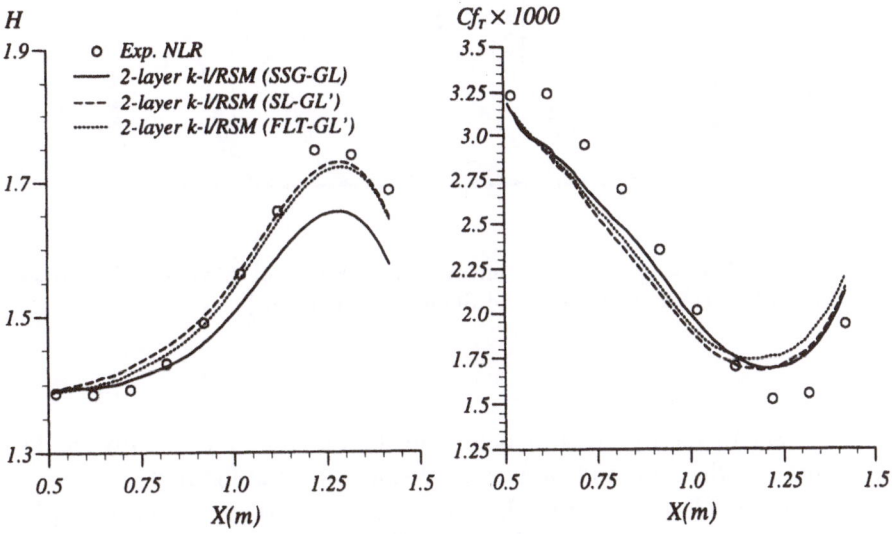

Fig. 8. — van den Berg et al. experiment – Shape factor and skin friction predictions – Non-linear Reynolds stress models

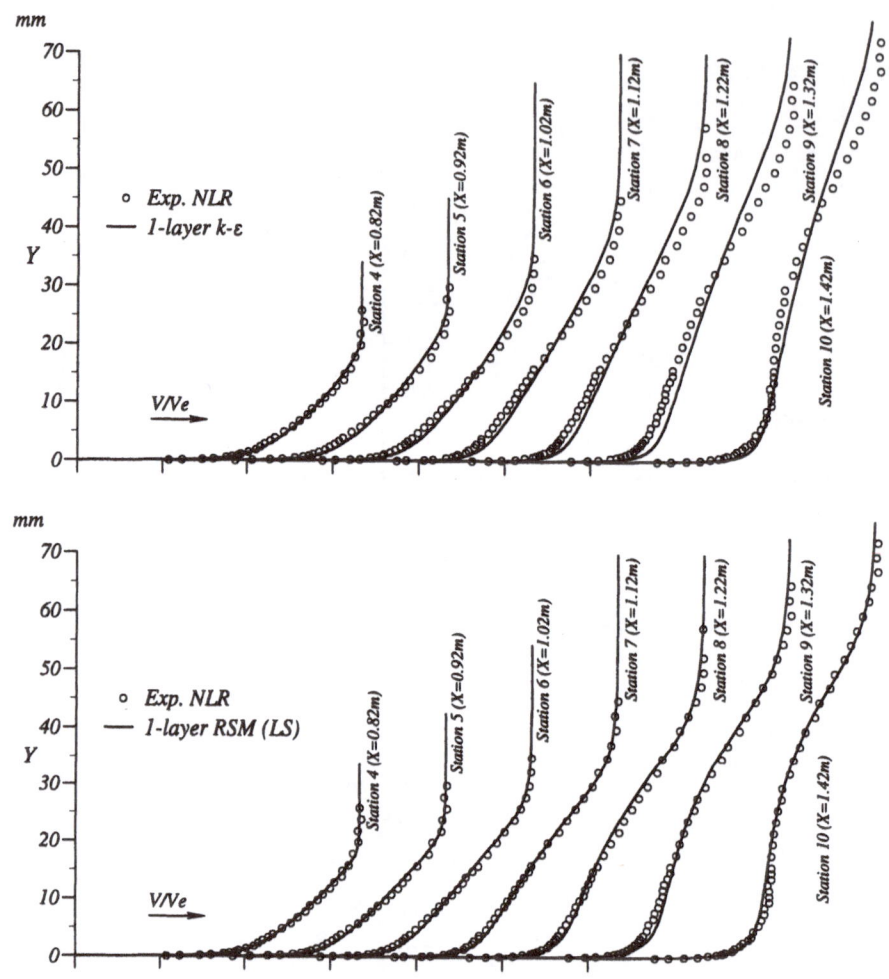

Fig. 9. — van den Berg et al. experiment – Comparison of mean velocity profiles –
$k - \varepsilon$ model (above) and Launder and Shima model (below)

Non linear Reynolds stress models yields better results, especially the Fu et
al. model and the Shih and Lumley model, as shown in figure 8.

As the Reynolds stress models better predict the turbulence anisotropy, they
give a better prediction of the velocity profile in the outer part of the boundary
layer, as shown in figure 9 where $k-\varepsilon$ and Launder and Shima model predictions
are compared. It must be reminded that the external velocity is not prescribed
in inverse mode.

However, even the Reynolds stress models are not able to predict the correct level of the main Reynolds stresses $(- < u'v' >, - < v'w' >)$ nor the correct direction of the "Reynolds stress vector" formed with these two stresses. The discrepancies on the level can be attributed to errors in the length scale prediction. Both errors are also due to the pressure/strain term and to the use of two-layer approaches as the peak of Reynolds stress production is in the wall region which is treated with a simpler model.

6. CONCLUSIONS

One of the key defects of the present modelling approaches is the lack of knowledge about the buffer layer where most of the turbulence production takes place. The models are tuned on an idealized description of the logarithmic region for zero pressure gradient flows but are unable to predict correctly the behaviour of this logarithmic region in other cases (compressible flows, but also relaminarization or adverse pressure gradient flows).

Two-equation models, which are very popular and less expensive than Reynolds stress models suffer from the double rôle the length scale has to play. It must describe at the same time the dissipation $(\varepsilon \sim k^{3/2}/L)$ and the Reynolds stress level $(\mu_t \sim \rho\sqrt{k}L)$. Anisotropic eddy viscosity models should help improving models. Moreover, improvements in the length scale prediction are still required.

At last DNS, and at a lesser extend LES, are a convenient tool to improve turbulence models. It must however be reminded that they only provide insights into the physics and numbers and that the model form has still to be developed by modellers to reproduce the physics and fit the numbers.

References

[1] Aupoix, B., Viala S., In *Symposium on Transitional and Compressible Turbulent Flows*. The Westin Resort, Hilton Head Island, South Carolina, USA (1995).

[2] Chevalier G., Prévision du Décrochage de Profils d'Aile par Résolution des Équations de Navier–Stokes Moyennées. *PhD thesis*, École Nationale Supérieure de l'Aéronautique et de l'Espace, Toulouse (1994).

[3] Chevalier G., Houdeville R., Cousteix J., *La Recherche Aérospatiale* 4 (1995) 277.

[4] Chien K.Y., *AIAA Journal* **20(1)** (1982) 33.

[5] Cousteix J., Houdeville R., *AIAA Journal* **19(8)** (1981) 976.

[6] Gleyzes C., *Rapport Technique 0A 71/2259AYD DERAT 55/5004-22* ONERA (1988).

[7] Gleyzes C., *Rapport Technique 0A 72/2259AYD DERAT 57/5004-22* ONERA (1989).

[8] Huang P.G., Bradshaw P., *AIAA Journal* **33** (1995) 624.

[9] Huang P.G., Bradshaw P., Coakley T.J., *AIAA Journal* **32(4)** (1994) 735.

[10] Launder B.E., Reece G.J., Rodi W., *J. Fluid Mech.* **68(2)** (1975) 537.

[11] Launder B.E., Sharma B.I., *Letters in Heat and Mass Transfer* **1** (1974) 131.

[12] Launder B.E., Shima N., *AIAA Journal* **27(10)** (1989) 1319.

[13] Malecki P., Étude de Modèles de Turbulence pour les Couches Limites Tridimensionnelles. *PhD thesis*, École Nationale Supérieure de l'Aéronautique et de l'Espace, Toulouse (1994).

[14] Malecki P., Cousteix J., Houdeville R., In *Refined Flow Modelling and Turbulence Measurements*, IAHR, ENPS Éditions (1993) 133.

[15] Menter F.R., *AIAA Journal* **30(6)** (1992) 1657.

[16] Menter F.R., *AIAA Journal* **32(8)** (1994) 1598.

[17] Michel R., Quémard C., Durand R., *Note Technique 154* ONERA (1969).

[18] Patel V.G., Rodi W., Scheuerer G., *AIAA Journal* **23(9)** (1985) 1308.

[19] Viala S., Étude Théorique et Numérique des Effets de Compressibilité et d'un Gradient de Pression Négatif sur la Couche Limite Turbulente. *PhD thesis*, École Nationale Supérieure de l'Aéronautique et de l'Espace, Toulouse (1995).

[20] Wilcox D.C., *AIAA Journal* **26(11)** (1988) 1299.

[21] Wilcox D.C., *AIAA Paper 91-1785*, 22nd Fluid Dynamics, Plasmadynamics and Laser Conference, Honolulu, Hawai (1991).

Fluid Instabilities in Inertial Confinement Fusion

D. Besnard[1][2], F. Ducros[2], D. Galmiche[2], Ph. Loreaux[2], B. Meyer[2], M. Naudy[2]

[1]*Commissariat à l'Énergie Atomique*
Centre d'Études de Buyères le Châtel
91680 Bruyères Le Châtel, France
[2]*Commissariat à l'Énergie Atomique*
Centre d'Études de Limeil-Valenton,
94195 Villeneuve St Geoges Cédex, France

Abstract

The design of optimized targets in the Inertial Confinement Fusion context requires instabilities to be estimated and mitigated whenever possible. Indeed, these processes might lead to shell break-up, therefore to a large decrease in yield. To address this problem, an experimental, theoretical as well as numerical research program was set up, which we partially describe in this paper. Theoretical questions pertain to ablative Rayleigh-Taylor instability, transition to turbulence, and mix modeling. As for numerical results, we present 3D Rayleigh-Taylor instability simulations. Current experiments are also discussed.

1. INTRODUCTION

For the past twenty years, a large effort has been put on ICF research. The scope of this paper is to offer some comments about current developments related to the problem of instabilities and turbulence in this context [1]. A typical example of these problems is encountered in the design of laser targets for the new Megajoule laser planned at DAM.

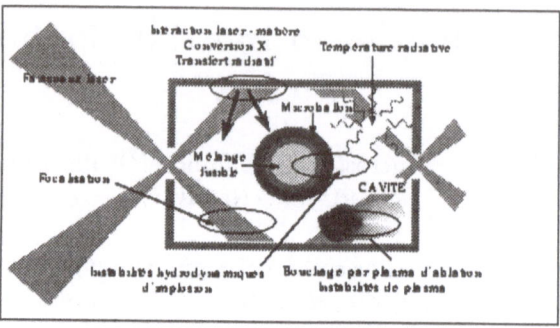

Fig. 1. — Schematic configuration of Megajoule laser target (revue Chocs n° 13, 1995).

A typical Megajoule laser target (Figure 1) consists of an outer cylindrical case (usually in gold) containing the capsule itself, which is surrounded by a gas. The overall dimensions of the case are of the order of 1cm. There is an aperture on both sides, closed by transparent windows, to retain the gas and allow the laser beam to penetrate in the case. The beams energy is absorbed by the gold case, and converted to X-ray energy. This energy is deposited in the capsule and drives the implosion. The target itself is composed of an outer shell (e.g. carbon foam), and the inner fuel is composed of cryogenic DT and DT gas. Once compressed, the DT generates fusion reactions, producing neutrons and alpha particles. Alpha particles help combustion to propagate; neutrons escape the target and provide the energy, which is eventually recovered. The goal is of course to design a target that leads to a gain in energy as large as possible. Because thin shell capsules require less energy to implode, and energy is extremely expensive to bring to the target, it is crucial to know what is the most optimized target one can safely use. We first deal, in Part 2 of this paper, with the linear phase of ablative Rayleigh-Taylor instability in ICF targets. Fully developed mix is obviously not welcome and is not expected at present time in Megajoule laser designs, but it is worthwhile to be able to predict the phenomena associated with such a possibility. Therefore, we address in Part 3 the problem of turbulent mix prediction in Rayleigh-Taylor instability induced circumstances, a major cause of target instability in laser targets.

2.   INSTABILITIES IN THE ICF CONTEXT

We consider a typical ICF target, with respectively the ablator, the cryogenic DT, and the DT gas. In figure 2, a typical implosion is shown through the plotting of the capsule radius as a fonction of time, as well as interface radii for each existing interface.

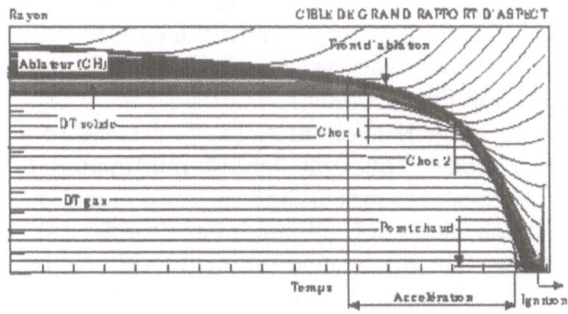

Fig. 2. — Implosion dynamics for a Megajoule laser target (revue Chocs, n° 13, 1995). The investigated target is constituted by a spherical ablator layer surrounding a cryogenic DT layer filled with DT gas.

One first notices the compression of the part of the ablator which is not ablated, with a decrease in radius for the ablation front location. A shock successively accelerates the CH/DT interface and the cryogenic DT/DT gas interface. Meanwhile, the transmitted shock reaches the capsule's center and reflects back. This pattern is duplicated through successive reflections on interfaces, until maximum compression is reached, and the first fusion reactions occur.

Fluid instabilities occur at the ablation front, at each shock passage (Richtmyer-Meshkov instability), and when the light DT gas pushes the heavy cryogenic DT (same thing as at the ablator/cryogenic DT interface) (Rayleigh-Taylor instability).

2.1.   Ablative Rayleigh-Taylor Instability

Let us consider the radial density, pressure and energy profiles of a typical laser capsule. Part of the ablator is progressively heated out, providing an acceleration, denoted g. The location of the limit at which the material is ablated is called the ablation surface. This surface is not fully spherical; this is due to defects in X-ray irradiation (smaller, however, than in the case of the direct irradiation by laser beams). This combination of a low density material (the ablated material) pushing a high density material (the non ablated part of the shell) in an overall accelerated fluid leads to the amplification of existing perturbations, that takes place in the ablation region around the ablation front.

One can explore the linear phase of this phenomenon. One starts with Navier-Stokes equations, coupled to radiation, in an accelerated frame. Small amplitude perturbations at the ablation front are shown to grow exponentially at a rate which is smaller than the theoretical Rayleigh-Taylor instability growth rate. The correction is due to a smoothing effect provided by the ablation process itself, characterized by the ablation mass flux. An approximate formula was derived by Takabe [2], showing a wavelength cut-off, therefore a stabilizing effect of ablation. However, very recent results seem to question the current estimate for this cut-off [3].

The corresponding numerical simulations we present here were performed with the DAM FCI2 code. This code handles multimaterial, multicomponent flows, with separate equations for ions and electrons energies. Thermal conduction, together with ion/electron Coulomb interaction are taken into account in the energy equations. The mass equation includes source and sink terms due to nuclear reactions. Momentum and energy equations are coupled to radiation, fast ions, and suprathermal electrons equations. Laser/matter interaction itself is treated using the classical optics approximation. Together with these equations, evolution equations for photons, neutrons, alpha particles, and fast electrons are solved. This code provides the time history of the primary variables describing the target evolution. It is then convoluted with the modeled transfer function of some experimental diagnostics, such as backlighting (by which a laser irradiates the target from behind, and the transmitted light is analyzed).

The actual simulations correspond to experiments performed with the PHEBUS laser, which is located at the Centre d'Etudes de Limeil-Valenton. The target itself consists of a sandwich composed of a 5 μm aluminum planar disk, and 10-50 μm of CH. 5-150 μm wavelengths, .1 μm in initial amplitude modulations are superimposed on the rear side of the target.

One can plot the instability growth rate at the ablation front as a function of the perturbation wavelength for a given target thickness and irradiance, and for different theories: classical RT, Munro theory, Takabe theory. Numerical simulations show that the growth rate is close to Takabe formula, provided that the mitigation coefficient β is taken to be equal to 5.5 (to be compared to the 2-3 value proposed by Takabe); it is also close to .6 times the classical RT growth rate.

In addition, Takabe formula shows a small wavelength cut-off that should be recovered by numerical simulations. Two difficulties may prevent this: on one hand, more physics is included in the code than in the theory; on the other hand, small wavelength perturbations lead to very fine meshes (several tens of cells are required to adequately describe one half wavelength). Demonstrating this threshold may be currently out of reach.

Therefore, one can safely conclude at the present time that:

- shell break-up can be predicted for large wavelengths;
- Takabe formula is fine for most wavelength provided β is taken to be 5.5;
- the cut-off value is smaller than currently available;
- more specific experiments are required, with a submicron spatial resolution.

2.2. Rayleigh-Taylor Instability

The second occurrence of fluid instabilities is at the ablator/fuel and cryogenic DT/DT gas interfaces. It is easier to study this type of instability in planar geometry; however, cylindrical designs are physically closer to laser targets that will be used at the Megajoule facility, and more easily diagnosed than spherical ones.

Consider the tri-layer planar experiment performed at CEA/LV [5]. It consists of a sandwich of Si/Al/Au. The Octal laser (8 beams) is used to generate a shock that propagates through the layers. The interface Al/Au is the one that is studied, using a laser probe to ablate a given thickness of the Au layer and possibly some of the mix region of Al and Au. The signal is analyzed using spectroscopy. If there is some mix, the ablated region contains some Al plasma, and the corresponding spectral lines appear.

Cylindrical experiments are performed at the Nova facility(Lawrence Livermore National Laboratory) as a joint LLNL, LANL and CEA/LV program. Eight Nova beams are converted into X-rays in a hohlraum. The cylindrical target is mounted transverse to this hohlraum, and is axially radiographed. The central part of the cylinder is composed of a low density foam, surrounded by a chlorinated plastic layer (called marker layer as it is opaque to the backlighter) a polystyrene layer (pusher) and a brominated ablator. Pertubations are machined on the outside surface of the ablator. This experiment allows for the study of instability growth in a convergent geometry. Feedtrough and growth of the Rayleigh-Taylor instability from imposed perturbations from the outer surface to the inner surface are observed.

Simulating the experimental X-ray radiographs is a multistep process. First, from experimental laser pulses, a temporal shape of the radiation drive temperature is obtained with a view factor code (such a code solves the stationary radiative transfer equation in the holraum -= case + external part of the cylinder-). The simulation of the implosion itself can then be performed with the FCI2 2D code. Finally, a diagnostic code is run to simulate X-ray backlighting (see figure 3). This last step is a requisite to compare numerical results to experimental radiographs. A good match has been observed for the position of the internal edge of the marker layer, which is not much perturbed before deceleration. The experimental outer edge of the marker is soon very perturbed, and presents significant components of several modes (mainly mode 4, due to the discrete number of beams). Up to now, only a qualitative agreement has been obtained for the outer edge location. More recents shots, with

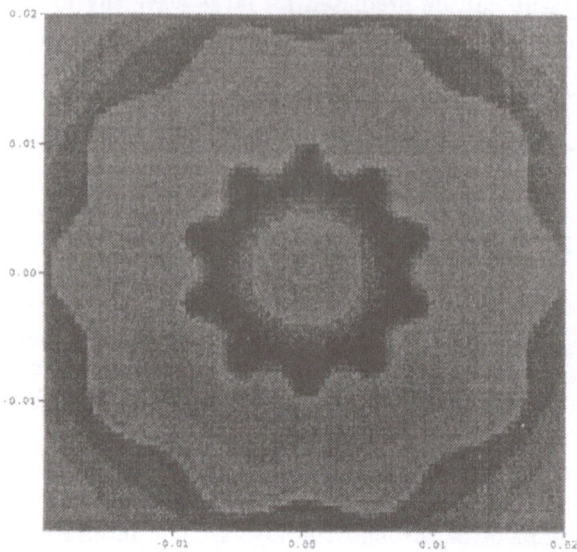

Fig. 3. — Simulation of X-ray radiagraph for a cylinder experiment. Central foam and polystyrene layer are transparent. The outer edge of the marker layer exhibits a well developed mode 10 perturbations.

better experimental radiographs, have to be analyzed. Once this is obtained, one can safely state that we have a predictive tool for RT growth rates in convergent geometries in the ICF context. This will be true, of course, provided that we can confirm this for more complex initial configurations, clear into the nonlinear phase of this instability.

3.   TURBULENCE AND MIX

Instability growth, as stated in the introduction, may lead to shell break-up, at worst. It may also induce some mixing between shell and fuel, and prevent burn. Quite a lot of effort was put into designing capsules immune, as much as possible, from mix, and more especially from turbulent mix.

Estimates of the Reynolds numbers occurring in a laser targets are difficult to obtain but they do not forbid the possibility of turbulence at high laser energy. To study this issue, it is important to develop adequate numerical tools. We note that the flows we are interested in are transient; we have therefore to cover the entire range of processes, from instability development to fully developed flows, including the transition process.

To explore the transition problem under Rayleigh-Taylor circumstances, we turned to LES techniques, and first considered idealized circumstances, describing the flow with Navier-Stokes, multifluid equations. The Richtmyer-Meshkov instability case was first explored by Besnard et al. [6]; The Rayleigh-Taylor case was then studied by Ducros [7], and Besnard et al. [8]. The difficulty is to develop a numerical scheme that allows for shocks and contact discontinuities, and that still preserves turbulence characteristics. Indeed, numerical schemes degenerate at shock location, since some amount of artificial viscosity must be introduced; they tend to damp wavelengths of the order of the cell size. On the other end, transition (numerically) occurs only if instable modes can grow.

To address this problem, we started [8] with 3D Navier–Stokes equations, together with concentration equations for the different fluids that are considered. The LES model we used is based on the work by Ducros [7]. The corresponding equations read:

$$\frac{\partial \overline{U}}{\partial t} + \frac{\partial (\overline{F}_1 - \overline{D}_1)}{\partial x_1} + \frac{\partial (\overline{F}_2 - \overline{D}_2)}{\partial x_2} + \frac{\partial (\overline{F}_3 - \overline{D}_3)}{\partial x_3} = \bar{\rho} g \quad,$$

where

$$\overline{U} = {}^T(\bar{\rho}, \overline{\rho u_1}, \overline{\rho u_2}, \overline{\rho u_3}, \overline{\rho e}, \overline{\rho c_n}) \quad,$$

with

$$\overline{F}_i = \begin{pmatrix} \bar{\rho}\tilde{u}_i \\ \varpi + \bar{\rho}\widetilde{u_i u_1} \\ \varpi + \bar{\rho}\widetilde{u_i u_2} \\ \varpi + \bar{\rho}\widetilde{u_i u_3} \\ (\bar{\rho}\tilde{e} + \varpi)\tilde{u}_i \\ \bar{\rho}\tilde{c}_n \tilde{u}_i \end{pmatrix} \quad, \quad \overline{D}_i = \begin{pmatrix} 0 \\ (\frac{\mu}{Re} + \rho\nu_t)\tilde{S}_{i1} \\ (\frac{\mu}{Re} + \rho\nu_t)\tilde{S}_{i2} \\ (\frac{\mu}{Re} + \rho\nu_t)\tilde{S}_{i3} \left(\frac{\mu}{Re} + \rho\nu_t\right)\tilde{u}_i\tilde{S}_{ij} \\ +\frac{\gamma}{Re}(\frac{\mu}{Pr} + \frac{\bar{\rho}\nu_t}{Pr_t})(\frac{\partial \tilde{e}}{\partial x_i}) \\ (\frac{\mu}{ReSci} + \frac{\bar{\rho}\nu_t}{ReSci_t})(\frac{\partial \tilde{c}_i}{\partial x_i}) \end{pmatrix} \quad,$$

$$\widetilde{S}_{ij} = \left[\frac{\partial \tilde{u}_j}{\partial x_i} + \frac{\partial \tilde{u}_i}{\partial x_j} - \frac{2}{3}(\vec{\nabla}.\vec{\tilde{u}})\delta_{ij}\right] \quad, \qquad \varpi = \bar{p} - \frac{1}{3}T_{ll} \quad,$$

where ν_t is modeled as in [7].

We used a predictor–corrector scheme, which was rendered TVD. This type of scheme is well adapted to multifluid, compressible flow.

When the flow turns turbulent, it is crucial to check that this scheme does not alter the properties of the flow. We first considered Homogeneous Isotropic Turbulence circumstances. Starting with an initial turbulent spectrum ($k^2 e^{-k^2}$), in a quasi–incompressible flow, we let this flow evolve.

If the Reynolds number is high, the kinetic energy should stay basically constant until the enstrophy catastrophy occurs. To allow for this, we had to introduce a limitation of the flux limiter, based on the value of the local Mach number, in order to prevent excessive numerical dissipation. After this

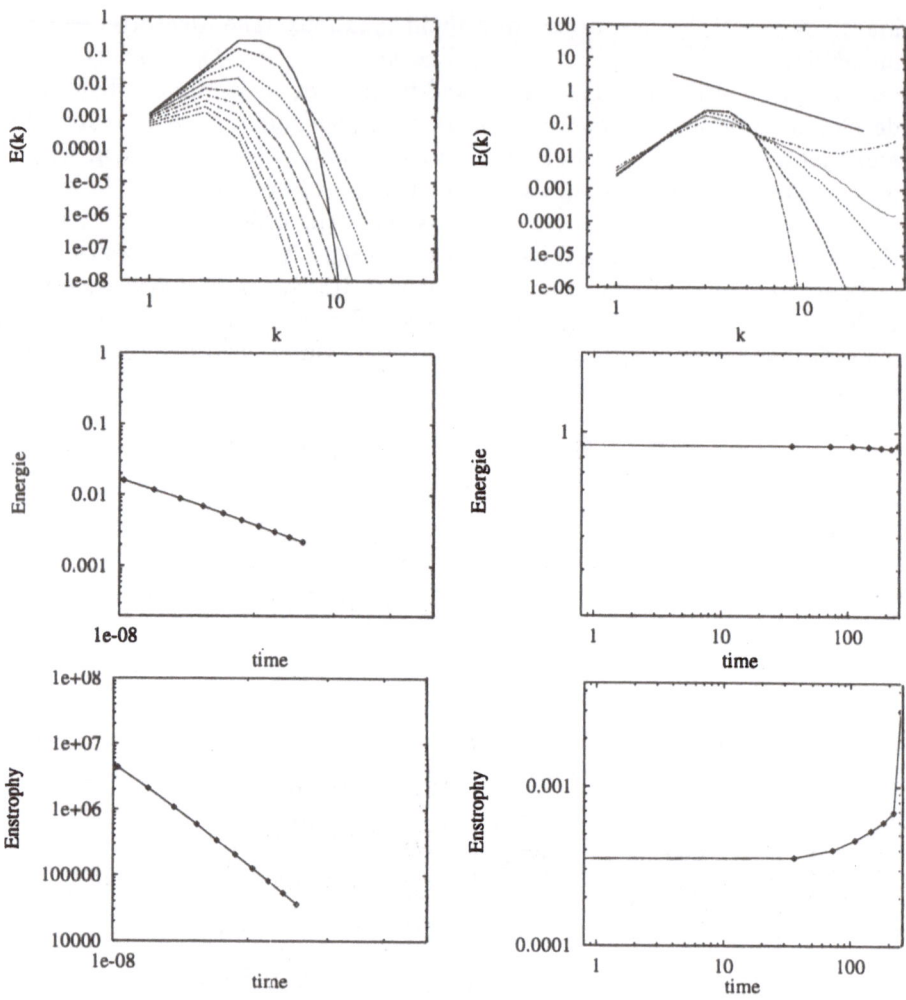

Fig. 4. — Homogeneous Isotropic Turbulence energy spectra, kinetic energy vs time, and enstrophy vs time. $Re_t = 3$ and $Re_t = 10^{10}$

correction, the code showed the expected increase in enstrophy around a characteristic time that is smaller than a theoretical value based on EDQNM theory, and larger than the value obtained by a spectral code, taken here as reference in terms of numerical dissipation (this critical time is inversely proportional to the square root of the enstrophy; less numerical dissipation induces more enstrophy, therefore a shorter critical time).

If the Reynolds number is small, the kinetic energy decreases due to viscous dissipation.

Figure 4 shows the two cases of $Re = 10^{10}$, and $Re = 3$, where $Re = \frac{Urms L0}{\nu}$, and $Urms$ is the root mean square of velocity fluctuations, $L0$ the size of the computational box, ν the kinetic viscosity; the turbulent Mach number is equal to 0.03.

We now consider circumstances under which the mixing is induced by Rayleigh-Taylor instability. As we mentioned above, Rayleigh-Taylor instability is one of the main source of mixing.

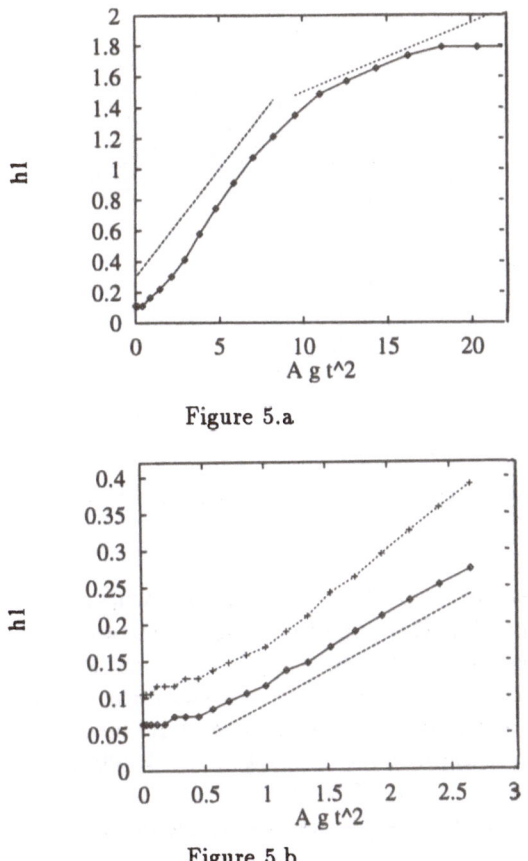

Figure 5.a

Figure 5.b

Fig. 5. — Mixing zone thickness vs time. Figure 5.a Stratified case; $g = 2.5\ 10^5 cms^{-2}$, $a_0/\lambda \approx 1$. Figure 5.b Non-stratified; $g = 5$. $a_0/\lambda \approx 1$; a_0 is the initial amplitude of the perturbation, λ an average wavelength.

The main results obtained in the RT case is that, after an initial period that depends upon initial conditions, another regime appears, that produces an interpenetration that depends only on the parameters of the flow, i.e. Mach, Atwood, and acceleration g. In other words, the interpenetration thickness is proportional to gt^2, which is the only length scale that can be built out of the available variables when the flow is fully developed.

In figure 5, the mixing thickness as a function of time is plotted for two values of the acceleration. After an initial phase, the caracteristics of which depend on the initial conditions, the curve exhibits an inertial regime, which eventually stops when the heavy fluid reaches the bottom of the computational box. The same type of simulation is repeated for a much lower value of g.

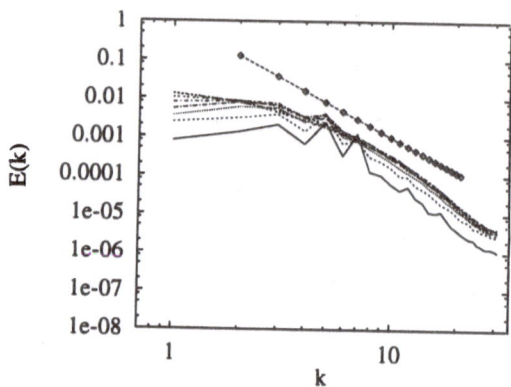

Fig. 6. — Energy spectra under Rayleigh-Taylor circumstances, $g = 2.510^5 cms^{-2}$.

The results we obtained [8] are in agreement with those of Linden [9], Youngs [10], and Glimm et al. [11]. The energy spectrum was obtained, demonstrating an inertial range close to the Kolmogorov one (see Figure 6).

More simulations are to be done, to explore the entire range of parameters occuring in the ICF context, but we already have here a powerful tool to explore the primary physical processes at work in the Rayleigh-Taylor and Richtmyer-Meshkov regimes.

For fully developed, highly non stationary and turbulent flows, we also developed a non spectral equilibrium model (see Besnard, Harlow, Zemach, Rauenzahn [12]). This model helps evaluate and improve simpler one-point models that can be in turn used in parametric studies.

4.   CONCLUSION

The results presented here are the result of on-going work. At the present time, a few conclusions may be drawn from them:

- Predictive capabilities are crucial for the Megajoule laser design;

- One major uncertainty is due to fluid instabilities occuring in large initial (and in-flight) aspect ratio targets;

- These instabilities may induce a break-up of the target's shell, and mix at interfaces;

- Fully developped turbulence is not very likely to exist, but if so, one might expect a large increase in mix, and a corresponding decrease in yield;

- Numerical simulation, as well as spectral modeling is crucial for a better understanding of critical phenomena leading to mix, since only indirect measurements are available;

- Current 2D Laser codes provide a satisfactory qualitative (and often quantitative) description of linear instability development;

- 3D simulations are necessary for non linear and turbulent phases;

- Modeling sheds additional light on mix formation, and helps validate simpler models used in general purpose ICF codes.

References

[1] Lindl J.D., McCrory R.L., Campbell E.M., *Physics Today* 32 (September 1992).

[2] Takabe H., et al., *Phys. Fluids* **28** (1985) 3676.

[3] Lafitte O., *private communication.*

[4] Meyer K.A., Blewett P.J., *Phys. Fluids* **15** (1972) 753.

[5] REVUE CHOCS, n° 14, 25 (1995).

[6] Besnard D., Haas J.F., Rauenzahn R.M., *Physica D* **37** (1989) 227.

[7] Ducros F., *CEA Report CEA-N-2811* (1996).

[8] Besnard D., et al., *Proceedings of the 5th International Workshop on the Physics of Turbulent Compressible Mixing*, Stony Brook (July 1995).

[9] Linden P.F., In *Proceedings of the 2nd International Workshop on the Physics of Compressible Turbulent Mixing*, Pleasanton, USA (November 1989).

[10] Youngs D.L., *Phys. Fluids A* **3(5)** (1991) 1312.

[11] Glimm J., Li X.L., Menikoff R., Sharp D.H., Zhang Q., *Phys. Fluids A* **2** (1990) 2046.

[12] Besnard D., et al., *J. Theo. & Comp. Physics* **8 (1)** (1996) 1.

Large Eddy Simulations in Nuclear Reactors Thermal-Hydraulics

D. Grand[1] and G. Urbin[1][2]

(1) *CEA- GRENOBLE DRN/DTP/STR*
17, rue des Martyrs 38054 Grenoble Cedex 9, FRANCE
e-mail: grand@dtp.cea.fr and urbin@pelvoux.alpes.cea.fr
(2) *LEGI Institut de Mécanique de Grenoble*
BP 53, 38041 Grenoble Cedex 9, FRANCE

---

1. INTRODUCTION

Nuclear industry requires detailed computations of thermal-hydraulic phenomena in different regions of nuclear reactors and in various components. The physical situations encountered can cover a wide range of flow regimes in single phase, two-phase and multiphase flows. Single phase flows only will be considered hereafter.

The coolant in the reactor system extracts the heat from the core and transfers it to the turbines and condensers through different circuits (primary and secondary) and components (steam generators, heat exchangers). With the high power involved, temperature differences of some 100°C can be generated in the fluid, and are either stationnary or transient. These temperature differences generate thermal stresses in the neighbouring solid structures, due to the differential thermal dilatation. They must be known for the design of the structures. In some situations of low flowrates, buoyancy forces can influence the flow field and induce combined convection (natural and forced).

Local velocity distributions must also be known for example in order to optimize the flow distribution inside the core or to minimize the risk of flow-induced vibrations in components like steam generators. Thus the knowledge of local velocity and temperature distributions (averages and fluctuations) is of great importance, for the performances and the safety aspects.

The current practice requires to use codes validated on experiments and to apply conservative margins to the results of reactor calculations. Experimental qualifications are always required and are based on two types of experiments:

- separate effects tests, where a key phenomenon is studied in a generic configuration. They provide the physical basis of the models implemented in the codes,

- integral experiments, where coupled phenomena are approximated with similarity criteria in realistic geometries. They contribute to the verification of the codes results in complex cases.

The increased performances of codes and computers should allow greater confidence in codes extrapolation to new geometries and operating conditions. This can authorize a reduction of the conservative margins adopted in the transposition of codes results to reactor conception. Another benefit will be an extension of the domain of prediction of computer codes to new physical variables like temperature fluctuations.

Turbulence modelling is of course a major challenge in the improvement of the codes predictions. The three major approaches available for computation of turbulent flows can be used at different levels in nuclear applications:

- Resolution of the Reynolds averaged Navier-Stokes equations (approximation by turbulent viscosity (k,ϵ) or resolution of PDE for Reynolds stresses)

- Large Eddy Simulation

- Direct Numerical Simulation

In any case the flows considered are confined and marked by strong spatial inhomogeneities. The simulation must be done in physical space.

It is interesting to know how these different simulations can be applied to the computation of flows inside a reactor vessel. A major constraint is the size of the computational mesh required for a given type of simulation. Consider the reactor vessel of a PWR. Its internal diameter (D) is of the order of 4m. The hydraulic diameter inside the core (d_h) is of the order of 1 cm. The microscale of turbulence (Kolmogorov length scale η) can be estimated of the order of 30μm. Thus the following estimation can be made for the number (N) of elements of the mesh required for the computation of the flowfield for the whole vessel:

- for a Direct Numerical Simulation N= 10^{15}
- for a Large Eddy Simulation N= 10^{10}

Considering the evolution of computer performances, application of LES to this calculation could be foreseen around 2000. Today, simulations with 10^6 elements become common practice in advanced industrial computations and the possibility will expand with parallel computers. Due to the time schedule of all research activity, the future use of these methods must be prepared today on academic cases which present the main features of these industrial flows. The methods should also be tested on some industrial applications where the spectrum of length scales is narrower than the ambitious case presented above.

2.   NUMERICAL TOOL CHARACTERISTICS

In the large-eddy simulation of turbulence approach, one gets rid of the scales of wavelength smaller than the grid mesh by using a subgrid-model. We have chosen the *structure function model* proposed by Métais and Lesieur [1] in its selective version (see Lesieur and Métais, [2]), which is well adapted for transitional flows and accepts non uniform grids [3]. Its main characteristic is that it switches off the eddy-viscosity when the flow is not three−dimensional enough. The LES filtered Navier-Stokes equations are solved using the TRIO-VF code. It is an industrial software developed for thermal-hydraulics applications at the Commissariat a l'Energie Atomique de Grenoble [4] . It has been thoroughly validated in many LES of various flows . It uses the finite volume element method on a structured mesh. The variables are located on a staggered grid, with pressure and scalar quantities located at the centre of the control volume and the velocity components on the side. Temporal discretisation is a first order Euler scheme; spatial discretisation is a third order Quick Sharp scheme. The Poisson equation for pressure uses an iterative method of conjugated gradient.

3.   STRATIFIED RECIRCULATING FLOWS

3.1.   Flow Downstream a Backward Facing Step

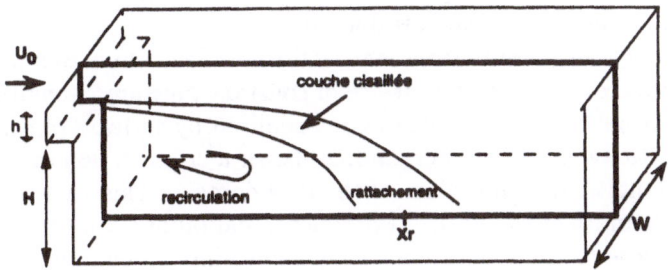

Fig. 1. — Geometry of the backward-facing step

The flow downstream a backward facing step combines complex flow phenomena in a simple geometry (Fig. 1). At the inlet, the boundary layer developped upstream in the channel, separates at the edge of the step produced by the sudden expansion. The free shear layer detached from the edge is curved by the recirculation flow in the wake of the step. It further interacts with the two walls bounding the channel and impinges the lower one.

This case was studied by Silveira [5] and Fallon [3]. The computed results are compared to experimental ones for isothermal flows. The comparison deals with the statistics of the flow (average velocity and Reynolds stresses), the energy spectrum and the visualization of the flow structures. The dynamics of the flow and the instabilities are analyzed and compared to experimental visualizations.

Different experimental and numerical studies have been conducted by other authors in this configuration and an updated biograph can be found in Fallon [3]. For brievety, we will mention the experimental results of Eaton [6] for a step with an aspect ratio $R = (H + h)/H = 2.5$, Ötügen [7] for different aspect ratios and more recently Jovic [8] for R=6. Large Eddy Simulations of this configuration have been conducted by Arnal and Friedrich [9] and Lee and Moin [10].

The boundary conditions in our studies are the following [5] [3]:

- lateral sides: free slip condition

- upper and lower walls: no slip condition and the shear stress is computed accordingly to the formulation given in Grötzbach

- outlet: advection equation for pressure

- inlet: profile of the longitudinal velocity measured in the experiment and superposition of a random pertubation of intensity 10^{-6}

Initially, the fluid is at rest. For the computation of statistics quantities, it is verified that the flow becomes statistically time independent before a characteristic time 160 H/Uo (approximatively four times the delay for complete renewal of the fluid inside the flow domain).

Fig. 2 displays an instantaneous view of the flow field: three Kelvin-Helmhotz (K-H) billows are visualized by surface of constant pressure. Longitudinal vortices streched between these billows are visualized by surfaces of constant longitudinal vorticity. In the background, is shown a cross section of the pressure field in the longitudinal mid-plane of the flow domain. This view is charateristic of the dynamics of the flow. Two dimensional billows are generated in the shear layer close to the separation. They evolve into three-dimensional structures downstream at the approach of the reattachment point. Downstream of the reattachment, longitudinal vortices are advected. They are similar to Λ shaped structures observed in the computations of the compressible boundary layer [11]. At subsequent time, it can be observed helical pairing of the K-H billows. These visualizations made evident the mechanism of forcing of the instabilities in the shear layer by turbulent structures advected in the recirculation region back to the step from the reattachement point.

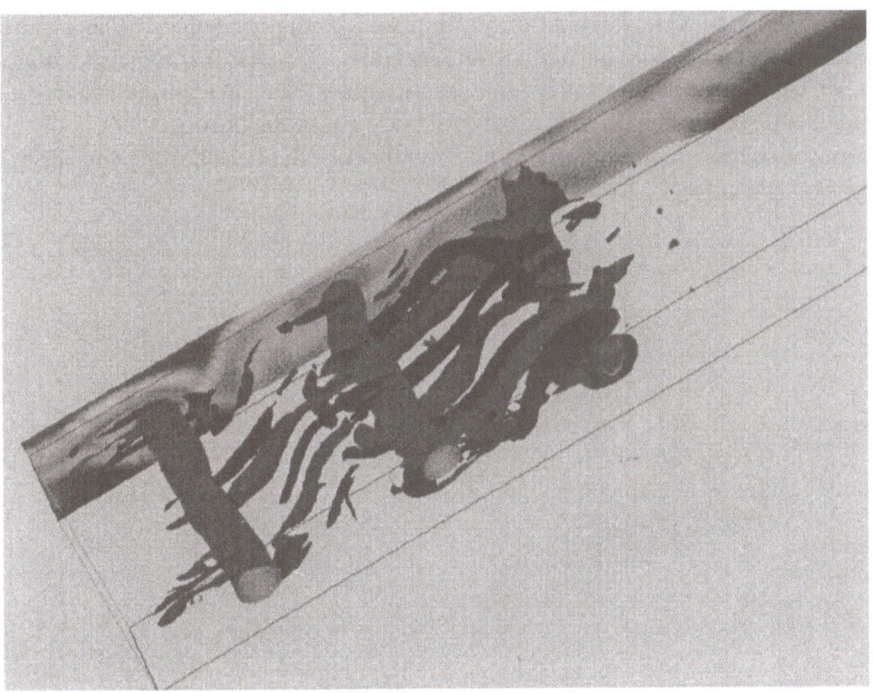

Fig. 2. — Backward-facing step. Visualization of eddies : low pressure isosurface, positive and negative longitudinal vorticity.

The statistics of the flow have been compared with experimental results [6]. The mesh used for this calculations is 100 * 40 *20 and follows previous recommendations [9]. The statistics compared are the mean velocity, turbulent intensity and Reynolds stresses. The position of the reattachment point is overestimated by 30% in the simulations with the structure function model. The selective structure function model overestimates it by only 10%. The error is attributed to the coarse mesh.

The results on the transverse profiles of turbulent quantities are qualitatively in good agreement with the experimental profiles and display the same positions of the maxima. Fig. 3 show the profiles drawn at the reattachment point of each experiment or simulation in order to remove the influence of different locations. All simulations overestimate the turbulent statistics. However, filtering of the experimental results can explain that the measured values are lower than real values of these quantities. Quite surprisingly, the results with the selective structure function model depart more from experimental results than the regular model. The coarsness of the mesh or the inlet boundary conditions may be responsible and this requires further analysis which is underway.

This study is extended in Fallon [3] to the stratified flow obtained when a

vertical temperature distribution is imposed in the inlet flow. The evolution
of the instabilities in the free shear layer are analyzed for different values of
the Richardson number. With increasing values, the three dimensional instabilities are progressively inhibited: first they remain bidimensional, then the
pairing mechanism is suppressed and finally turbulence collapses and energy is
dispersed by gravity waves.

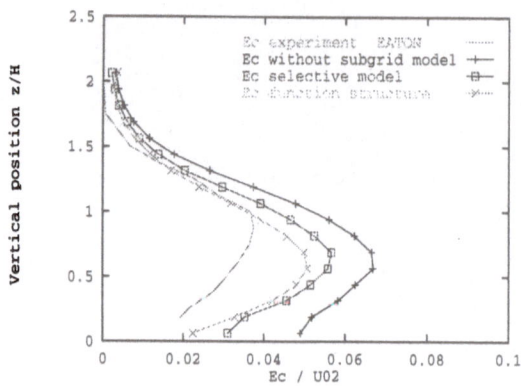

Fig. 3. — Vertical profile of kinetic energy at the reattachment station.

3.2. Recirculating Flow and Thermal Stratification in a Rectangular Cavity.

In the design of the pool type Liquid Metal Reactors, the primary circuit is
enclosed in a main vessel is divided into two separate pools or plenums:

- a cold plenum which feeds the core inlet with cold sodium forced by
  pumps,
- a hot plenum which collects the hot sodium at core outlet.

The two plenums are separated by a thin structure (redan) which transfers
heat by conduction. The conductive heat flux from the hot pool to the cold
pool generates a thermally stratified region in the lower part of the hot pool.
Temperature gradients, oscillations of the stratification interface and inherent
local temperature fluctuations must be known to assess the thermomechanical
behaviour of the redan. Time-variations of the temperature due to turbulent fluctuations can be as critical as spatial-variations. In this case, the LES
provides an attractive method of predicting instantaneous flow topology, time-
dependent turbulent interactions and therefore the temperature fluctuations.

Moreover, the results of a numerical simulation can be compared to an experiment Cormoran which has been carried out to investigate the thermal-

hydraulics in a rectangular cavity filled with liquid sodium [12]. The test section is a rectangular cavity 1m high, 0.5 m wide and 0.2 m deep. The hot incoming sodium flow is introduced on the left vertical side through a rectangular slot (0.1 meter high). It leaves the cavity through a rectangular opening at the top. The opposite vertical wall of the cavity simulates the redan, which is cooled externally by a forced sodium flow. The measurements are made with thermocouples are located at the mid-plane of the cavity along three vertical profiles.

Several tests were conducted, covering a range of values of the Richardson number from 0.5 to 10. The Richardson number is based on the maximum temperature difference between the two flows and on the bulk velocity in the inlet slot. A value of 9.76 is chosen for the present calculation.

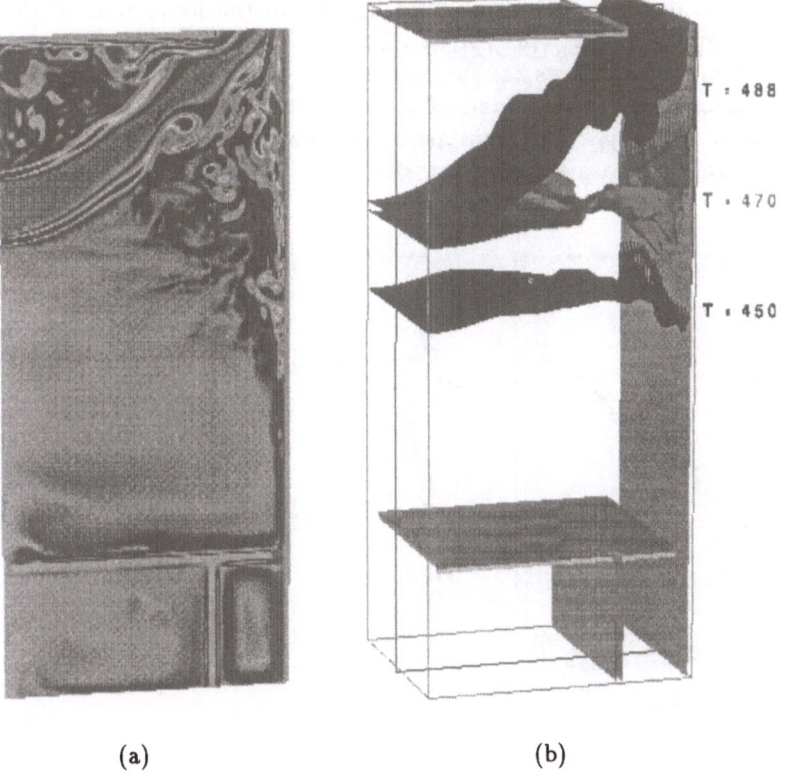

(a)   (b)

Fig. 4. — Cormoran: 3D flow topology. a) Cut at mid-plane of the vorticity modulus field. b) Temperature isosurfaces: T= 488°C, T= 470°C,T= 450°C.

The study and the analysis were performed by Fallon [3] using the selective structure function model. The boundary layers at the solid walls (including internal structures) are not explicitly resolved but replaced by wall functions. The wall shear stress is computed by integration across the turbulent boundary layers and an universal logarithmic velocity profil is assumed. Conductive heat transfer is computed inside the redan, separating the cavity from the cooling duct. The cooling duct is treated with a two-dimensional description. Temperature and velocity are specified at both inlets. From industrial point of view, the mesh size is a compromise between the available computational resources (or CPU time billing) and the need of accuracy for the description of turbulent fields.

Figure 4 displays the temperature field at a given time with 3 isotherms. The upper one shows how the temperature rolls up around coherents vortices in the lower shear layer of the incoming jet. This roll up is quasi-homogeneous along the spanwise direction. The middle one visualizes the flow becoming tridimensional at mid-distance betwen the inlet and the redan. The lower one (horizontal) displays the strong stratification in the lower part of the cavity. The same figure displays the vorticity field in the mid-plane of the cavity where Kelvin-Helmholtz billows can be visualized. The recirculating flow is highly tridimensional and carries small coherent vortices which interact with the lower shear layer of the jet. In a thin area along the cold redan vortex interactions with the wall increase the heat transfer.

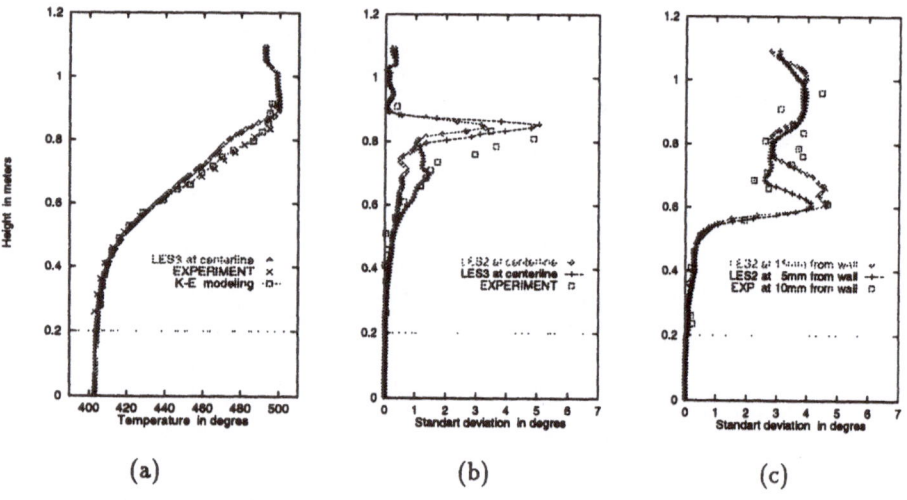

(a)   (b)   (c)

Fig. 5. — Comparison between computed temperature and measurements. a) b) at centerline, c) near the redan

In Fig. 5 the mean temperature and standard deviation of temperature fluctuations are compared with experimental measurements along a vertical profile at cavity mid-plane and 1 cm from the redan. Previous results based on the classical (k,ϵ) model (computed with the same code [12]) are added for the mean temperature (Fig. 5.a). For the mean temperature, the results given by Large-Eddy Simulation are globally better (the two profiles surrounding the exact location of the experimental profile are drawn). The standart deviations of temperature fluctuations are obtained by time integration over 20 s of the instantaneous LES predictions. At centerline (Fig. 5b), maxima standard deviations are located in the lower shear layer of the jet. The numerical results indicate that the position of the curved jet is slightly overestimated but the maximum intensity is in very good agreement with the experimental measurements. Near the wall (Figs 5c), the predictions agree very well in particular at the maximum in the stratification position.

4.   TURBULENT JETS

4.1.   Introduction

Axisymmetric jets are a prototype of free shear flows of vital importance from both a fundamental as well as a more applied point of view. Indeed, a better understanding of the jet vortex structures should make possible the active control of the jet (spreading rate, mixing enhancement...) for engineering applications [13] [14]. We present hereafter numerical simulations investigating the three-dimensional coherent vortices in the near region of a round jet, the role of the inflow perturbation on these structures, and the jet statistics. Our purpose is to demonstrate that a simulation, both three-dimensional and in a spatial box, is feasible with reasonable computational cost for engineering applications. Through Large Eddy Simulations (LES based upon the selective structure function model), we first describe the different coherent structures growing in a free round jet at high Reynolds number $Re = 25000$ based upon the jet diameter D and its bulk velocity W_o. We then show how the jet dynamics can be controlled by imposing the inflow excitation.

4.2.   Flow Configuration

We consider the domain starting at the nozzle and extending up to a dozen of diameters downstream. This domain, including the potential core, is characterized by strong vorticity and coherent vortex structures generation. Cartesian coordinates were preferred to cylindrical coordinates since the former allow to consider more complex geometries of industrial interest like multi-jets for instance. The inlet velocity profile is close to a top-hat shape. The grid is three-dimensional, non-uniform, with 204000 cells. The Z axis is along the jet axis. The origin of axes is located at the nozzle center. This relatively low number of cells makes possible long time integration, which is compulsory to reach statistical convergence: over one hundred adimensional time (D/W_o) are represented.

4.3. The Natural Jet

In order to validate the numerical approach, we first compare the computed statistics with experimental results. Figure 6a shows the downstream evolution of the mean axial velocity on the jet axis. The fall around $Z/D = 5.5$ indicates the end of the potential core. The numerical value stays within the experimental range. The downstream evolution of the r.m.s. fluctuating axial velocity on the jet axis is displayed on Figure 6b. The characteristic peak is well reproduced both in terms of its maximal value and its axial position.

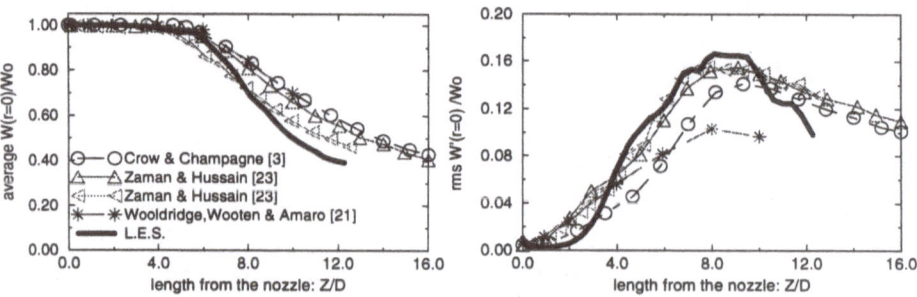

Fig. 6. — Numerical and experimental results comparisons. Centerline axial velocity
a) mean ; b) r.m.s. fluctuations

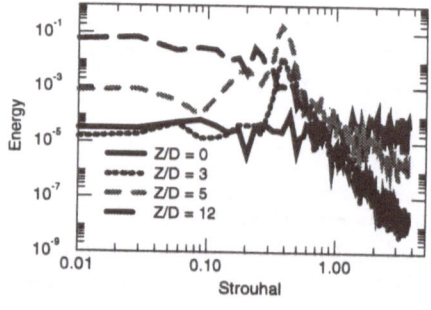

We then turn to the frequential aspect by considering the frequency spectra (see Figure 7). From the white noise ($Z = 0$) emerges a predominant vortex-shedding Strouhal number, $Str_D = 0.35$ (based on the diameter) in good correspondance with the experimental value (Str_D between 0.3 and 0.5; see Hussain and Zaman [15].

Fig. 7. — Frequency spectra of the fluctuating axial velocity at different downstream locations

The Kelvin-Helmholtz instability along the border of the jet yields, further downstream,to vortex structures, mainly axisymmetric tores. However, these structures are not always present and alternate with vortices of helicoidal shape (see figures 8a and 8b taken at two different times). The spatial jet variations of the present jet could be at the origin of the alternance between the two structures. We then follow the structures further downstream. The 3D visualisation exhibits an original vortex arrangement subsequent to the varicose mode growth: the "alternated pairing". Experimental evidence of "alternated pairing" was recently shown by Broze and Hussain [16]. This mode will be further discussed in the next section.

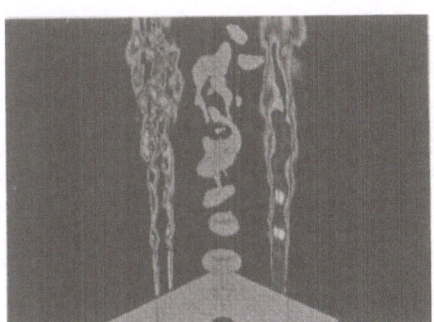 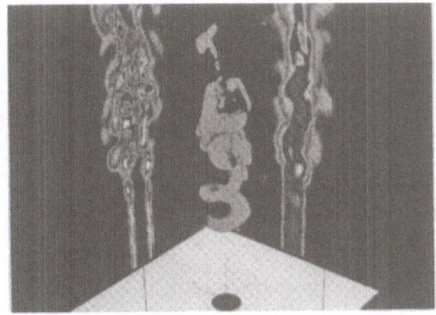

Fig. 8. — Natural jet three-dimensional visualization at two different times. Low pressure isosurface in light gray; cross-sections at $X = 0$ of the vorticity modulus and at $Y = 0$ of the velocity modulus.

4.4.   The Excited Jet

We then show how the jet dynamics can be controlled by imposing the inflow excitation. As mentioned above, when the toroidal structures are advected downstream, they display alternated localized pairings. Since this vortex arrangement seems to be characteristic of the jet topology, we next propose to preferentialy excite it from the nozzle. The excitation method is based on the principle of imposing a periodic perturbation to the amplitude of the axial velocity W at the nozzle:

$$W(r) + \epsilon \, W_o \, sin \left(\frac{Str_D W_o}{D} t \right) \, sin \left(\frac{Y}{D} \right) \qquad (1)$$

where $r = \sqrt{X^2 + Y^2}$ is the radius

The periodic fluctuations level is $\epsilon = 5\%$, and its imposed frequency corresponds to $Str_D = 0.35$. Half of the jet presents a speed excess, while a speed defect is imposed on the other half, and this alternatively. Note that this perturbation has a preferred direction, chosen along the Y axis. The resulting structures are shown on Figure 9. Inclined vortex rings exhibiting localized pairing persist far downstream till $Z/D = 10$. The striking fact is the very distinct spreading rates in the Y and X directions. In the YZ plane, the spreading rate reaches nearly 25° and even higher values near the exit. Conversely, it is close to zero in the XZ plane. Recently, the experiment performed by Longmire and Duong [17] displays similar results by using a special geometry of the nozzle (made with two half nozzles). One important technological application of this excitation resides in the ability to polarize the jet in a preferential direction.

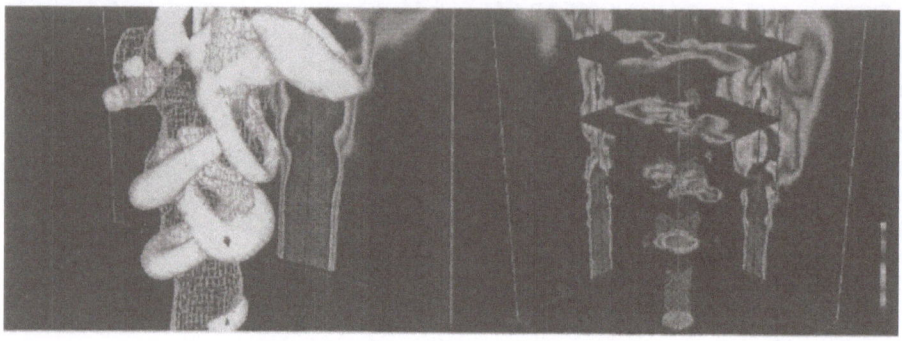

Fig. 9. — Jet with "alternate pairing" excitation: low-pressure isosurfaces in white; vertical and horizontal cross-sections of the axial velocity W

Exciting the primary instabilities is not the only way to control the flow. We remember that jet contains streamwise vortices too [18] [19][20]. The next excitation is designed to obtain a preferential development of the jet in three particular azimuthal directions, positionned each 120 degrees. Three side jets are excited with their couples of counter-wise longitudinal vortices by mean of this new perturbation.

$$W(r) + \epsilon\, W_o\, sin\left(\frac{Str_D W_o}{D}t\right)\, sin\,(3\Theta) \qquad (2)$$

where Θ is the azimuthal angle

The same intensity and frequency are used as before. We actually observe, fig. 10 and 11, the growing of such side jets and the development up to a three branch star. This results can be compared to experimentals studies of Zaman [13], Reeder and Samimy [14] in the case of turboreactor jet. Indeed, by the help of vortex generators in the form of small tabs at the nozzle exit, they obtain a similar evolution. The main interest of this control is that it creates furthermore a strong amplification of the entrainment rate, as a super mixing.

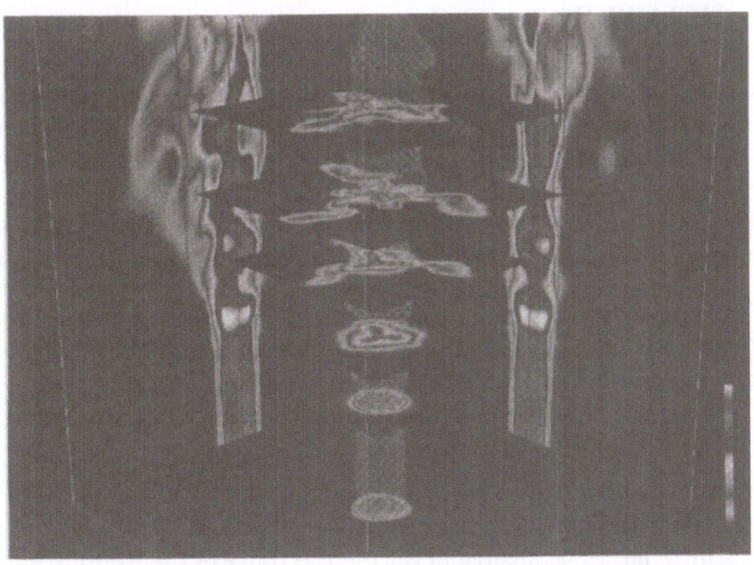

Fig. 10. — Jet with three azimuthal mode excitation: similar to Figure 9b

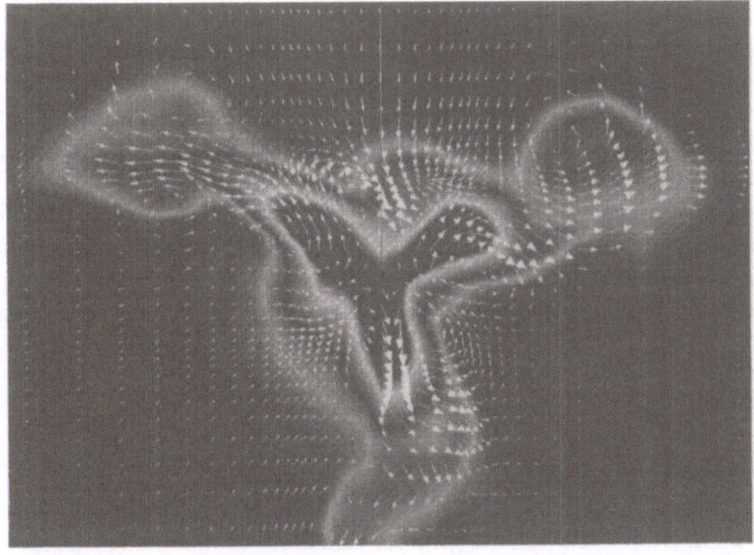

Fig. 11. — Extract of fig 10

5.   CONCLUSION

In this paper is presented a review of recent results on numerical simulations of prototypic flows in nuclear reactors. These simulations are obtained with Large Eddies Simulations (LES) and two subgrid models based on the structure function.

Let us sumarize the advantages of LES for the future of computational fluid Dynamics applied to industrial flows and in particular in nuclear reactors:

- The large scales of the flow, the most influenced by boundary conditions (geometry, inlet flow conditions or heat exchange with the surroundings) are directly solved from the transient Navier-Stokes spatially filtered. They do not require models for the large scales which are generally difficult to extrapolate from one case to the other.

- The modelling is limited to the small scales of the flow which depend mostly of the local and instantaneous conditions. The cutoff of the spatial filter has to be located in the inertial range of the Kolmogorov cascade where the eddies are directly influenced by local and instantaneous variables. Thus the subgrid models developped tend toward an universal behaviour for a wide range of flow conditions.

Thus LES provides a qualitative gap toward models with increased extrapolation capabilities.

The examples listed above show that physical complex flows become accessible to the simulation: confined mixing layers, stratified shear flows, externally excited turbulent jets. In all cases, correct predictions were obained for velocity and thermal fields statistics (mean values, mean standard deviations, Strouhal numbers,..) and also for the dynamics of the flow (onset of instabilities, dynamics of vortices). Furthermore a better understanding of the vortex structures should make possible the active control of the flow (jet spreading rate, mixing enhancement...) and develops better engineering applications. Still limited to simple geometrical configurations, LES possesses a great potential for more complex geometries. With the development of computers, larger meshes will become accessible (toward 10^9 cells in 2000) and in return this will open new challenges on physical models, numerical methods and codes structures.

References

[1] Métais O., Lesieur M., *J.Fluid Mech.* **239** (1992) 157.

[2] Lesieur M., Métais O., *Annu. Rev. Fluid. Mech.* **28** (1996) 45.

[3] Fallon B., Simulation des grandes échelles d'écoulements turbulents stratifiés en densité, *PhD thesis*, Natl. Polytech. Inst., Grenoble (1994).

[4] Grand D., Magnaud J.P., Pages J.R., Villand M., In *Advances in mathematics, computations and reactor physics*, Pittsburgh (1991).

[5] Silveira-Neto A., Grand D., Métais O., Lesieur M., *J.Fluid Mech.* **256** (1993) 1.

[6] Eaton J.K., Johnston J.P., *Standford University, Rep. MD-39* (1980).

[7] Ötügen M.V., *Experiments in fluids* **10** (1991) 273.

[8] Jovic S., Driver M., *NASA Ames Research Center*, unpublished (1992).

[9] Arnal M., Friedrich R., *Winter GAMM Conference on Numerical Methods in Fluid Mechanics*, Lausanne (1991).

[10] Lee H., Moin P., *Annu. Res. Brief.*, Stanford University (1992) .

[11] Normand X., Lesieur M., *Theor. Comput. Fluid Dyn.* **3** (1992) 231.

[12] Surle F., Berger R., Menant B., Grand D., Comparison between sodium stratification tests on the Cormoran model and TRIO-VF computations. NURETH 6 Paper 141 (1993), Grenoble.

[13] Zaman K.B.M.Q., Reeder M.F., Samimy M., *Phys. Fluids* **6(2)** (1994) 778.

[14] Reeder M.F., Samimy M., *J.Fluid Mech.* **311** (1996) 73.

[15] Hussain F., Zaman K.B.M.Q., *J.Fluid Mech.* **110** (1981) 39.

[16] Broze G., Hussain F., *J.Fluid Mech.* **311** (1996) 37.

[17] Longmire E.K., Duong L.H., *Phys. Fluids* **8(4)** (1996) 978.

[18] Lasheras J.C., Lecuona A., Rodriguez P., In *The Global Geometry of Turbulence*, edited by J.Jimenez (Plenum Press, New-York) (1991).

[19] Monkewitz P.A., Pfizenmaier E., *Phys. Fluids A* **3(5)** (1991) 1356.

[20] Liepmann D., Gharib M., *J.Fluid Mech.* **245** (1992) 643.

Coherent Vortices in Rotating Flows

O. Métais([1])and E. Lamballais([1])([2])

([1]) LEGI/IMG
BP53, 38041 Grenoble-Cedex 09, France.
([2]) LEA/CEAT
43 Route de l'Aérodrome, 86036 Poitiers Cédex, France.

———————

1. INTRODUCTION

Turbulent or transitional shear flows in a rotating frame have been extensively studied due to their importance in many geophysical and engineering applications. Within these flows, the local Rossby number, which characterizes the relative importance of inertial and Coriolis forces, can vary significantly. Typical values of the Rossby number $R_o = U/2\Omega D$, where U is a characteristic velocity associated with the eddy and D its diameter, are on the order of 0.05 in mesoscale oceanic eddies and in Jupiter's Great Red Spot, 0.3 for large synoptic-scale atmospheric perturbations, and 2.5 for the atmospheric wake of a small island. Turbulence in rotating fluids finds numerous industrial applications in turbo-machinery; e.g., the turbulent characteristics of the flow in blade passages of radial pumps and compressor impellers determine the efficiency of these devices. Turbulence is also of great importance for the cooling by the fluid inside the blades. Depending upon the magnitude of the radial velocity, the Rossby number within rotating machines can range from values close to unity to very small ones (of the order 0.05).

Laboratory experiments, theoretical works, numerical simulations and atmospheric and oceanic observations show that they are three basic effects associated with rotating bounded- or free-shear flows. (i) If the shear vorticity is parallel and of same sign as the rotation vector (cyclonic rotation), the flow is made more two-dimensional. (ii) If the two vectors are anti-parallel (anti-cyclonic rotation), destabilization is observed at moderate rotation rates (high Rossby numbers), while (iii) two-dimensionalization is recovered for fast rotation. The section 2 and 3 will be devoted to the dynamics of coherent vortices present in free and wall-bounded shear flows submitted to solid-body rotation. Our goal here is to demonstrate the ability for the LES to correctly reproduce the detailed vortex topology even in the presence of external forces.

Most of the eddies encountered in the atmosphere and the ocean originate from the development of instabilities resulting from the combined effects of both fast rotation and horizontal or vertical density gradients. This is the case of the baroclinic eddies as well as those resulting from strong convective events. Furthermore, the Reynolds numbers accessible with the DNS are several orders of magnitude smaller than those encountered in atmospheric and oceanic flows. For a realistic representation of these flows, LES are therefore compulsory. In section 4, LES techniques are applied to the computation of atmospheric and oceanic mesoscale eddies resulting from baroclinic instability.

Turbulent convection as a result of intense freezing and/or cooling is a common phenomenom taking place in the ocean. In polar regions, especially in wintertime, due to the ice formation high salinity water forms. It then sinks and gives rise to intense convective turbulent plumes. The fluid so generated is dense enough to penetrate the underlying stable density stratification and evenually reaches the ocean bottom. If deep enough, the convective plumes become affected by rotational effects. This mechanism of deep water formation is the subject of considerable interest due to its role in influencing the oceanic heat transport.

In the Golfe du Lion, fierce Mistral events induce intense cooling at the sea surface reaching $\sim 1000 W m^{-2}$: these comprise ascending and descending currents that reach speeds $\sim 10 cms^{-1}$. The horizontal scale of homogeneous convection layer is on order of tens of kilometers and violent mixing occurs in localized, intense plumes on scales of order $1km$.

The dynamics of rotating turbulent convection has been investigated through theories based upon dimensional arguments [1, 2], in situ measurements, laboratory experiments [2], and non-hydrostatic Boussinesq numerical simulations [1]. In [1] different (ad hoc) horizontal and vertical diffusion coefficients and simple Laplacian operator were used for the dissipative and diffusive terms. The goal of the present study is to show that more sophisticated subgrid-scale models like the structure-function model proposed in [3] (see also [4]) is able to reproduce the main features of rotating convection (see section 5).

2. ROTATING FREE-SHEAR FLOWS

Here, one considers a free-shear flow of basic velocity, $\vec{\bar{u}} = (\bar{u}(y), 0, 0)$ (x, y and z are respectively the longitudinal, shear and spanwise directions). One works in a frame rotating with a rotation vector $\vec{\Omega} = (0, 0, \Omega)$ oriented along the span (positive or negative). The vorticity vector associated with the basic velocity profile $\vec{\omega} = \left(0, 0, -\frac{d\bar{u}}{dy}\right)$ can be parallel or anti-parallel to $\vec{\Omega}$. We will refer to the first case as the cyclonic case, while the second will be called anticyclonic.

We here consider the rotating mixing layer and the rotating plane wake and will show, through Direct Numerical Simulations (DNS) and Large-Eddy Simulations (LES), how the rotation modifies the three-dimensional flow topology. The reader is referred to [5, 6, 7] for more details. A solid-body rotation does not influence a two-dimensional flow in a plane perpendicular to the rotation axis, in the sense that the Coriolis force is then proportional to the gradient of the stream function, and may be included into the pressure gradient. Therefore, the phenomena observed in the laboratory experiments can only be explained by considering the influence of rotation on the growth of three-dimensional perturbations.

The mixing layer is associated with an hyperbolic-tangent mean velocity profile of initial vorticity thickness δ_i, and the wake with a gaussian profile of half deficit velocity width r_m. The Rossby number is here based upon the maximum ambient vorticity of the mean profile, that is, the vorticity at the inflexion point(s), $-(d\bar{u}/dy)_i$, i.e.,

$$R_o^{(i)} = -(d\bar{u}/dy)_i/2\Omega \quad . \tag{1}$$

In the mixing layer, $R_o^{(i)}$ is positive for cyclonic rotation and negative for anticyclonic rotation. For the wake, one side is cyclonic, while the other is anticyclonic. In that case, one considers the modulus $|R_o^{(i)}|$ of the Rossby number.

In order to describe the early stage of the flow development, a three-dimensional linear-stability analysis of planar free-shear flows has been carried out in [8]. For cyclonic rotation and for strong anticyclonic rotation, it was found that the flow was two-dimensionalized and the instability diagram in the k_x, k_z plane concentrated around the Kelvin-Helmholtz mode (which is not affected by the rotation). For moderate anticyclonic rotation $R_o^{(i)} < -1$, and in addition of the Kelvin-Helmholtz instability, a new instability was discovered consisting of the strong amplification of a purely longitudinal mode (along the k_z-axis; $k_x = 0$), corresponding to a purely streamwise instability: the shear/Coriolis instability. For both the mixing layer and the wake, this shear/Coriolis instability has larger amplification rates than the co-existing Kelvin-Helmholtz instability for roughly the range $-8 < R_o^{(i)} < -1.5$, and its effect is maximum for $R_o^{(i)} \simeq -2.5$. This Rossby number corresponding to maximum destabilization will be called the *critical Rossby number*. It is important to note that the wavelength of the shear/Coriolis mode is much smaller than the Kelvin-Helmholtz's mode for large Reynolds number flows. It was shown in [8] that, for purely longitudinal modes ($k_x = 0$), and if the stability problem is reduced to perturbations such that $k_x = 0$, a necessary and sufficient conditions for inviscid instability is that the local Rossby number $R_o(y) = -d\bar{u}/dy/2\Omega$ should be smaller than -1 somewhere in the layer. This is essentially the result previously found in [9, 10]. Notice that, for a purely longitudinal perturbation, the eigenvalue equation is similar to the one governing the radial velocity when studying the inviscid linear centrifugal instability in the limit of axisymmetric disturbances.

The analysis performed in [8] allows to describe the early linear stage of the perturbations growth. Further insight in the nonlinear regime, can be obtained, firstly by examining the vorticity stretching mechanisms, and secondly by performing three-dimensional simulations of the full Navier-Stokes equation. In a former study [5], the importance of considering the absolute vorticity was emphasized, and not just the relative vorticity, since Kelvin's circulation theorem directly applies to the absolute vorticity. For example, if the relative vorticity is written as the sum of the ambient $-(d\bar{u}/dy)\vec{z}$ and fluctuating $\vec{\omega}'$ components. The absolute vorticity then writes $(2\Omega - d\bar{u}/dy)\,\vec{z} + \vec{\omega}'$. If the flow is locally cyclonic (i.e., Ω and $-(d\bar{u}/dy)$ have the same sign), then the absolute vortex lines are closer to the spanwise direction than the corresponding relative ones. Therefore, as compared to the non-rotating case, the effectiveness of vortex turning and stretching is reduced. Conversely, if the flow is locally anticyclonic, especially for the regions where 2Ω has a value close to $d\bar{u}/dy$ (weak absolute spanwise vorticity), absolute vortex lines are very convoluted, and will be very rapidly stretched out all over the flow, as a dye would do. It was thus predicted that in rotating shear flows, the vortex filaments of Rossby number -1 (hence anticyclonic) would be stretched into longitudinal alternate vortices. This phenomenological argument will be referred to as the weak-absolute vorticity stretching principle.

2.1. Coherent Vortices Topology

We here concentrate on the rotating and non-rotating plane wake coherent vortices obtained in LES based upon the *structure function model* proposed in [3] (see also Lesieur's lecture notes in the present volume). Temporal shear flows are here considered with periodicity in the streamwise direction. Initially, a low-amplitude three-dimensional random noise is superposed upon the ambient velocity profile.

As pointed out in [3], in large-eddy simulations, low-pressure centers are better tracers of the coherent structures than high-vorticity regions, which are encumbered by small-scale structures. Low pressures contours are displayed on Figure 1 for the nonrotating plane wake at $t = 65.5 r_m/|U_0|$. Pressure troughs vizualize not only the primary vortices of the Karman street, but also the longitudinal vortices between them: intense longitudinal vorticity stretching takes place implying strong depressions: here, $\omega'^{max}_x \approx \omega'^{max}_y \approx 4.8|\omega^{(i)}_{2D}|$ ($\omega^{(i)}_{2D}$ vorticity maximum associated with the initial mean velocity profile). The longitudinal vortices are located within the braids connecting consecutive Karman vortices of anti-parallel vorticity. They exhibit a lambda-shaped structure with a characteristic arrangement. Indeed, the wake consists of in-phase oscillations of the Karman rollers; this leads to the formation of aligned lambda-shaped

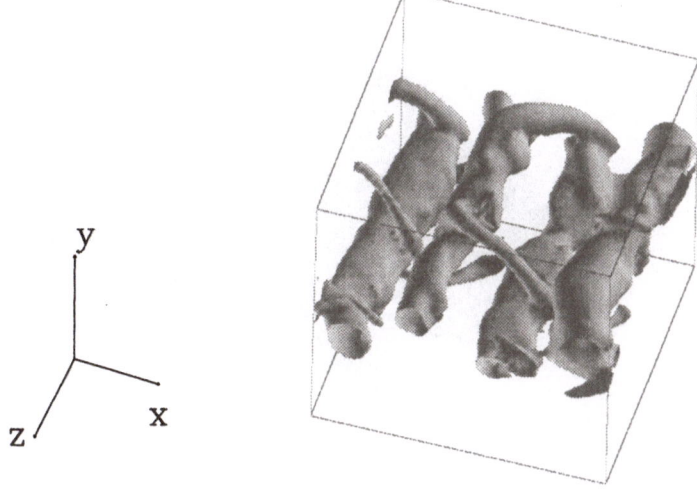

Fig. 1. — Plane wake ($|R_o^{(i)}| = \infty$), at $t = 65.5 \ r_m/|U_0|$. Low-pressure isosurfaces (from [7]).

vortices similar to those studied in [11]. Their spanwise wavelength is of the order of the size of the computational box.

In the rotating case, the simulations confirm the global trends observed in the experiments and predicted by the linear-stability analysis: the Karman vortices are two-dimensionalized by the rotation when these are cyclonic. We have checked that this is also true for the anti-cyclonic vortices when rapid rotation is applied. Conversely, the primary anti-cyclonic vortices are disrupted when a moderate rotation is applied. Figure 2 shows, for $|R_o^{(i)}| = 2.5$ (corresponding to a maximum anticyclonic destabilization), an isosurface of low "ageostrophic" pressure for which the contribution of the Coriolis term has been discarded. Here, the initial perturbation is quasi-twodimensional and favours the appearance of the Karman rollers (forced transition; see [7], for details). The wake exhibits a strong topological asymmetry. On the anticyclonic side of the wake, the Karman vortices are superposed with hairpin-shaped longitudinal vortices. The spanwise wavelength of the latter is approximately one-fourth of the computational domain width, in good agreeement with the prediction of the linear stability analysis [8].

2.2. Vorticity Stretching Mechanisms

A close examination of the time evolution of the *absolute* vortex lines has been performed [7] in the destabilized anticyclonic region. The initial velocity perturbation was purely three-dimensional (natural transition). It was showed

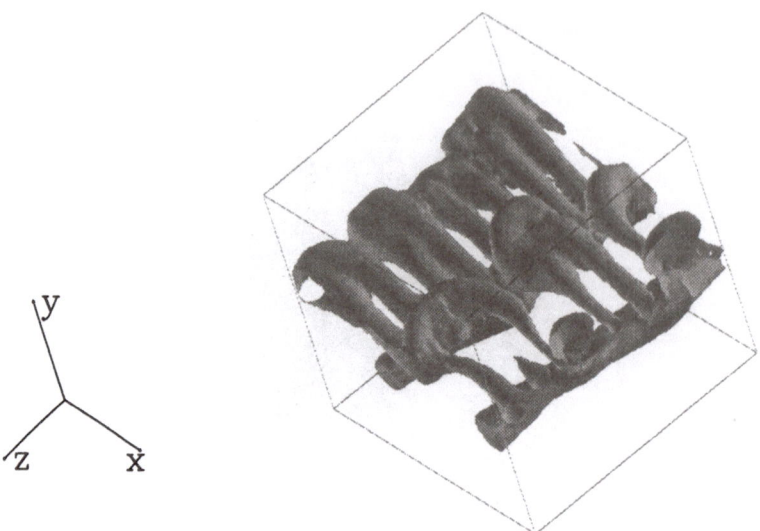

Fig. 2. — Plane wake at $|R_o^{(i)}| = 2.5$ and at $t = 32.3\ r_m/|U_0|$. Isosurfaces of low "ageostrophic" pressure (from [7]).

that the flow undergoes very distinct stages. In the first stage, the vorticity dynamics are dominated by quasi-linear mechanisms yielding absolute vortex lines inclined at 45° with respect to the horizontal plane. These are in phase in the longitudinal direction. It has been furthermore checked that maximum longitudinal vorticity stretching is achieved in the flow regions with a local Rossby number of approximately -2.8. In a second stage, nonlinear stretching mechanisms yield quasi-horizontal longitudinal hairpins of absolute vorticity. These absolute vortex tubes correspond to local Rossby number ≈ -1. The ω_x stretching terms become larger than for ω_y, and those terms are found to be maximum within the legs of the vortices. The dynamics are then dominated by a strong quasi-horizontal stretching of longitudinal *absolute* spanwise vorticity. During that stage, we observe, for both the mixing layer and the plane wake, that the mean velocity profile exhibits a well-defined range of nearly constant shear whose vorticity exactly compensates the solid-body rotation vorticity. Figure 3(a) shows the time evolution of the plane-wake mean velocity profile $\bar{u}(y,t)$ obtained in a DNS at Reynolds number of 200 and for a rotation rate corresponding to $|R_o^{(i)}| = 2.5$ (three-dimensional initial perturbation). Starting from the initial gaussian shape, the anticyclonic side of the mean profile exhibits, in the highly nonlinear state a long range of nearly constant shear, whose intensity $d\bar{u}(y)/dy$ is independent of time. This is confirmed by the profile of $R_o(y,t) = -\frac{d\bar{u}(y,t)/dy}{2\Omega}$ which displays at later time a very well defined plateau around the value of -1 (see Figure 3(b)).

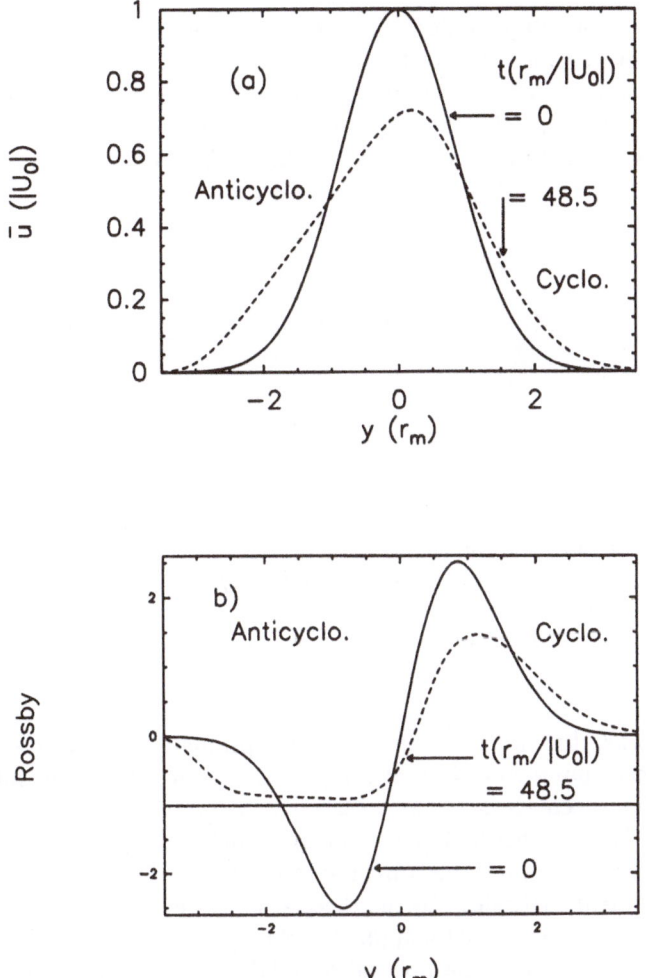

Fig. 3. — Plane wake $|R_o^{(i)}| = 2.5$ a) Mean velocity profile $\bar{u}(y,t)$ at $t = 0$ and $t = 48.5\ r_m/|U_0|$. b) Rossby number profile $R_o(y,t)$.

Thus a very efficient mechanism to create intense longitudinal vortices in rotating anticyclonic shear layers is provided, thanks to a linear longitudinal instability followed by a vigorous stretching of absolute vorticity taking place in regions of weak spanwise absolute vorticity in agreement with the phenomenological theory proposed in [5].

3. ROTATING CHANNEL FLOW

The results presented here are analogous to those presented in Lesieur's lecture notes of the present volume, except that we now impose a solid-body rotation orientated along the spanwise direction. The spectral-dynamic eddy viscosity model described in Section 3 of Lesieur's notes is used without any modification. The channel width is $2h$, and the macroscopic Reynolds number $R_e = 2hU_m/\nu$ is $14,000$. The flow configuration is presented on Figure 4.

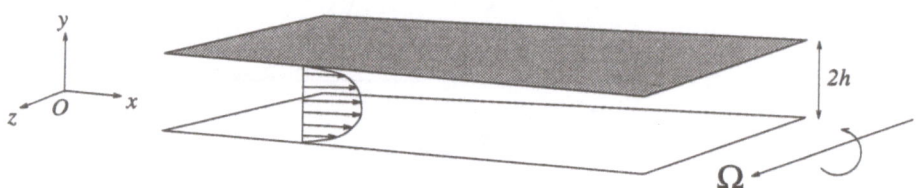

Fig. 4. — Schematic view of the rotating channel
h : half-channel height ; U_m : bulk velocity ; $Re = U_m 2h/\nu$: Reynolds number
$Ro_g = 3\, U_m/\Omega 2h$: global Rossby number
x, y, z are respectively the streamwise, transverse and spanwise directions (from [20]).

In previous works, the rotating channel flow problem has been studied by means of direct numerical simulation (DNS) at low Reynolds number [12, 13, 14]. It has been shown that rotation can modify considerably the vortex topology of the channel flow. Previous authors [15, 16, 17] have already used LES successfully to predict the principal modifications brought by a weak rotation upon turbulent statistics. Of the three studies, reference [17], where a variant of the dynamic procedure [18] applied to Smagorinsky's model was used, gave the best agreement with experimental data [19]. Here, we present results completing these previous works by a study which focuses upon a detailed study of vortex organization, and upon higher rotation rates and Reynolds numbers regimes. The global Rossby number $Ro_g = 3\, U_m/\Omega 2h$ is taken equal to 6. Detailed results are presented in [20].

3.1. Statistical Results

Figure 5 shows profiles of the r.m.s velocity fluctuations for the two simulations considered here (non-rotating and rotating cases). A tendency similar to what was observed previously in DNS [13, 14, 12, 7] is recovered : when rotation is applied, the cyclonic region (corresponding to $y > 0$) of the flow is stabilized while turbulence remains very active in the anticyclonic region ($y < 0$). Similarly to the previous section on free-shear flows, the term cyclonic (respectively

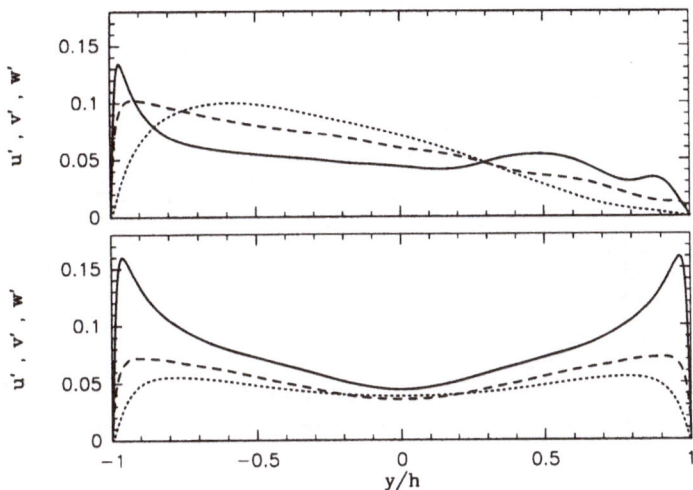

Fig. 5. — R.m.s of the velocity fluctuations in global units ($Re = 14000$) : —, $\sqrt{\langle \overline{u'^2} \rangle}$; ..., $\sqrt{\langle \overline{v'^2} \rangle}$; - - -, $\sqrt{\langle \overline{w'^2} \rangle}$; (a) $Ro_g = 6$; (b) $Ro_g = \infty$ (from [20]).

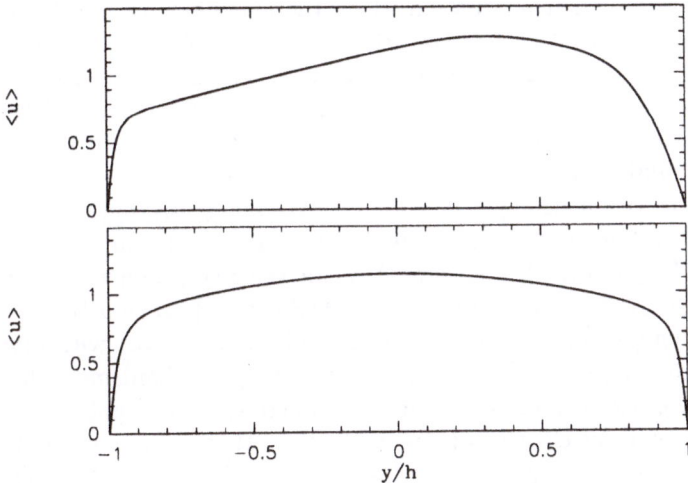

Fig. 6. — Mean velocity profiles in global units ($Re = 14000$) : (a) $Ro_g = 6$; (b) $Ro_g = \infty$ (from [20]).

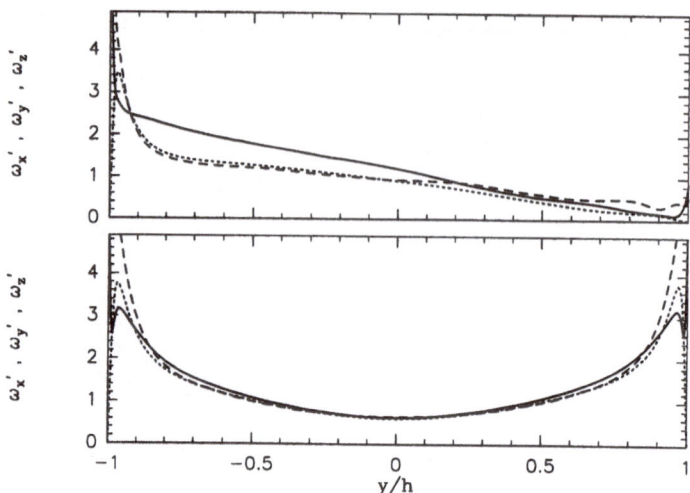

Fig. 7. — R.m.s of the vorticity fluctuations in global units ($Re = 14000$) :   —,
$\sqrt{\langle \overline{\omega'_x}{}^2 \rangle}$; $\cdots$, $\sqrt{\langle \overline{\omega'_y}{}^2 \rangle}$; - - -, $\sqrt{\langle \overline{\omega'_z}{}^2 \rangle}$; (a) $Ro_g = 6$; (b) $Ro_g = \infty$ (from [20]).

anticyclonic) designates a situation where the mean vorticity vector is paralell (respectively antiparallel) to the rotation vector $\vec{\Omega} = (0,0,\Omega)$. One sees that, in the anticyclonic region, rotation reduces the longitudinal velocity fluctuations and increases the transverse ones, which indicates a tendency for the low-and high-speed streaks of the non-rotating case to transform into transverse and spanwise oscillations.

Similarly to the shear-layer and the plane wake, a striking characteristic of the rotating turbulent channel flow is the existence of a linear zone in the mean velocity profile, where the (constant) mean velocity gradient is equal to 2Ω (see figure 6). As low Reynolds number DNS [13, 14] previously showed, such a range is dynamically very important since it is associated with longitudinal vorticity production by absolute-vortex stretching mechanisms. The effect of this stretching can be observed in figure 7, where the x-component of the r.m.s vorticity fluctuations clearly dominates the two other components in the region where the mean velocity profile is linear. The ability for LES technique to account for such phenomena already observed via DNS is an important result.

3.2.   Coherent-Structure Dynamics

In figure 8, we compare isosurfaces of the vorticity modulus $\omega = 4.5 \, U_m/h$ for the non-rotating and rotating channel flow. Note that the coherence of motion

$Ro_g = \infty$ (Non-rotating case)

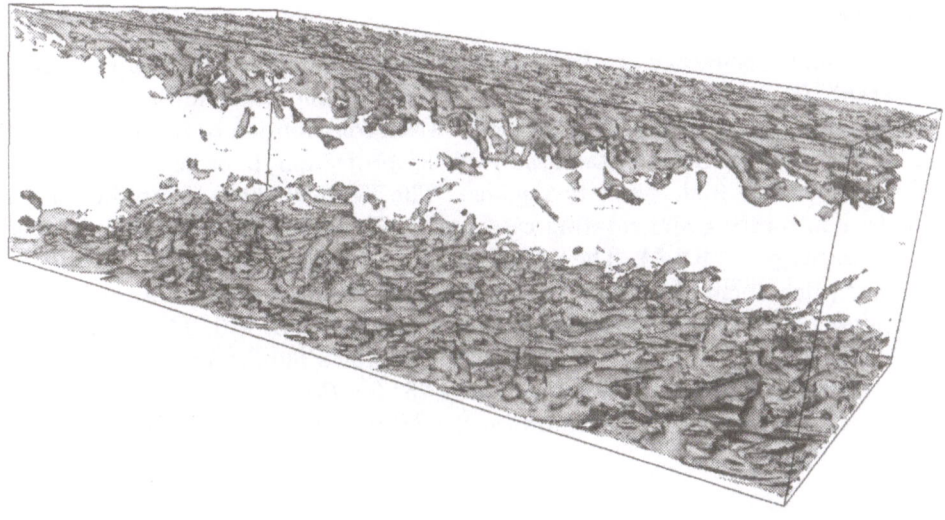

$Ro_g = 6$ (Rotating case)

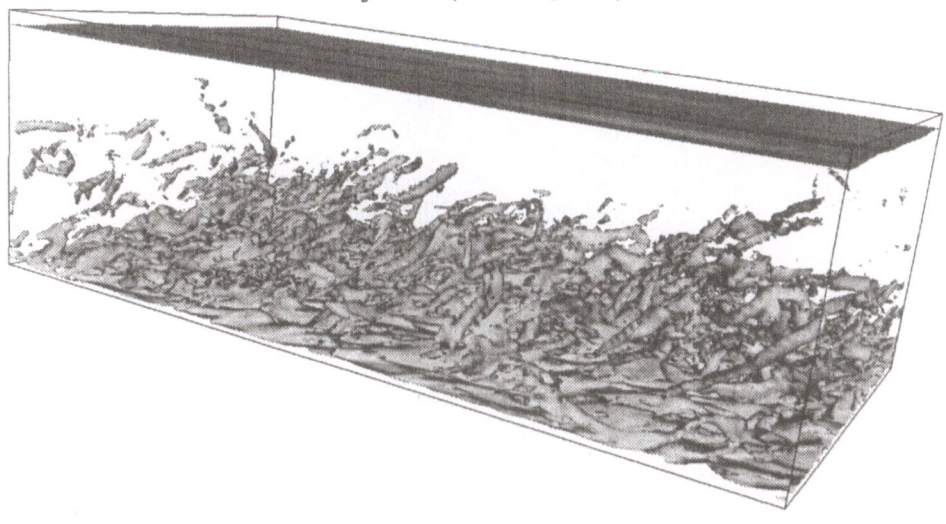

Fig. 8. — Isosurfaces of vorticity modulus $\omega = 4.5\ U_m/h$ ($Re = 14000$)
Size of the computational domain : $(L_x, L_y, L_z) = (2\pi h, 2h, \pi h)$
Grid : $128 \times 97 \times 64$ (from [20]).

is preserved in both LES, and well organized vortex structures can clearly be observed. The vortex topology modification in the anticyclonic region is similar to lower Reynolds number DNS [13, 14]: as compared with the non-rotating case, the vortical structures are more inclined with respect to the wall and more organized.

This double tendency can be quantified by examining the p.d.f of the angle $\theta = tan^{-1}(\omega_y/\omega_x)$, representing the angle between the wall and the projection of the vorticity vector on the (x, y)-plane. Each contribution to the distribution is weighted by the magnitude of the vorticity projection, in order to emphasize the part due to high vorticity regions. The figure 9 shows θ's p.d.f. both in the non-rotating and rotating cases, at a y-location where hairpin vortices are present ($y = -0.6\, h$). The p.d.f is shifted to lower values when rotation is applied (more inclined vortex structure) and becomes narrower (more organized motions). This was previously noticed in the DNS by [13, 14]. The most probable angles obtained in DNS (at low Reynolds number) and in LES are identical (45° without rotation and nearly 25° for $Ro_g = 6$). This confirms the fact than the LES correctly reproduce the detailed vortex topology.

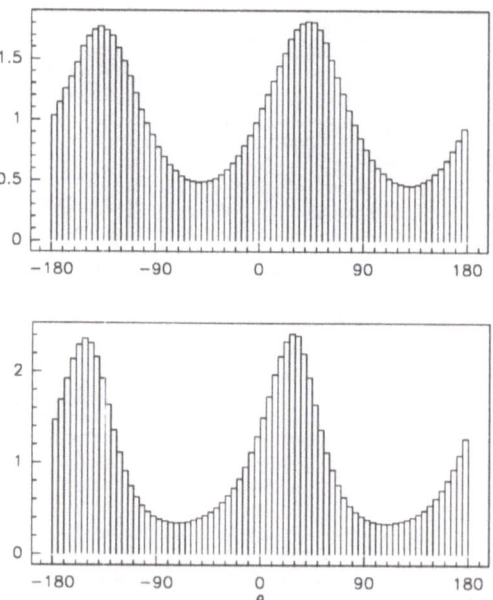

Fig. 9. — P.d.f of θ ($Re = 14000$) at $y = -0.6\, h$: $Ro_g = \infty$ (top) ; $Ro_g = 6$ (bottom) (from [20]).

4. NUMERICAL SIMULATION OF BAROCLINIC EDDIES

A subgrid-scale model which is widely used in the study of geophysical flows because of its simplicity is the hyper-viscosity model. The hyper-viscosity consists in replacing, the molecular dissipative operator $\nu \nabla^2$ of the Navier-Stokes equations by $(-1)^{\alpha-1}\nu_\alpha(\nabla^2)^\alpha$, where α is a positive integer. One of the drawback of this model is that ν_α is a constant (positive) coefficient which has to be empirically adjusted. This has been widely used in two-dimensional isotropic turbulence (see [21]), with $\alpha = 2$ or $\alpha = 8$, as a way to shift the dissipation to the neighbourhood of k_c. This allows for a reduction of the number of scales strongly affected by viscous effects, and has rendered possible in the case of two-dimensional turbulence to demonstrate the existence of coherent vortices. In three-dimensional turbulence, it was used [22] to study the influence of a solid-body rotation, with surprisingly good results.

We here use the generalized hyper-viscosity model based upon the structure function subgrid-scale model presented in Lesieur's lectures notes of the present volume (see his equation 50, and [4]), which is more physically justified than hyper-viscosity models. This model will be designated as the "cusp" structure function model, since it allows to take into account the cusp-like behaviour near the cut-off wavenumber k_C. One of the characteristic of geophysical flows resides in their large aspect ratios with horizontal scales much larger than the vertical ones. In non-isotropic grids such that $\Delta x \neq \Delta y \neq \Delta z$, the second-order velocity structure function may however be evaluated in each point with the help of the six closest, using the Kolmogorov's law $F_2 \propto (\epsilon \, r)^{2/3}$. Is is obtained :

$$
\begin{aligned}
\bar{F}_2(\vec{x}, \Delta c) \quad &= 1/6 \sum_{i=1}^3 \{ \| \, \vec{u}(\vec{x}) - \vec{u}(\vec{x} + \Delta x_i \vec{e_i}) \, \|^2 \\
&+ \| \, \vec{u}(\vec{x}) - \vec{u}(\vec{x} - \Delta x_i \vec{e_i}) \, \|^2 \} \, (\Delta c / \Delta x_i)^{2/3}
\end{aligned}
$$

where $\Delta c = (\Delta x \, \Delta y \, \Delta z)^{1/3}$

The baroclinic instability results from the combined effects of horizontal temperature gradients and fast rotation on a stably-stratified fluid. It corresponds to a very efficient mechanism of conversion of potential energy into horizontal kinetic energy. When one considers horizontal scales of the order of the internal Rossby radius of deformation ($\approx 1000 km$ in the atmosphere and $\approx 50 km$ in the ocean, at mid-latitude), this instability becomes very active and gives rise to "baroclinic" eddies. Our goal is here to study in the case of a baroclinic jet submitted to a rapid rotation the nature of the coherent vortices and in particular the asymmetry between cyclonic and anticyclonic eddies. LES are be used to study frontal secondary instabilities at high Reynolds numbers. The following results are described in details in [23, 24, 25].

We then consider a stably-stratified medium associated with a constant vertical mean density gradient $\bar{\rho}(z)$ characterized by a constant Brunt-Vaissala frequency N:

$$N = \sqrt{-\left(\frac{g}{\rho_0}\right)\frac{d\bar{\rho}}{dz}} \quad . \tag{2}$$

The initial basic state is an horizontal density front oriented in the meridional direction $\vec{x}$:

$$\rho(x, z) = \Delta\rho_H \, tanh\left(\frac{2x}{\delta}\right) + \bar{\rho}(z) \quad , \tag{3}$$

where $2\Delta\rho_H$ is the density difference imposed between the two meridional boundaries of the computational domain and δ the front steepness.

The rotation vector $\vec{\Omega}$ is oriented in the z direction. Let $f = 2\Omega$ be the Coriolis parameter. In the limit of fast rotation and strong stratification, it can be shown that the density front has to be associated with a basic velocity profile. Indeed, the geostrophic equilibrium corresponding to a balance between the Coriolis force and the pressure gradient and the hydrostatic balance imply that this basic state has to satisfy the thermal wind equation (see [26]):

$$\frac{\partial \vec{u}_H}{\partial z} = -\frac{g}{\rho_0 f}\vec{z} \times \vec{\nabla}_H \rho \quad , \tag{4}$$

where $\vec{a}_H$ stands for the horizontal projection of the vector $\vec{a}$ on the horizontal plane. This gives the following velocity profile:

$$\bar{V}(x, z)\vec{y} = -\frac{g}{\rho_0 f}\frac{2\Delta\rho_H}{\delta}\frac{1}{ch^2(2x/\delta)} \, (z - z_0)\vec{y} = V(z)\frac{1}{ch^2(2x/\delta)}\vec{y} \quad , \tag{5}$$

corresponding to a Bickley jet directed along $\vec{y}$ of vorticity thickness δ, sheared (with a constant shear) in the z direction.

When a small random perturbation is superposed to this basic state the flow becomes unstable. The nature of the instability is however very different depending upon the characteristic parameters. The two characteristic length scales for instabilities are the barotropic length scale δ and the baroclinic length scale $r_I = NH/f$ the internal Rossby radius of deformation. z varies between z_B and z_T with $z_T - z_B = H$ the height of the computational domain. The two non-dimensionalized parameters are the Rossby (Ro) and Froude (Fr) numbers defined as:

$$Ro = \frac{\omega_{2D}^i}{f} = \frac{V(z_T)}{\delta f}; \quad Fr = \frac{V(z_T) - V(z_B)}{2NH} \quad . \tag{6}$$

where ω_{2D}^i stands for the vorticity maximum associated with the mean velocity profile. We then have :

$$\frac{Ro}{Fr} = \frac{r_I}{\delta} \quad . \tag{7}$$

Two regimes have to be distinguished :

1) $\delta < r_I$ (or $Ro/Fr > 1$): the instability is weak and mainly barotropic: it is of Kelvin-Helmholtz type and is associated with the inflexional nature of the mean velocity profile.

2) $\delta \geq r_I$ (or $Ro/Fr \leq 1$): the baroclinic instability corresponding to a conversion of potential energy associated with the horizontal density gradient into horizontal kinetic energy can develop. The amplification of the perturbations is much stronger than in the barotropic case.

We now concentrate on the second regime $Ro/Fr \leq 1$: here $Ro/Fr = 0.5$. The Rossby number is fixed to 0.1. The numerical code is similar to the channel flow study of the previous section except that compact differences schemes are here used into two spatial directions. The resolution is $193 \times 64 \times 20$ points both for the DNS and the LES. The Reynolds number $Re = V(z_T)\delta/\nu$ is 400 in the DNS 10,000 in the LES.

4.1. Direct Numerical Simulations

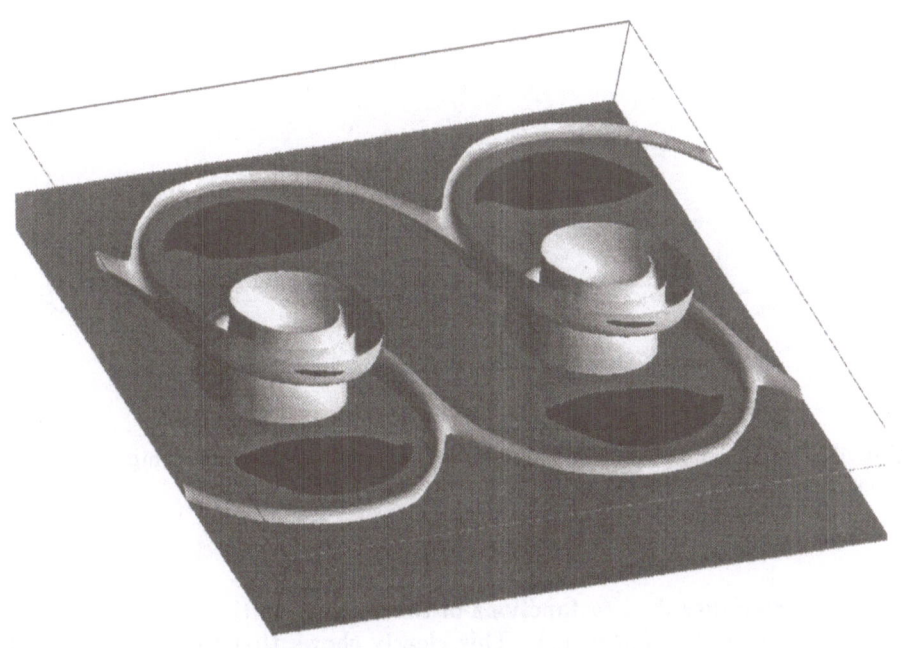

Fig. 10. — Iso-surfaces of vertical vorticity corresponding to $|\omega_z| = 2\omega_{2D}^i$; light-gray: cyclonic vorticity, black: anticyclonic vorticity. $t = 47r_I/V(z_T)$ (from [23]).

Figure 10 shows the vorticity structure once the instability has fully developed for $Ro/Fr = 1$. We observe the formation of cyclonic eddies of strong intensity, composed of nearly two-dimensional cores between which braids of very high vorticity are formed. The vorticity maxima are observed within these braids and correspond here to ≈ 8 times the vorticity maximum of the initial mean velocity profile. The vorticity intensification in the anticyclonic eddies is weaker (3 times the initial vorticity): we have checked that those are far more three-dimensional than the cyclonic eddies and strongy stretched by them. The asymmetry cyclones/anticyclones is well illustrated when consider-

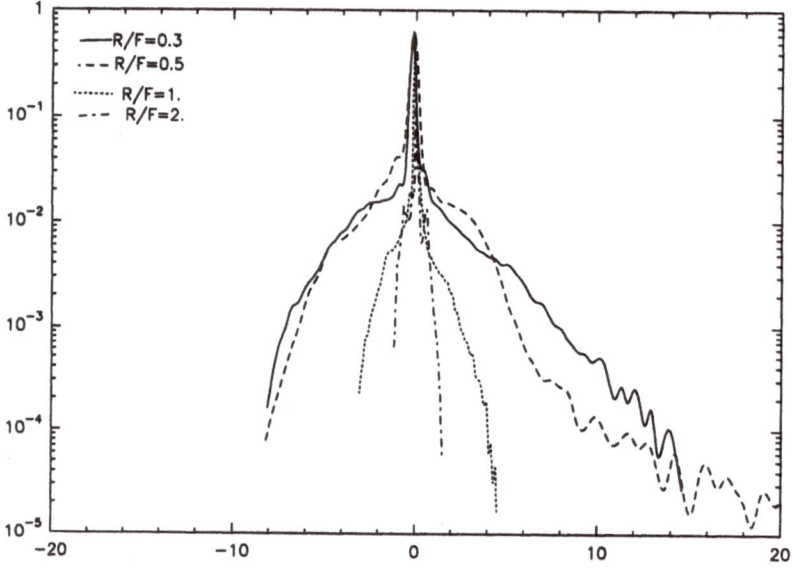

Fig. 11. — Probability density functions of the vertical vorticity component ω_z for several values of R_o/Fr (from [23]).

ing the probability density functions of the vertical vorticity for several value of the ratio Ro/Fr (figure 11). This clearly shows that for $Ro/Fr \leq 1$ the vorticity maxima are cyclonic and are localized in very concentrated regions of the space (low probability). Contrarily, the anticyclonic vorticity exhibit a quasi-gaussian behaviour indicating that, for moderate vorticity intensity, the most probable values are negative.

4.1.1. Large-Eddy Simulations

In order to reach Reynolds number closer to those encountered in atmospheric or oceanic flows, we mainly focus here on results from LES at $Re = 10,000$. In order to represent the effects of the subgrid-scales on the large scale field we use the "cusp" structure-function model.

Figure 12 shows a time evolution of the vorticity contours. As compared with the DNS presented in the preceeding section, one notice that the spiralling of the vorticity contours inside the core of the cyclonic eddies is much more pronounced. Due to viscous effects, the vorticity was indeed homogenized in the DNS. We have checked that the frontal region are much steeper in the LES indicating more energy near the wavenumber cut-off. The steepening of the fronts is associated with the appearance of a secondary instability resulting in a local intensification of the vertical vorticity. This instability takes place in regions where the local values of the Rossby and Froude number $Ro(\vec{x})$ and $Fr(\vec{x})$ verify the criterion $Ro(\vec{x})/Fr(\vec{x}) \leq 1$. The potential energy associated with local horizontal fronts is then converted into horizontal kinetic energy and gives rise to vertical vorticity intensification. It is important to notice that if the structure model without cusp is used excessive accumulation of energy is observed at the smallest scales eventually leading to numerical divergence. This clearly demonstrates the importance of the cusp-like behaviour and the feasibility of the subgrid-scales described in section 2 for the LES of geophysical flows with quasi-twodimensional regions and sharp frontal regions.

5. TURBULENT CONVECTION INTO A ROTATING FLUID

The numerical model is identical to the previous section on baroclinic instability. The LES based upon the structure-function subgrid-scale model are carried out in a doubly periodic box $32km$ square ($L_x = L_y = 32km$) by 2 km deep ($H = 2km$) on a $128 \times 128 \times 20$ grid. Convection is induced over a disk of radius $r = 8km$, depth $h = 200m$ by the application of localized buoyancy flux B_0. The fluid below the disk is initially of uniform density ρ_0. B_0 is here chosen to correspond to a heat flux $\mathcal{H} = 800Wm^{-2}$. The forcing is stopped after 2 days. The rotation vector is oriented along the vertical direction: here we have taken $f = 4.\ 10^{-4}s^{-1}$.

5.1. Dimensional Arguments

We first recall a dimensional analysis proposed in [1, 2]. Let $\Delta\rho$ be the density of the source fluid above ambient and Q its flow rate per unit area. The buoyancy flux writes:

$$B_0 = Q\ \frac{\Delta\rho}{\rho_0}\ g = Q\ g^* \tag{8}$$

where g^* is the reduced gravity. Several stages have to be distinguished: t is the period of time of application of the forcing.

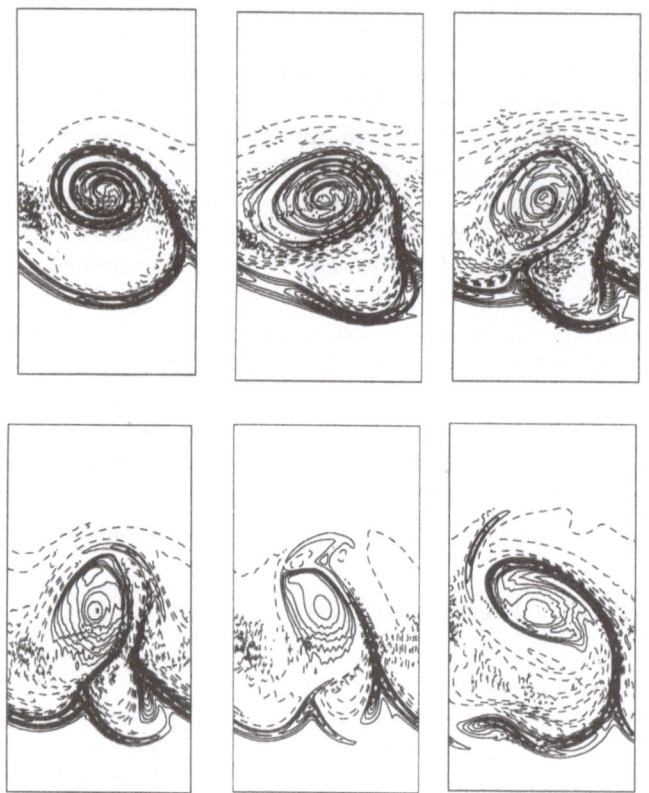

Fig. 12. — LES with the "cusp" structure-function model Hyperviscosity simulation: time evolution of the vertical vorticity isocontours at $z = z_T$. Solid-lines cyclonic vorticity; dashed lines anticyclonic vorticity (from [23]).

● $t < \frac{1}{f}$ The turbulence generated by the negative buoyancy flux is three-dimensional and unaffected by rotation, and progresses along a front of depth z_f. The relevant parameters are then B_0 and z_f. Dimensional arguments suggest:

$$Q \propto (B_0 \, z_f)^{1/3} \, ; g^* \propto \left(\frac{B_0^2}{z_f} \right)^{1/3} \qquad (9)$$

From these one can construct a Rossby number characteristic of the front:

$$Ro_f \sim \frac{Q}{fz_f} \propto \frac{B_0^{1/3}}{fz_f^{2/3}} \qquad (10)$$

One would expect that when the front has reached a depth z_c, at which Ro_f is of order one, then the turbulent fluctuations would be affected by rotation and a regime of rotationnaly dominated convection would result. At this point:

$$z_c \sim B_0^{1/2} f^{3/2} \qquad (11)$$

• $t \geq \frac{1}{f}$. The relevant parameters are now B_0 and f. If $z_c < H$, vortices of characteristic diameter

$$D_v \sim \frac{B_0^{1/2}}{f^{3/2}} \qquad (12)$$

form for $z > z_c$, and propagate downward. Once these have reached the bottom, they create a water column of dense fluid which becomes unstable to baroclinic instability. The resulting vortices should have a diameter of the order of the local Rossby radius of deformation:

$$r_I = \sqrt{g^* H}/f = B_0^{1/4} H^{1/2}/f^{3/4} \qquad (13)$$

5.2. Large-Eddy Simulations

Figure 13 shows a time evolution of a density isosurface corresponding to the dense surface fluid being cooled ($f = 4. \, 10^{-4} s^{-1}$). The convection front begins to propagate down at $t \approx 6 hours$. We then assist to the formation of plumes of well defined length scales $t \approx 18 hours$. We have checked that the vertically accelerating fluid induces an horizontal convergence near the surface. Due to the Coriolis force, this generates a cyclonic circulation in the upper part of the computational domain. When the plumes have reached the bottom, they form a water column of fluid denser than the outer fluid. The horizontal and vertical density gradients combined with rotation trigger baroclinic instability mechanisms and the isosurface exhibits well defined oscillations ($t \approx 48 hours$). At the bottom the horizontal flow divergence is associated with an anticyclonic circulation. In a final stage, the baroclinically generated vortex columns tend to propagate away from the source.

6.   CONCLUSION

Our goal here was to demonstrate the ability for the LES with a proper subgrid-scale model to correctly reproduce, in high Reynolds number flows, the detailed

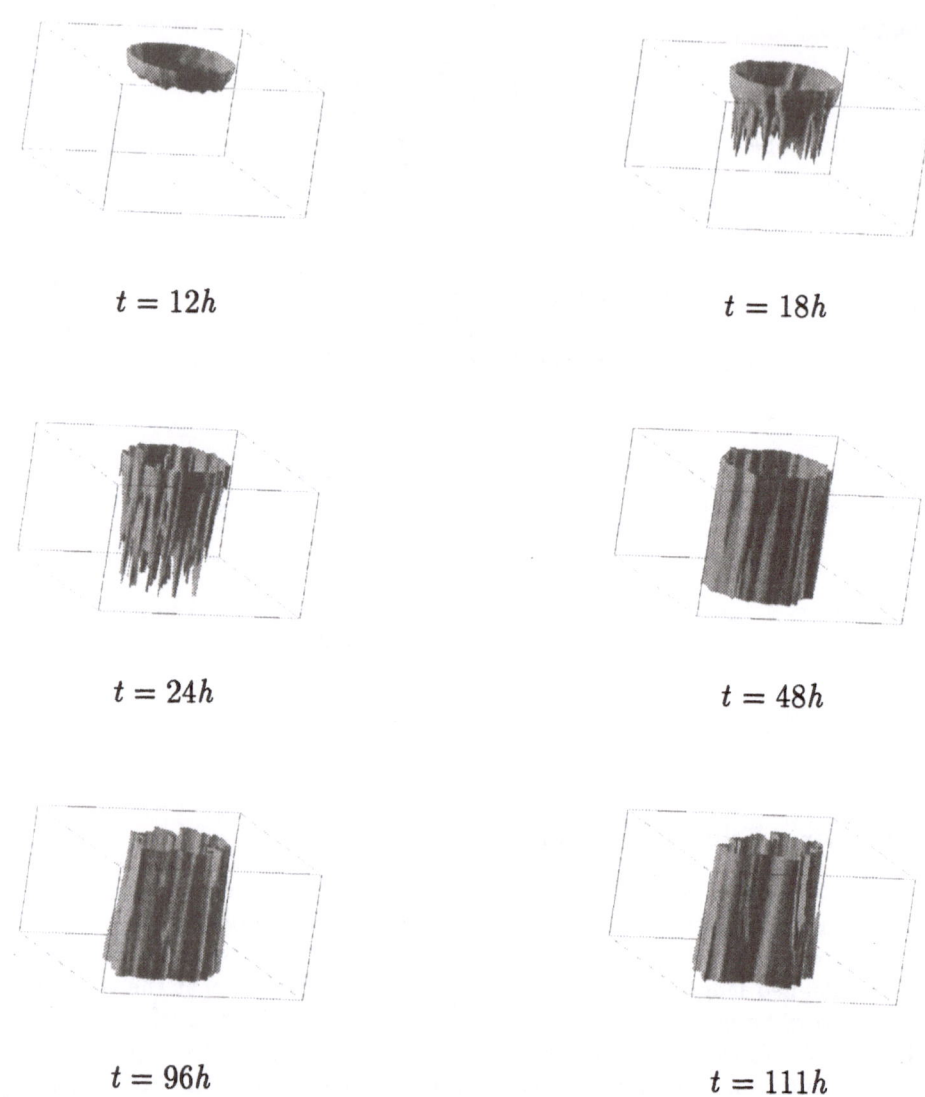

$t = 12h$ $t = 18h$

$t = 24h$ $t = 48h$

$t = 96h$ $t = 111h$

Fig. 13. — LES using the structure function subgrid-scale model of turbulent rotating convection. Time evolution of a density isosurface of the dense surface fluid. $f = 4.\ 10^{-4}s^{-1}$.

vortex topology even in the presence of external forces. We have here focused on the influence of solid-body rotation alone or combined with density gradients. The structure function subgrid-scale model [3] was successfully applied to the rotating temporal wake. The spectral-dynamic model [4, 20] has been proved able to capture the principal modifications of the channel flow behaviour under the effect of rotation. Similarly to the DNS, at moderate anticyclonic rotation and for all the simulated shear-flows, the mean velocity profile obtained through the LES exhibits a characteristic linear region of slope 2Ω. In such a range of near-zero absolute spanwise vorticity, the concentration and production by stretching of longitudinal vorticity is highly favored and leads to the formation of longitudinal hairpin vortices.

In order to study flows of geophysical interest, we have next focused on results of LES aimed at investigating the effects of stable density stratification and/or solid-body rotation on turbulence and coherent vortices. We have shown how LES techniques can be applied to the computation of atmospheric and oceanic mesoscale eddies resulting from different instability mechanisms such as baroclinic instability and convective instability. We have demonstrated the importance of the cusp-like behaviour for the eddy-viscosity and the feasibility of the generalized hyper-viscosity subgrid-scale models for the LES of geophysical flows with quasi-twodimensional regions and sharp frontal regions.

Acknowledgments

We are indebted to E. Briand and E. Garnier who contributed greatly to the numerical simulations of turbulence presented here, Computations were carried out at the IDRIS (Institut du Développement et des Ressources en Informatique Scientifique, Paris).

References

[1] Jones H., Marshall J., *J. Phys. Oceanography* **23** (1993) 1009.

[2] Maxworthy T., Narimousa S., *J. Phys. Oceanography* **24** (1994) 865.

[3] Métais O., Lesieur M., *J. Fluid Mech* **239** (1992) 157.

[4] Lesieur M., Métais.O., *Ann. Rev. Fluid Mech.* **28** (1996) 45.

[5] Lesieur M., Yanase S., Métais O., *Phys. Fluids A* **3** (1991) 403.

[6] Métais O., Yanase S., Flores C., Bartello P., Lesieur M., Turbulent Shear Flows VIII (Springer-Verlag, Berlin, 1992) p. 415.

[7] Métais O., Flores C., Yanase S., Riley J.J., Lesieur M., *J. Fluid Mech.* **293** (1995) 47.

[8] Yanase S., Flores C., Métais O., Riley J.J., *Phys. of Fluids. A.* **5 (11)** (1993) 2725.

[9] Pedley T.J., *J. Fluid Mech.* **35** (1969) 97.

[10] Hart J.E., *J. Fluid Mech.* **45** (1971) 341.

[11] Meiburg E., Lasheras J.C., *J. Fluid Mech.* **190** (1988) 1.

[12] Kristoffersen R., Andersson H. I., *J. Fluid Mech.* **256** (1993) 163.

[13] Lamballais E., Lesieur M., Métais O., *Int. J. Heat and Fluid Flow* **17** (1996) 324.

[14] Lamballais E., Lesieur M., Métais O., *C. R. Acad. Sci.* Série II b **323** (1996) 95.

[15] Kim K., In *Proc. 4th Symp. on Turbulent Shear Flows*, Karlsruhe (1983) 6.14.

[16] Tafti D.K., Vanka S. P., *Phys. Fluids A* **3(4)** (1991) 642.

[17] Piomelli U., Liu J., *Phys. Fluids A* **7(4)** (1995) 839.

[18] Germano M., Piomelli U., Moin P., Cabot W., *Phys. Fluids A.* **3 (7)** (1991) 1760.

[19] Johnston J.P., Halleen R. M., Lezius D. K., *J. Fluid Mech.* **56** (1972) 533.

[20] Lamballais E., Métais O., Lesieur M., Second ERCOFTAC Workshop on Direct and Large Eddy Simulation (Kluwer Academic Publisher, 1996), *in press.*

[21] Basdevant C., Sadourny R., *J. Mec. Theor. et Appl.*, Numéro Spécial (1983) 243.

[22] Bartello P., Métais O., Lesieur M., *J. Fluid Mech.* **273** (1994) 1.

[23] Garnier E., Etude numérique des instabilités de jets baroclines, PhD thesis, Natl. Polytech. Inst., Grenoble (1996).

[24] Garnier E., Métais O., Lesieur M., *C. R. Acad. Sci.* Série II b **323** (1996) 161.

[25] Garnier E., Métais O., Lesieur M., Synoptic and frontal-cyclone scale instabilities in baroclinic jet flows. *submitted to J. Atmos. Sci.* (1996)

[26] Lesieur M. *Turbulence in fluids, third edition* Kluwer Academic Publishers (1997).

Large-Eddy Simulation of Air Pollution Dispersion: a Review

F.T.M. Nieuwstadt and J.P. Meeder

J.M. Burgers Centre, Delft University of Technology
Rotterdamseweg 145, 2628 AL Delft

1. INTRODUCTION

Large-eddy Simulation (LES) has become an established technique to study turbulent flows, in particular turbulence at high Reynolds numbers. In contrast with so-called Direct Numerical Simulation (DNS) where all turbulent flow scales are resolved numerically, one computes in LES only a fraction of the turbulent flow field. The advantage of LES with respect to DNS when applied to the same turbulent flow, is that with LES one needs fewer grid point and thus less computational resources than with DNS.

This distinction between flow scales in LES is accomplished by applying a filter with a cut-off length ℓ_f to the flow field, so that all flow scales below ℓ_f are removed. As a result only the "large-eddies" remain. By applying the filter operation to the Navier-Stokes equations a set of "filtered" equations result. These equations are to be solved numerically to simulate the large-scale turbulent motions and this explains the term LES. The usefulness of this approach lies in the fact that the large-scales are in general characteristic for a particular turbulent flow and thus carry important information about various flow processes. So, for instance the turbulent transport of momentum, temperature, contaminants etc. is dominated by the large flow scales.

Before we turn to the application of LES to atmospheric turbulence, we first consider some more details of this simulation technique. It will be clear that LES will lead only to a successful simulation of the "large-eddies" when ℓ_f is chosen such that $\ell_f \ll \ell$ where ℓ represents the size of the large scales in the turbulence. Another parameter to be considered is the flow domain h, which later shall be set equal to the boundary-layer height. Apart from solving for the "large-eddies", our LES model must also resolve the flow in the domain with size h. In some cases, in particular those where $\ell \ll h$, the computational requirements to perform a LES become very large and need resources even beyond the capacity of present-day computers. This will be illustrated below with some examples of the atmospheric boundary layer.

Although the advantages of filtering out the large eddies are clear, the application of the filter operation to produce the LES equations comes leads also to disadvantages. Namely, filtering of the non-linear terms which e.g. appear in the Navier-Stokes equations, results in an additional unknown term which is known as the subgrid term. This term must be parameterized by a so-called subgrid model. Physically, the subgrid term describes the interaction between the resolved, large scales and the unresolved, small scales which have been removed by the filtering operation. This interaction can be best visualized in terms of the well-known cascade process which is the classical picture of turbulence dynamics [1]. In this process turbulent energy is passed down a cascade of flow scales from the large-scale motions where turbulent energy is created to the smallest scales where energy is dissipated. The subgrid model must provide for an uninterrupted energy transfer across the cut-off filter length, ℓ_f so that the cascade process is not disturbed. This means that the subgrid model must be able to handle the non-stationary and complicated flow dynamics near ℓ_f. In view of this, it will be no surprise that a fully satisfactory subgrid model is not yet available. This fact should be kept in mind when one considers the results of LES especially in those cases where the effect of the subgrid model can not be ignored.

Table I. — Summary of the performance of LES for the three prototypes of the atmospheric boundary layer; ℓ is the size of the "large-eddies" to be simulated and h represents the flow domain which in this case is the boundary-layer height; Resolution stands for capability of present-day computers to resolve the flow problem: ++ excellent, -/+ adequate and - in sufficient; Subgrid model denotes the importance or influence of the subgrid model: - small, + large and ? not clear; Realistic simulation gives the reliability and accuracy of the simulation: ++ very realistic simulation; + realistic simulation (e.g. away from areas where subgrid model is important) and - not (yet) realistic simulation.

		Resolution	Subgrid Model	Realistic Simulation
CBL	$\ell \sim h$	++	−	++
NBL	$\ell \sim 0.1h$	-/+	+	+
SBL	$\ell < 0.1h$	-	?	-

As mentioned at the start of this introduction, LES is primarily applied to high-Reynolds-number turbulence which is characterized by such a large range of flow scales that in view of computational resources a full simulation is impossible. As a result, LES is employed to simulate the turbulent flow fields at the very high Reynolds numbers which occur in the atmosphere. For instance, LES has been used with success to simulate the convective boundary layer (CBL) [2] and the neutral boundary layer [3]. For the stable boundary layer (SBL) the situation is less cleat and it seems that this type of boundary layer is still more or less beyond the capabilities of LES [4] and [5]. The present state of the art for the application of LES to the atmospheric boundary layer is summarized in table 1 and for a recent review of the LES-technique in atmospheric applications we may refer to [6].

We have argued that LES leads to a realistic turbulent velocity field as function of space and time, in particular for the CBL and NBL. This velocity field can be used to compute the transport of contaminants by the turbulent flow and from this we can obtain information on the statistics of turbulent dispersion. In this paper we aim to give an introduction and some results of this so-called numerical simulation of dispersion. We shall limit ourselves here to only a few examples. For more information and results we refer to the literature such as [7], [8], [9], [10] and [11].

In next section we will first discuss the techniques that can be used to simulate turbulent dispersion by means of LES. In the following section we discuss some examples of the LES of turbulent dispersion with emphasis on the dispersion from localized sources.

2. TECHNIQUES FOR LES OF TURBULENT DISPERSION

The LES of turbulent dispersion can be carried out with help of two methods which we shall designate as Eulerian and Lagrangian. In the Eulerian method an equation for the conservation of mass for the contaminant is solved alongside with the other conservation equations for the velocity and temperature field. In the Lagrangian method the contaminant is represented by a so-called "marked" fluid particle which is advected by the flow field. The number of "marked" fluid particles in a given volume then becomes a measure for the concentration. In the following subsections we shall discuss both methods in somewhat more detail together with their advantages and disadvantages.

2.1. Eulerian Method

The conservation equation for the instantaneous contaminant concentration c which is subjected to transport by a velocity field u_i and to molecular diffusion with a coefficient κ, reads

$$\frac{\partial c}{\partial t} + u_i \frac{\partial c}{\partial x_i} = \kappa \frac{\partial^2 c}{\partial x_i^2}. \tag{1}$$

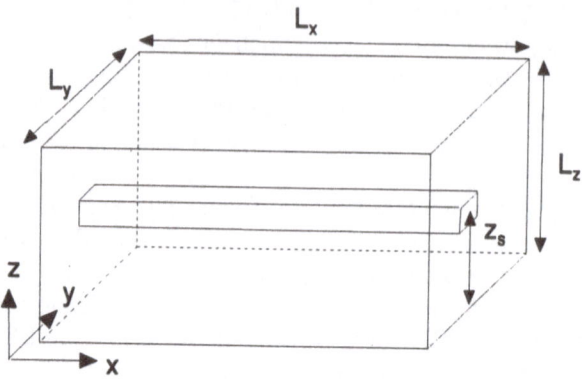

Fig. 1. — Illustration of the Eulerian method for the initial condition of a line source.

To isolate the effect of the large eddies, we apply to this equation a filter operation which is denoted by an overbar. The result reads

$$\frac{\partial \bar{c}}{\partial t} + \bar{u}_i \frac{\partial \bar{c}}{\partial x_i} = \kappa \frac{\partial^2 \bar{c}}{\partial x_i^2} + \frac{\partial f_i}{\partial x_i}. \tag{2}$$

The term f_i represents the subgrid term which formally is defined as

$$f_i = -(\overline{u_i c} - \bar{u}_i \, \bar{c}). \tag{3}$$

For atmospheric applications, the molecular term in (2) can be neglected with respect to the subgrid term.

For the subgrid term (3) a model must be adopted as we discussed in the introduction. In most applications to date a very simple quasi-molecular parameterization has been chosen. It reads

$$f_i = K_s \frac{\partial \bar{c}}{\partial x_i} \tag{4}$$

where K_s is the subgrid exchange coefficient which can be connected to the subgrid model for the filtered temperature and velocity field. It appears that the simple subgrid model (4) is sufficient for most cases.

The equation (2) must be solved simultaneously with the other equations of our LES in given computational domain. Apart from boundary conditions for $\bar{c}$, one must also supply an initial concentration field. One possibility is illustrated in Fig. 1 which denotes a line source configuration of which the dispersion is simulated as a function of time. Another initial condition is the point source where concentration is introduced at a point in the flow. This is also called a puff. Another related configuration is the case of a point source from which the concentration is introduced continuously in the flow field. The resulting concentration field in this case is known as a plume.

For the numerical simulation of dispersion, (2) must be discretized on the same numerical grid that is used for the velocity and temperature field. This leads in some cases to problems because the discretization of the concentration puts other demands on the numerical methods than the discretization of the velocity and temperature field.

The discretization scheme of the concentration field should preferably satisfy the following requirements:

1. conservation of mass

2. positive definiteness

3. maintenance of sharp gradients

Conservation of mass

The importance of this requirement is quite obvious. Namely, when one is interested in computing dispersion e.g. in order to estimate concentration levels, either creation or annihilation of contaminant mass is not acceptable.

Positive definiteness

It is clear that the concentration is a positive definite variable, i.e. the value of the concentration can never go below zero. It is important and for some applications even essential that the numerical solution of the concentration field satisfies the same requirement. First-order schemes, such as the standard upwind method satisfy this criterion but at the expense of failing the third requirement as we shall see below. Linear second-order schemes produce in some conditions spurious oscillations which may lead to negative concentrations. Therefore, to satisfy this criterion and at the same time maintain at least second-order accuracy, other usually non-linear methods must be used. An example is the algorithm proposed by [12]

Maintenance of sharp gradients

In some cases, in particular the simulation of plumes and puffs, the instantaneous concentration fields are characterized by sharp concentration gradients. These gradients coincide with the boundary of the instantaneous plume or puff. They are caused by the fact that in particular in the initial phase of dispersion, diffusion either by resolved or subgrid motion is small with respect to translation of the plume or puff in its entirety by the resolved large-scale turbulent motions. This latter process is also known as meandering. It is clear that the translation by resolved motion should not be accompanied by extra (numerical) diffusion and this rules for instance the use of first-order upwind schemes out.

An additional complexity in this case is that in the initial phase, the size of the plume or puff is by definition small and frequently not more than a few grid sizes. In other words, the initial resolution of the plume or puff is rather coarse. To maintain a sharp gradient in combination with such a coarse

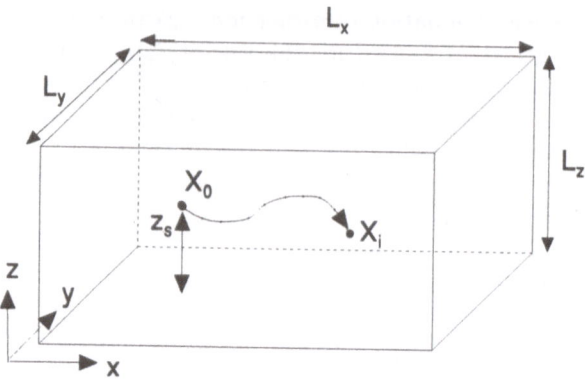

Fig. 2. — Illustration of the Lagrangian method based on the computation of particle trajectories.

resolution is far from trivial from a numerical point of view. Nevertheless various numerical algorithms have been developed which can handle this problem. Examples are the total variance diminishing (TVD) schemes, e.g. [13] or the flux-corrected transport (FCT) schemes, e.g. [14]. For more information on numerical treatment of this so-called advection diffusion equation we refer to [15].

Solution of (2) simultaneously with the other LES equations results in an instantaneous resolved concentration field $\bar{c}(x,t)$. From $\bar{c}(x,t)$ we can obtain statistics such as the mean concentration $<c>$ and the concentration fluctuations $<c'^2> \equiv <c^2> - <c>^2$ as a function of space and time. Here, the angular brackets $<>$ denote an ensemble average. For (quasi)-stationary conditions the ensemble mean can be computed by means of a time average. For non-stationary problems such as a concentration pattern that develops as a function of time, one usually employs a spatial average over the homogeneous direction of the flow to compute the ensemble average. An example is the slab average for the horizontally homogeneous case. In addition, one usually also performs an average over different realizations of the dispersion problem. Different averaging procedures can be combined to obtain results with smaller statistical errors.

In conclusion, we may summarize that by means of the Eulerian method the instantaneous concentration $\bar{c}(x,t)$ is simulated. From this concentration field all concentration statistics can in principle be obtained. This is one of the main advantages of applying the Eulerian method

2.2.   Lagrangian Method

The Lagrangian method is based on the equivalence between the mean concentration field of e.g. a plume from a point source and the probability density of the position of a particle that is released from the same point source (see e.g. [16]). This implies that we can describe the concentration field by following the trajectories of "marked" fluid particles which are introduced in the flow field at a given release point as illustrated in Fig. 2. In terms of the velocity field generated by our LES, the particle trajectory given by the coordinates X_i can be computed from

$$\frac{dX_i}{dt} = \overline{u}_i + u_i'$$ (5)

where the term u_i' stands for the contribution to the particle displacement by subgrid motions and molecular effects (the latter is usually neglected in atmospheric applications). This equation must be solved subject to the initial condition $X_i = X_o$ at $t = 0$.

The subgrid contribution in (5) requires again a parameterization. A choice which is equivalent with (4) is the random walk model given by

$$\Delta X_i' \equiv \int_t^{t+\Delta t} u_i' \, dt = \sqrt{2K_s \Delta t} \, \Omega_i.$$ (6)

where X_i' stands for the subgrid contribution to the particle displacement. The Ω_i in (6) represents a stochastic process known as the Wiener process, which satisfies $<\Omega_i> = 0$ and $<\Omega_i \Omega_j> = \delta_{ij}$. The square brackets $< >$ in this case denote again an ensemble average, i.e. an average over many realizations of the particle trajectory.

The procedure is now to perform a number of N_p independent realizations of the trajectory of a single particle released at a given point. The total number of particles in a given subvolume of the computation domain and at the given time instant with respect to the release time is designated as N. Based on the equivalence between particle probability and mean concentration discussed above, we have in the limit for $N_p \rightarrow \infty$.

$$<c(x_i, t)> = \frac{N}{N_p} \equiv p(x_i, t; x_o, 0).$$ (7)

where $p(x_i, t; x_o, 0)$ is the probability density at position x_i and time t of a particle released in x_o at $t = 0$. This definition of the ensemble averaged concentration is complete equivalent with the Eulerian ensemble averaged concentration introduced on page 270.

The main advantage of the Lagrangian method is that the three requirements mentioned on page 269 are automatically satisfied. The disadvantage, however, is that by releasing single particles and computing their trajectories one can only compute the mean concentration field. Other statistics such as concentration fluctuations require a computation in which multiple particles are released simultaneously. In other words, to compute the concentration variance $<c^2>$ one needs a two-particle probability density. Another disadvantage of the Lagrangian method is the fact that is primarily suitable to describe the dispersion of so-called passive contaminants which we shall consider in the next section.

3. DISPERSION OF PASSIVE CONTAMINANTS

A passive contaminant is defined as a substance which follows all flow motions and which, apart from transport and molecular diffusion, does not interact with the flow or any other contaminants present in the flow. As a result of these restrictions, all dispersion problems have been omitted where the contaminant satisfies another dynamic equation than the flow. An example is the dispersion of say heavy particles which due to inertia or drag and lift forces do not follow the path of a fluid particle. We also do not consider the dispersion of substances which modify the flow field and through this their dispersion. An example is the dispersion of fluids which are buoyant or heavy with respect to the fluid in which they disperse. These are all cases which are characterized as active contaminants and which will not be discussed in this paper.

In the following subsections we shall discuss three examples of plume dispersion in the atmospheric boundary layer. For the atmospheric boundary layer we shall consider only the CBL and NBL which both, as we have seen above, can be adequately simulated with LES. At the same time these examples will also illustrate the two simulation techniques discussed in section 2.

3.1. Dispersion in the CBL

The CBL is characterized by large-scale flow motions which consists of strong updrafts of hot air, also known as thermals, and more weak downdrafts surrounding these thermals. We can consider these thermals as the "large eddies". The size of these thermals is more or less equal to the boundary-layer height and this explains why the CBL is very suitable for LES (see also table 1). The thermals and the surrounding downdraft have different strengths, i.e. the mean upward velocity in an updraft is larger that the mean downward velocity in a downdraft. For further details and information on the structure and dynamics of the CBL we refer to [17].

The difference in strength between updrafts and downdrafts causes an asymmetry of the turbulent flow field in the CBL. Upward vertical velocity fluctuations which are connected to updrafts, are larger than the downward fluctuations caused by the downdrafts. At the same time the mean vertical velocity must be zero. The consequence is that the probability distribution of the vertical velocity fluctuations has a positive skewness. This skewness has a direct influence on the dispersion of contaminants. For instance, the dispersion from a source at the surface, which is primarily caused by positive velocity fluctuations, is different from the dispersion from a source at the top of the boundary layer. In [7] these two cases have been denoted as bottom-up and top-down diffusion.

We have thus found that in a CBL the dispersion characteristics depend on the location of the source. This implies that in the CBL dispersion cannot be described in terms of standard gradient diffusion theory because in this theory diffusion is independent of the location of the source. Other theories must be

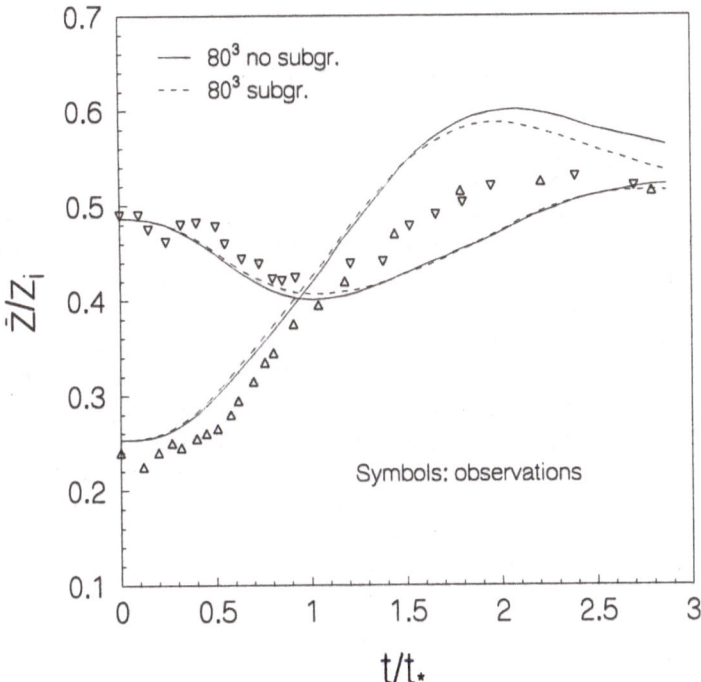

Fig. 3. — The mean height non-dimensionalized with the boundary-layer height of a line source released at a time $t = 0$ and at a height z_s in a CBL as a function of non-dimensional time t/t_*; release heights are: $z_s/h = 0.25$ and 0.5; the solid line denotes the LES results for the resolved motions only, i.e. without contribution of a subgrid model; the dashed curves are the LES data obtained by including the subgrid contributions; the symbols denote laboratory experiments.

developed to describe dispersion in the CBL. To formulate and validate these theories one can use LES.

Let us consider a line source of passive contaminants as illustrated in Fig. 1. At $t = 0$ the source is introduced in the CBL at a height z_s. In terms of the Lagrangian method described in section 2, the dispersion of the line source can be computed by releasing particles on a horizontal plane at the height z_s above the surface (note that the CBL is horizontally homogeneous so that the horizontal location on this plane does not matter and can be chosen freely subject to the condition of independency of the computed trajectory). The mean concentration field $< c >$ follows from these particle trajectories according to (7). Based on this mean concentration we can compute several statistics. One of these is the mean plume height which is defined as

$$<z>(t) = \frac{1}{N_p} \sum_{i=1}^{N_p} z^{(i)}(t) \equiv \frac{1}{h} \int_0^h z <c>(z,t)\,dz \qquad (8)$$

where $z^{(i)}$ stands for the height of the ith-particle trajectory and where we have assumed $\int_0^h <c>\,dz = 1$. In Fig. 3 we shown the data for $<z>$ which have been non-dimensionalized with the boundary layer height h, as function of time t which has been scaled with the convective time scale t_* (for the definition of t_* and the background of scaling in the CBL we refer to [17]).

The two curves shown in Fig. 3 give results for the release heights $z_s/h = 0.25$ and 0.5. The difference between the dashed and solid curve is the contribution of the subgrid model (6). It is clear that the influence of the subgrid model is very small. Also in other statistics we find that the subgrid contribution is small. The background is the excellent resolution that can be obtained in a LES of the CBL as already discussed in the introduction. In Fig. 3 we also show some results of laboratory experiments. The agreement between the LES and experimental data is quite acceptable.

The results shown in Fig. 3 give also a good illustration of the non-trivial influence of the asymmetric turbulence structure on dispersion. The curves for $<z>$ start by definition at $t = 0$ at the release height. For $t \to \infty$ the value of $<z>/h$ should approach 0.5. This latter value follows from a uniformly distributed concentration across the boundary layer, i.e. the concentration of the initial line source has filled for $t \to \infty$ the complete boundary layer, $0 < z/h < 1$ and the concentration has become independent of height. For the lowest release heights ($z_s/h = 0.25$) we find that the plume height overshoots its asymptotic value before it settles back to 0.5. For the largest release height, i.e. $z_s/h = 0.5$, the opposite occurs and the plume initially moves downward before it approaches the asymptotic value of $<z>/h = 0.5$. These results can be explained given the structure of the CBL in terms of updrafts and downdrafts. They have been also confirmed in atmospheric field experiments and this once more underlines the realism of LES where it must be stressed that the LES was done before the atmospheric experiments.

3.2.   Dispersion in the NBL

In this subsection we turn to the simulation of dispersion in the NBL computed by the Eulerian technique of section 2. In this technique we compute the instantaneous concentration field from which all dispersion statistics can be obtained. For the source configuration we choose a continuous point source from which a plume develops. An illustration of the instantaneous concentration field in given in Fig. 4 and this clearly shows that the LES of the plume appears quite realistic.

The LES computation was carried out for a case for which data from an atmospheric dispersion experiments are available [18]. The details of the experiments and the LES computation have been collected in Table II. One

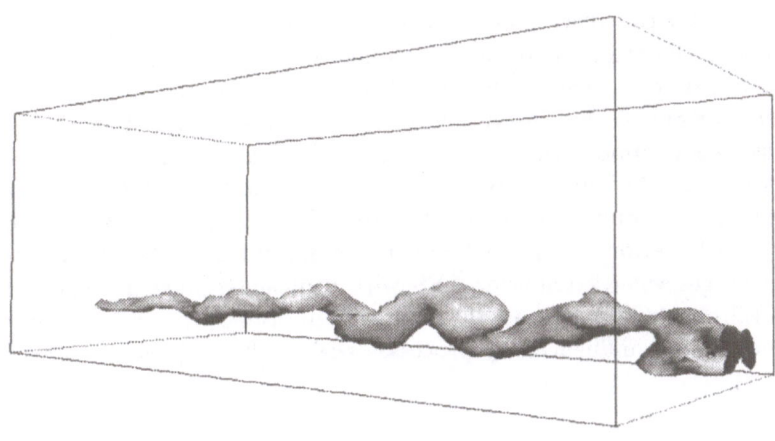

Fig. 4. — Instantaneous concentration field of a plume dispersing in the NBL.

should notice that the initial source size d which in our case has been taken equal 10m, is obviously much larger than the source size used in the experiments. However, the source size in the computations was the result of the numerical grid size, the consequences of which we have discussed in section 2.

Table II. — Details of the plume dispersion experiment in the NBL

Boundary-layer height	$h = 1000$ m
Computational domain	
L_x	8000 m
L_y	2000 m
L_z	1000 m
Resolution	$200 \times 100 \times 100$
Source height	$z_s = 115$ m
Source size	$d = 10$ m

Some of the computational results are illustrated in Fig. 5 together with experimental data. In Fig. 5 (upper figure), we show the concentration distribution at ground level for a given distance from the source and averaged over a period of 60-minutes. The observations indicated by the symbols are substantially smaller than the LES results which are indicated by the solid line. The reason for this discrepancy is the fact that due to low frequency wind oscillations in the atmosphere the experimental plume meanders on a time scale of several minutes. This is illustrated in Fig. 5 (lower figure) where concentration distributions are shown obtained during three separate runs within the total averaging time period of 60 minutes. The averaging time of these runs is 20 minutes. The meandering motion of the plume is clearly visible. This meandering motion is not present in our LES and this explains why the computed concentrations are larger than the observations. It also shows one of the disadvantages when comparing LES with atmospheric experiments. Namely, in a LES one only simulates the turbulent motions whereas the atmosphere flow motions are influenced by processes with a much larger spectrum of time scales.

3.3. Dispersion in Combination with Chemistry

In this subsection we will extend the dispersion problem that we have considered in the previous two subsections, with chemistry. Let us consider two chemical species A and B of which the concentrations are denoted as c_A and c_B. These substances can undergo a chemical reaction given by

$$A + B \xrightarrow{k_s} C \tag{9}$$

where k_s is the rate constant of the reaction.

The conservation equation for the concentrations c_A and c_B must be modified by an additional term which describes the change in the concentration due to the chemical reaction. Let us restrict ourselves to c_A because the equation for c_B is completely equivalent. The conservation equation for instantaneous value c_A then becomes

$$\frac{\partial c_A}{\partial t} + u_i \frac{\partial c_A}{\partial x_i} = \kappa \frac{\partial^2 c_A}{\partial x_i^2} - k_s c_A c_B. \tag{10}$$

In comparison with (1) we find an additional term on the right hand which is known as the chemical production term. As both c_A and c_B are > 0, this term is clearly a sink of concentration c_A which is consistent with rate equation (9).

Next we apply our filter procedure to (10) in order to derive an equation for the resolved concentration. The results reads

$$\frac{\partial \bar{c}}{\partial t} + \bar{u}_i \frac{\partial \bar{c}}{\partial x_i} = \kappa \frac{\partial^2 \bar{c}}{\partial x_i^2} + \frac{\partial f_i}{\partial x_i} - \underbrace{k_s \left(\bar{c}_A \bar{c}_B + \overline{c_A c_B} \right)}_{P_C}. \tag{11}$$

We find that the chemical production term in (10) gives rise to two additional terms in the resolved concentration equation. The first is the chemical produc-

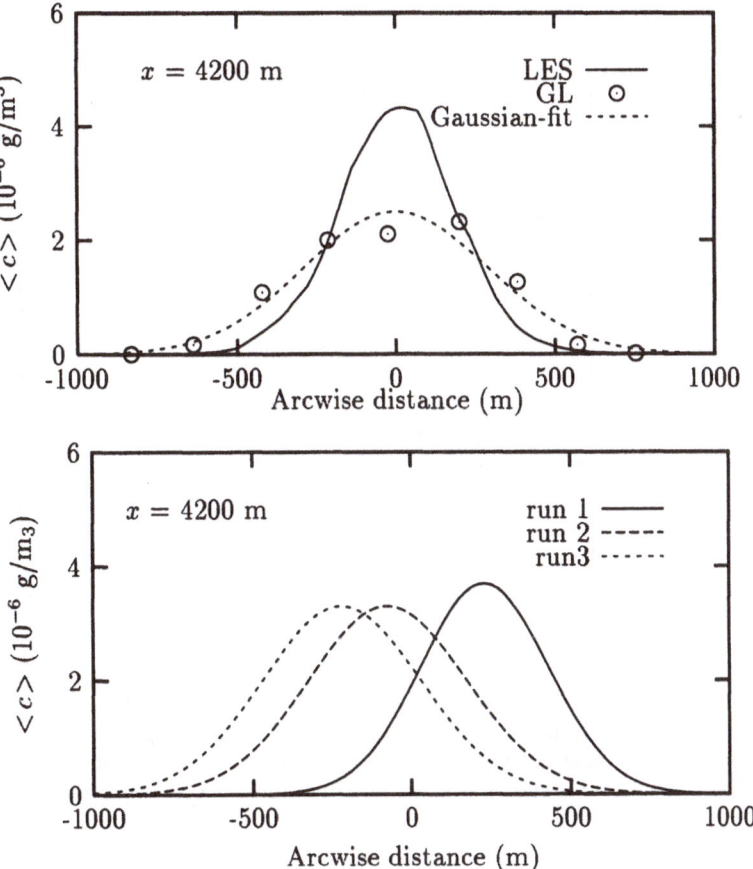

Fig. 5. — Upper figure: The mean concentration distribution at ground levels as a function of the distance to the plume axis; solid line: LES results; dashed line: Gaussian fit to the experimental data; symbols: experimental data; Lower figure: Three individual experimental runs which averaged together, produce the dashed line in the upper figure

tion in terms of the resolved concentrations and it can be directly incorporated in the solution procedure of (11). This is not the case for the second term which forms a new term in the equation. Its appearance has the same background as the subgrid term in (2) and in order to solve (11) one needs again a subgrid parameterization.

Frequently, the total chemical production term, P_C, in (11) is written in the

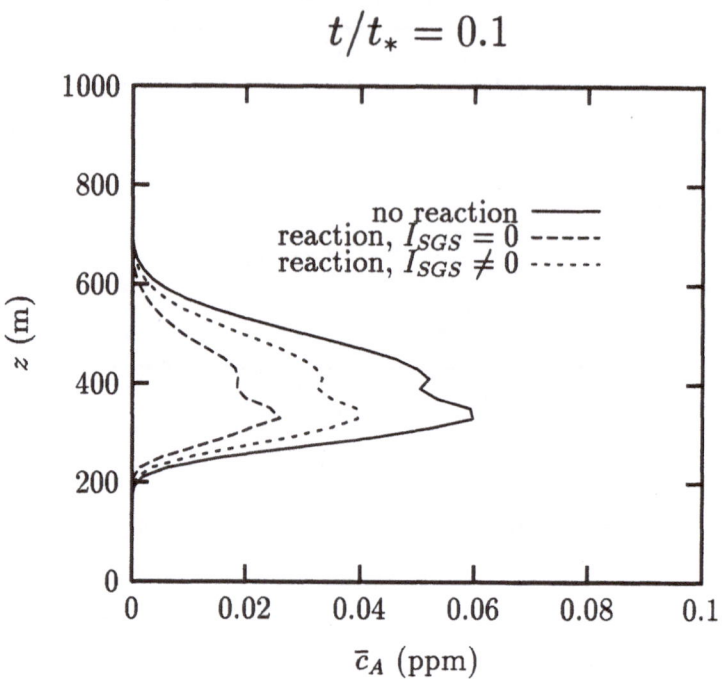

Fig. 6. — Vertical concentration distribution at a given time from the initial line source release for a inert substance and for a substance which undergoes a chemical reaction; the latter case is shown with and without inclusion of the subgrid segregation term I_{SGS}.

following form

$$P_C = k_s \bar{c}_A \bar{c}_B \left(1 + \frac{\overline{c_A c_B}}{\bar{c}_A \bar{c}_B} \right) = k_s \bar{c}_A \bar{c}_B \left(1 + I_{SGS} \right) \tag{12}$$

where the term I_{SGS} is denoted as the subgrid segregation coefficient. It describes how well the concentrations are mixed on a subgrid level. Complete mixing would imply $I_{SGS} = 0$ whereas total spatial separation of the two components c_A and c_B results in $I_{SGS} = -1$. The subgrid parameterization that we have mentioned above, can be formulated in terms of I_{SGS}. It depends not only on the subgrid mixing but also on how fast the reaction takes place.

Given a specific subgrid model for the I_{SGS} we can perform a LES with the Eulerian method according to the procedure as discussed in section 2. It will be clear that the Lagrangian method will not be very suitable for this particular problem because it assumes that the marked fluid particles, of which the trajectory are computed, remain unchanged with respect to their composition. This is clearly not the case here.

We have performed a computation for a dispersing line source as a function of time in a CBL. The results for the vertical concentration distribution at a given time after the initial release are shown in Fig. 6. We find that the chemical reaction has a large influence on the concentration. More important is that the effect of the subgrid segregation coefficient I_{SGS} is non-negligible. Therefore, a satisfactory subgrid model for the I_{SGS} is mandatory if one wants an accurate simulation of plume in combination with a chemical reaction

4. SUMMARY AND CONCLUSIONS

In this paper we have given a review of the application of LES to the atmospheric boundary layer for the simulation of turbulent dispersion of contaminants. Simulation of the dispersion can at best be as good as the simulation of the turbulent velocity field. At this stage only the convective (CBL) and the neutral (NBL) boundary layer can be adequately simulated with LES. Therefore we have restricted our discussion to these two cases.

The simulation of contaminant dispersion can be carried out with two techniques, i.e. the Eulerian and the Lagrangian method. Each of these methods has its specific advantages and disadvantages. So for each individual dispersion problem the most suitable model has to be selected separately.

Some examples of dispersion simulations are discussed. The first is the dispersion of an instantaneous line source in the CBL. Here we have found that the non-trivial characteristics of dispersion in the CBL are adequately captured by the LES. As a matter of fact, LES in this case can be used to generate databases based on which more simple dispersion models can be developed to be used for practical application.

The second example is the dispersion of a plume in the NBL. Despite its realistic simulation of turbulence, a LES can only produce the large-scale flow motions that it can resolve. In the real atmosphere usually a much larger range of flow motions is present and as a result the agreement between LES and observations must be considered in the light of the importance of these unresolved large-scale flow motions.

The third and last example treats dispersion in combination with a chemical reaction. The chemical production results in a non-linear term in the conservation equation for the concentration and as result one obtains an additional subgrid term when the LES filter operation is applied. This subgrid effect can be expressed in terms of a segregation coefficient which is measure for the spatial segregation of the reactive concentrations. For the case of dispersing line source we find that the influence of this segregation term on the concentration level is in general large in the neighbourhood of the source. Therefore, the simulation of chemically reactive plumes requires an acceptable subgrid model for this subgrid segregation term.

Acknowledgments

The second author is indebted to NWO-GOA for financial support towards carrying out this study.

References

[1] Tennekes H., Lumley J.L., A first course in turbulence (The MIT press, Cambridge Mass, 1972) p. 256.

[2] Nieuwstadt F.T.M., Mason P.J., Moeng C.-H., Schumann U., Large-eddy simulation of the convective boundary layer: a comparison of four computer codes, Turbulent Shear Flows 8 (Springer-Verlag, Berlin, 1992) p. 343.

[3] Andren A., Brown A.R., Graf J., Mason P.J., Moeng C.-H., Nieuwstadt F.T.M., Schumann U., *Quart. J. Roy. Meteor. Soc.* **120** (1994) 1457.

[4] Mason P.J., Derbyshire S.H., *Boundary-Layer Meteor.* **53** (1990) 117.

[5] Andren A., *Quart. J. Roy. Meteor. Soc.* **121** (1995) 961.

[6] Mason P.J., *Quart. J. Roy. Meteor. Soc.* **120** (1994) 1.

[7] Wyngaard J.C., Brost R.A., *J. Atmos. Sci.* **41** (1984) 102.

[8] Nieuwstadt F.T.M., *Atmos. Environment* **26A** (1992) 485.

[9] Mason P.J., *Atmos. Environment* **26A** (1992) 1561.

[10] Sykes R.I., Henn D.S., *Atmos. Environment* **26A** (1992) 3145.

[11] Klemp J.R., Thomson D.J., *Atmos. Environment* **30** (1996) 2911.

[12] Smolarkiewitz P.K., *J. Comput. Phys.* **76** (1986) 325.

[13] Leer B. van, *J. Comput. Phys.* **23** (1977) 276.

[14] Zalezak S.T., *J. Comput. Phys.* **31** (1979) 335.

[15] Vreugdenhil C.B., Koren B. (eds.), Numerical methods for advection-diffusion problems Notes on Numerical Fluid Mechanics Vol. 45 (Friedr. Vieweg & Sohn, Braunschweig/Wiesbaden, 1993) p. 373.

[16] Csanady G.T., Turbulent diffusion in the environment. Geophysics and Astrophysics Monographs Vol. III (D. Reidel Publishing Company, Dordrecht, 1973) p. 87.

[17] Nieuwstadt F.T.M., Duynkerke P.G., *Atmos. Research* **40** (1996) 111.

[18] Gryning S.-E., Lyck E., *J. Clim. and Appl. Meteor.* **23** (1984) 651.

Numerical Simulations of Compressible Convection

D. H. Porter and P. R. Woodward

Department of Astronomy and
Laboratory for Computational Science and Engineering
University of Minnesota, Minneapolis, U.S.A.

1. INTRODUCTION

We outline the results of a series of three–dimensional computations of thermally driven, compressible, and turbulent convection in deep atmospheres[1, 2, 3, 4, 5, 6, 7] performed with the PPM algorithm[8, 9, 10]. We use these numerical simulations to examine the nature of turbulence which is driven by convection in deep atmospheres. Boundary conditions are free slip impenetrable walls at top and bottom, and periodic in the two horizontal directions. The top boundary is kept at a constant temperature, and a uniform heat flux is imposed along the lower boundary. The flow follows a γ–law with an adiabatic index of $\gamma = 5/3$. A uniform gravitational acceleration is imposed in the vertical, which is normal to the impenetrable walls. The entire layer is convectively unstable and the resulting convection is efficient. Hence the layer is well mixed and nearly adiabatic. The total heat content of the system is chosen so that the resulting gravitationally stratified atmosphere has a density contrast of 11. Runs are performed on uniform grids ranging from $64 \times 64 \times 32$ to $512 \times 512 \times 256$ computational cells. All of the simulations have aspect ratios of 2 by 2 by 1, with the short dimension being in the vertical. Effective large scale Rayleigh and Prandtl numbers range from $Ra = 4.354 \times 10^{12}$ and $Pr = 3.865 \times 10^{-2}$ to $Ra = 2.229 \times 10^{15}$ and $Pr = 7.549 \times 10^{-5}$. Stellar convection, in stars like the Sun, is characterized by high Rayleigh number and low Prandtl number. All four of these simulations are run long enough to be in convective equilibrium with the imposed heat flux. The computation on the largest grid is run for 40 large eddy turnover times.

These simulations of compressible convection are characterized by a network of downflow lanes near the top boundary which merge together to form two intersecting and perpendicular downflow lanes at mid depths, which then merge into one large downflow plume near the lower boundary. Typically, there are 4.5 pressure scale heights across a the vertical extent of these simulations, which leads to buoyancy driving over a wide range of length scales. Below, we analyze the resulting fluid turbulence in terms of velocity spectra and correlations between the small scale vorticity and the large scale flow.

2. DISCUSSION

2.1. Velocity Power Spectra

Power spectra of the vertical velocity give us one measure of the flow field over a range of scales. In these vertically stratified simulations we anticipate that there will be systematic changes with depth, so it is appropriate to take the power spectra on horizontal cuts. Figure 1a shows 2–D Fourier power spectra of the vertical velocity field taken on the horizontal plane at mid–layer in each of the four simulations. We plot the average spectral energy per mode, $A(u_z)$, as opposed to the total energy in a spherical shell, or circular annulus, in wave number space. The four curves show the power spectra for this series of runs, which are all identical except for their mesh resolutions, and which range from $64 \times 64 \times 32$ to $512 \times 512 \times 256$ in mesh resolution. Wavenumber k has units of inverse distance and is consistent with the unit of distance being the depth of the layer. Since the aspect ratio is 2x2x1 the minimum wavenumber is $k_{min} = \pi$. The vertical velocity spectra are nearly the same at each k, independent of mesh resolution, from $k = k_{min}$ to $k = k_{max}/32$, where k_{max} is the Nyquist wavenumber of the mesh in each case. All seven spectra shown Figures 1a have their peak amplitudes at $k = k_{min}$ and decrease strongly with k, which is consistent with the flow being dominated be a single convection cell which spans the depth and breadth of the simulation region, as discussed above. Numerical dissipation is seen to have a strong effect up to wavelengths of about $10\Delta x$, where the curves drop off sharply. A reasonable power–law fit to $A(u_z)$ at long wavelength is $k^{-8/3}$, For comparison the slope corresponding to a Kolmogorov inertial range, which here is the mean energy per mode $\sim k^{-11/3}$, is drawn in Figure 1a.

The shapes of the velocity power spectra in the dissipation range are very similar from one resolution to the next. In Figure 1b we compare the dissipation ranges between the four simulations by scaling the wavenumber by the maximum (i.e., Nyquist) wavenumber k_{max}, scaling the vertical velocity spectra $A(u_z)$ by their values at k_{max}, and then compensating each spectrum by

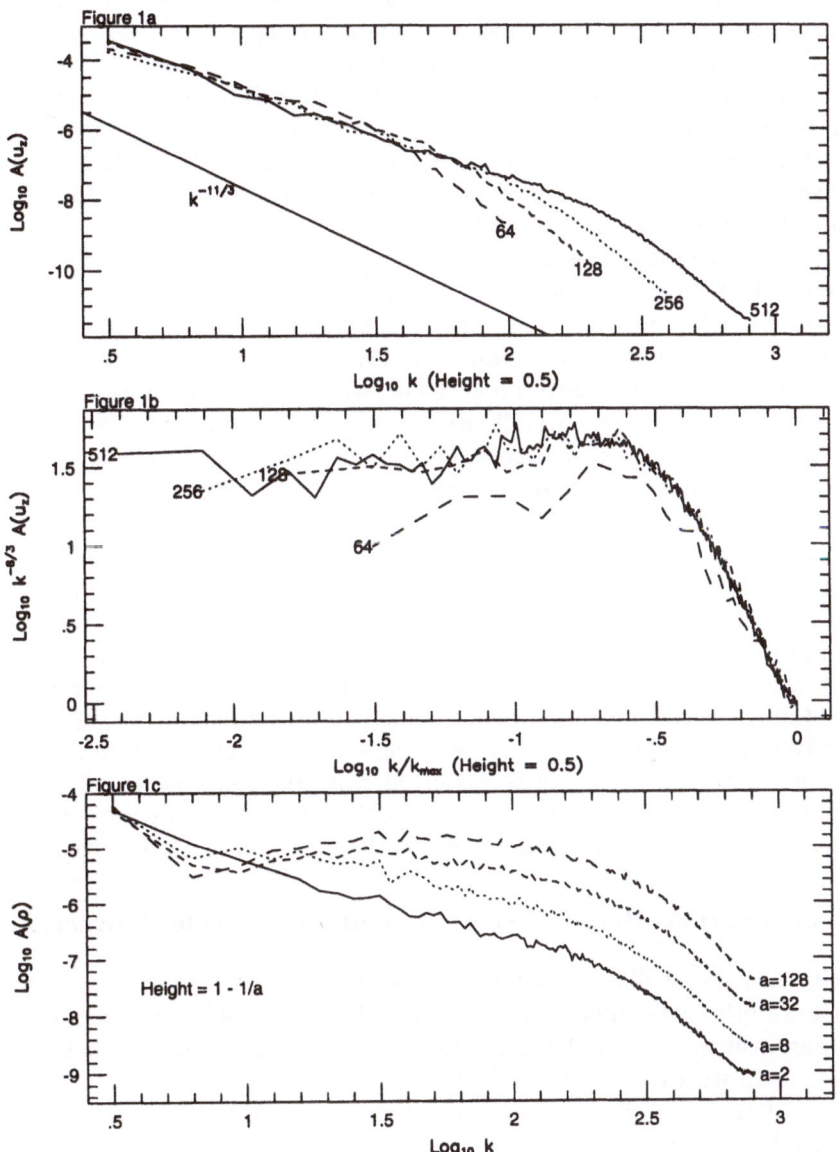

Figure 1

$(k/k_{max})^{-8/3}$ in order to remove the overall trend for $k < k_{max}/4$. We see that the dissipation ranges ($k > k_{max}/4$ or $\delta < 8\Delta x$) match quite well. Agreement for $k < k_{max}/4$ is also remarkably good. These velocity power spectra do not possess such an inertial range, in fact the spectra are consistent with a $k^{-8/3}$ power law for $k < k_{max}/8$, which we identify as a range of scales on which the flow is driven. There is, however, the characteristic excess in energy in the near dissipation range which is seen in both decaying flows[11] and in continuously driven flows[12, 13, 14]. These flows are driven on scales ranging from the size of the smallest cold drips forming in the thermal boundary layer along the top boundary to the scale of the large downflow lanes which can span the depth and breadth of the region simulated.

We can quantify the strength of buoyancy driving as a function of scale by examining the Fourier power spectra of the mass density ρ in 2-D horizontal cuts at various heights. Figure 1c shows the total Fourier power of ρ for the highest resolution run at heights $1 - 1/a$ for $a=2$, 8, 32, and 128. The buoyancy driving, which just goes as ρ, is clearly broad band, with significant contributions for $1 <= k/k_{min} <= 32$. The overall circulation in these convection simulations ensures that velocity perturbations which are driven at one height are quickly advected to all other heights. Hence, the range of scales which are substantially influenced by buoyancy driving extends to scales at least as small as $k = 32k_{min}$. This bouyancy driving over the full range of length scales is the most likely reason for the failure of the velocity spectra to conform to the Kolmogorov model.

We conclude that the spectrum of velocity perturbations in a convection zone is shallower than that expected in a Kolmogorov inertial range over a range of distance scales which covers the range of pressure scale heights in the convectively unstable layer. The largest scales of this energy containing range can be accurately modeled (in terms of quantitative convergence of velocity power spectra as mesh resolution is increased) with an Euler method, such as PPM, even with as few as 128 zones across the principal scale of convection.

2.2. Correlations Between Vorticity and Large–Scale Convection

Vorticity is a useful diagnostic for identifying turbulent regions within flows which are both three dimensional and at a high Reynolds number. Vortex stretching insures that the largest local rates of strain and Reynolds stresses are associated with vortex tubes, which therefore play an important role in the behavior of turbulent flows. Our simulations of convection generate narrow and turbulent downflow lanes. Vorticity is strong in these downflow lanes. Is this vorticity generated by the strain fields associated with the downflow lanes, or is it simply advected from the upper boundary where it is generated by the baroclinic terms of the convectively unstable cold drips in the upper boundary, or is the vorticity in the downflow lanes simply a result of the Kelvin–Helmholtz instability of the strong shear field at each side of a downflow lane? We characterize the contribution of each of these mechanisms. Vorticity is a vector quantity and the geometry of a gravitationally stratified convective flow is far from isotropic. There are preferred directions of vortex stretching in the strongly anisotropic strain fields, which leads to to alignment of vorticity with large-scale convective structures, as we show below.

There is a positive correlation between downflows and enstrophy. However, downflow lanes are sufficiently turbulent that at a given instant in time there can be positive vertical velocity embedded within the downflows. These small regions of counterflow are often due to very strong vortex tubes dominating the local flow and do not correspond to any long lived updraft. We can get a much more representative measure of the large scale circulation by filtering the data. For these simulations of compressible convection we use a Favre, or mass weighted, filter. Given any field quantity, Q, and any filter which produces $\overline{Q}$, the Favre filter of Q is defined by

$$\tilde{Q} = \frac{\overline{\rho Q}}{\bar{\rho}} \ .$$

We choose a Gaussian filter. Since there are systematic trends in the vertical direction, we only filter in the two horizontal directions. Hence, the filter we use is

$$\overline{Q}(x,y,z) = N \int\int e^{-((x_1-x)^2+(y_1-y)^2)/\delta_k^2} \, Q(x_1,y_1,z) \, dx_1 dy_1 \ ,$$

where $\delta_k = 2\pi k/L$ and L is the periodic horizontal width of the simulation region. We use the Favre filtered velocity, $\tilde{u}_z$, filtered at $k = 4$ (as shown in Figures 3e and 3f) to distinguish between large scale downflows and upflows. The filtered velocity is smooth in the vertical direction, despite the fact that no filtering was done in the vertical.

We can now quantify the correlation between downflows and enstrophy. In Figures 2a, 2b, and 2c we show the mean enstrophy (solid curve) as a function of $\tilde{u}_z$ in three narrow bands of depth centered near the top boundary (Figure 2a), at Z=0.68 (Figure 2b), and near the lower boundary (Figure 2c). Downflows correspond to $\tilde{u}_z < 0$. The varying ranges of $\tilde{u}_z$ in the different depth intervals reflect the dependence of the strength of the flow with depth. Everywhere there is a strong trend of increasing enstrophy with decreasing $\tilde{u}_z$. Away from the vertical boundaries (Figure 2b) the enstrophy contrast is about a factor of 20 from the strongest downflows to the strongest upflows.

The dotted curves in Figures 2a, 2b, and 2c show the dependence of the vertical component of enstrophy, ω_z^2, on $\tilde{u}_z$. We have scaled ω_z^2 by a factor of three for comparison with the total enstrophy. There is a systematic trend for vertical enstrophy to be preferred in downflows in the upper regions ($3\omega_z^2/\omega^2 = 1.20$) and upflows in the lower regions ($3\omega_z^2/\omega^2 = 1.29$). Similarly, the horizontal components of enstrophy are preferred in upflows in the upper regions ($3\omega_z^2/\omega^2 = 0.40$) and downflows in the lower regions ($3\omega_z^2/\omega^2 = 0.71$).

This trend corrosponds to horizontally oriented pairs of vortex tubes in the uppermost regions, as visualizations of the flow there clearly show.

We can understand the correlation of total enstrophy with downflow by examing the source terms in the enstrophy equation. By taking the curl of the momentum equation we can derive an equation for vorticity which looks like

$$\partial_t \omega + u \cdot \nabla \omega = \nabla P \times \nabla \frac{1}{\rho} + \omega \cdot \nabla u - (\nabla \cdot u)\omega \ .$$

The left hand side of this equation is the comoving time rate of change of the vector vorticity ω, the three terms on the right hand side of this equation are the baroclinic, vortex stretching, and divergence terms. The barioclinic term tells us where the convective instability generates vorticity, the vortex stretching term indicates where the strain field is amplifying vorticity, and the compressional term shows where vorticity is being concentrated or spread out. We can write a scalar equation for the comoving time rate of change of enstrophy

$$\frac{d\omega^2}{dt} = B + S + C \ ,$$

in terms of the baroclinic term

$$B = 2\omega \cdot \nabla P \times \nabla \frac{1}{\rho} \ ,$$

the vortex stretching term

$$S = 2\omega \cdot \omega \cdot \nabla u \ ,$$

and compressional term

$$C = -2\omega^2 \nabla \cdot u \ .$$

Figures 2d, 2e, and 2f show the stretching, compressional, and baroclinic terms as functions of the filtered vertical velocity $\tilde{u}_z$ in the same three depth ranges as used in Figures 1a, 2b, and 1c. Semi–log axes are used here in order to display the large dynamic range in the source terms of enstrophy. Everywhere, at all depths and independent of vertical velocity, the stretching term S (solid curve) is the strongest of the three terms. The contribution of vortex stretching to the vorticity is proportional to the vorticity itself. Hence, as soon as any vorticity is produced, vortex stretching quickly amplifies it. Vortex stretching is the strongest in downflows, with the contrast in S from the strongest downflows to the strongest upflows being as large as a factor of 100. The baroclinic term B (dashed curve) is strongest near the top boundary where temperature gradients are the steepest and the convective instability is the strongest. The baroclinic term is also enhanced near the top boundary by the low densities there. There is also a trend for the baroclinic term to be stronger in downflows at all depths, except along the bottom boundary. The compressional term C (dotted curve)

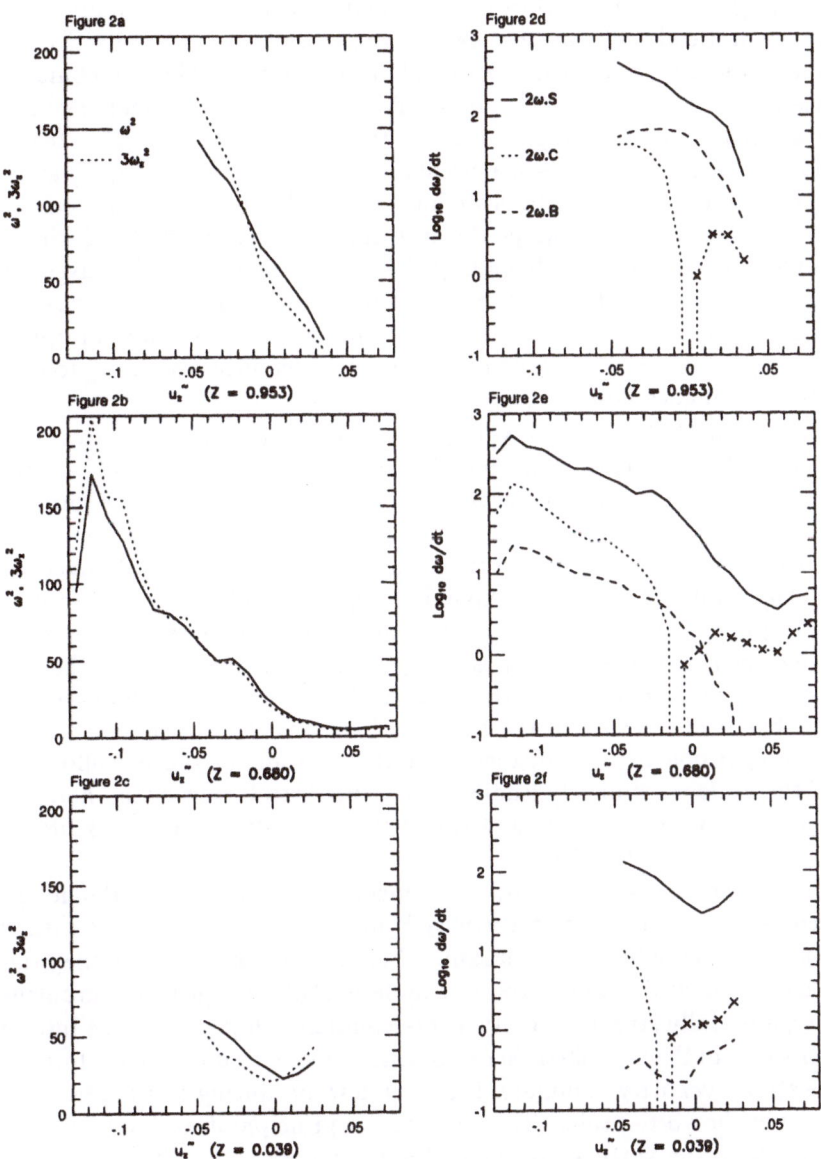

Figure 2

contributes to the enstrophy in downflows, where the flow is systematically compressing, and is negative (i.e., acts to decrease the enstrophy) in upflows, where there is systematic expansion of the fluid. In these semi–log plots we plot $log_{10}|C|$, and negative values of C are indicated with "x"s. Compressional effects systematically enhance the enstrophy in downflows, and diminish enstrophy in upflows. This data suggests the following scenario. The baroclinic term generates enstrophy in the very smooth flow near the upper boundary layer. Vortex stretching quickly amplifies the enstrophy, which is then advected into downflow lanes by the large–scale circulation. All three terms (S, C, and B) act to enhance the enstrophy in downflow lanes, only viscous dissipation of enstrophy, which scales as ω^2, keeps the enstrophy from exponentially increasing indefinitely. Along the lower boundary the large–scale circulation advects the strong enstrophy at the base of the downflows into the upflows. Viscous dissipation and decompression act together to diminish the enstrophy in upflows. Near the top boundary the effects of expansion are sufficiently strong to create an extremely smooth flow.

The preferential alignment of vorticity with the horizontal or vertical direction in identifiable regions of the flow can be understood in terms of the strain fields associated with the large–scale circulation. As was seen in Figures 2d–2f, the vortex stretching term S dominates the production of enstrophy. The stretching term is strongest for vorticity aligned with the principal direction of strain. The fluid systematically expands in upflows and compresses in downflows. At the tops of the downflows and bottoms of the upflows the vertical component of the divergence is positive (i.e., there is expansion in the vertical), leading to maximal vortex stretching in the vertical direction in these two regions. Further, the sum of the horizontal components of the divergence is positive (expansion in the horizontal) at the bottoms of the downflows and tops of the upflows, similarly leading to maximal stretching in the horizontal direction in these two regions, and thus with vorticity preferentially oriented in the horizontal direction there.

The alignments of vorticity with the horizontal and strain with the horizontal discussed above just reflect vorticity being preferentially aligned with the principal direction of large–scale strain. Alignment of vorticity with the principal directions of strain is also seen in simulations of homogeneous turbulence. Vorticity, especially strong vorticity, is preferentially aligned with the intermediate direction of the rate of strain tensor associated with unfiltered velocity[15] because strong vorticity dominates the local flow, producing a strain field from shear within the vortex tube. This generates a principal direction of strain at a right angle to the vorticity vector within the vortex tube. The intermediate direction of strain, then, is typically associated with the strain field on larger scales. In simulations of homogeneous turbulence, by constructing the rate of strain tensor from the filtered velocity, which takes out the self–induced strain field of the vortex tube, one finds that the local vorticity is, indeed, aligned with the principal direction of strain associated with the large–scale velocity field[9].

3. CONCLUSION

Buoyancy driving spans a wide range of scales in these simulations, leading to velocity spectra that obey a power law which is shallower than Kolmogorov over a broad range of scales. Downflow lanes are turbulent, while updrafts at the same altitude are relatively laminar. Vorticity is preferentially aligned with the principal direction of strain associated with the large scale convective motion. The very rapid horizontal expansion of gas in updrafts near the upper boundary leads to a horizontal network of very strong vortex tubes. These vortex tubes typically exist in counter–rotating pairs, which straddle downflow lanes. The corresponding velocity field consists of two strong and localized updrats, one on each side of a given downflow lane. The phenomenon of localized updrafts immediately next to downflows is seen in solar observations and may be due to the same mechanism of vortex stretching seen in the simulations presented here.

These simulations are of a single convectively unstable layer with impenetrable walls at top and bottom, and constant thermal diffusivity. Models of penetrative convection, where the lower boundary is replaced by a convectively stable layer, and thermal diffusivity is derived from a Kramers opacity, as well as simulations of spheroidal rotating convection are currently in progress.

Acknowledgments

The authors wish to thank J. Ferziger and O. Métais for organizing this school. We are also pleased to acknowledge grants of computer time from an NSF metacenter allocation grant through the Pittsburgh Supercomputing Center (PSC) and a grant of computer time at the Minnesota Supercomputer Institute at the University of Minnesota. Our $512 \times 512 \times 256$ computation was performed on the 512-PE Cray T3D of PSC. Visualization of these simulations was performed in the Laboratory for Computational Science and Engineering at the University of Minnesota. This work was supported by the National Science Foundation, through grand challenge grant ASC-9217394.

References

[1] Porter D.H., Woodward P.R., Yang W., Mei Q., Nonlinear Astrophysical Fluid Dynamics (Annals of the New York Academy of Science 617, R. Buchler S. T. Gottesman eds., 1990) p. 234.

[2] Malagoli A., Cattaneo F., Brummell N.H., *Astrophys. J.* **361** (1990) L33.

[3] Cattaneo F., Brummell N.H., Hurlburt N.E., Malagoli A., Toomre J., *Astrophys. J.* **370** (1990) 282.

[4] Porter D.H., Woodward P.R., Mei Q., *Video Journal of Engineering Research* **1** (1991) 1.

[5] Bogdan T.J., Cattaneo F., Malagoli A., *Astrophys. J.* **407** (1993) 316.

[6] Porter D.H., Woodward P.R., *Astrophys. J. Suppl.* **93** (1994) 309.

[7] Chan K.L., Sofia, *Astrophys. J.* **336** (1989) 1022.

[8] Woodward P.R., Colella P., *J. Comput. Phys.* **54** (1984) 115.

[9] Woodward P.R., Astrophysical Radiation Hydrodynamics (Reidel, K.-H. Winkler and M. L. Norman eds., 1986) p. 245.

[10] Colella P., Woodward P.R., *J. Comput. Phys.* **54** (1984) 174.

[11] Porter D.H., Pouquet A., Woodward P.R., *Phys. Fluids A* **6** (1994) 2133.

[12] Porter D.H., Pouquet A., Woodward P.R., Small-scale structures in fluids and MHD (Springer–Verlag Lecture Notes in Physics 462, M. Meneguzzi A. Pouquet P.L. Sulem Eds., 1995) p. 51.

[13] Borue V., Orszag S.A., *preprint Princeton University* (1994).

[14] She Z.S., Jackson E., *Phys. Fluids A* **5** (1993) 1526.

[15] Kerr R.M., *Phys. Rev. Lett.* **59** (1987) 783.

Dynamical Evolution of the Turbulent Interstellar Medium at the Kiloparsec Scale

A. Pouquet (¹), T. Passot (¹) and E. Vazquez–Semadeni (²)

(¹) *CNRS URA 1362, OCA, BP 4229, 06304 Nice Cedex 4, France,*
pouquet@obs-nice.fr, passot@obs-nice.fr
(²) *Instituto de Astronomía, UNAM, Apdo. Postal 70-264 ,*
México, D. F. 04510, México,
enro@astroscu.unam.mx

---

1. INTRODUCTION

The interstellar medium (ISM) is a highly compressible medium, with pervasive magnetic fields, abundant energy sources such as supernova blasts, expanding HII regions, bipolar outflows, large-scale shear due to differential rotation, etc., and a range of densities spanning at least nine orders of magnitude (see *e.g.* [1]), from the rare, hot diffuse medium with $n \sim 10^{-3}$ cm^{-3} to the densest clumps in molecular clouds with now $n \sim 10^6$ cm^{-3}. Thus a better understanding of turbulent motions within the interstellar medium is in demand. For example, shocks have been known for a long time to play an essential role in star and cloud formation, be it the density wave at the origin of the spiral structure of the galaxy, or the blast wave emanating from a supernova. Many approaches can be found in the literature (*e.g.* [2] [3] [4]) to tackle this problem. This paper is devoted to the description of the model developed in this context at the kiloparsec scale [5] [6] [7]. It incorporates magnetic fields, heating and cooling, self–gravity, thresholded and discrete star formation, rotation and shear; numerical computations using a pseudo–spectral code and hyper–viscosities have been performed, mostly in two dimensions; some of their results are described succinctly here.

The importance of the role of turbulence in the interstellar medium has

been advocated by numerous authors (see [1] [8] [9] [10] for recent reviews).
The intermittency of the flow (^{1}) has been observed and quantified in several
instances; for example, a reinterpretation of the wind tunnel data of Gagne
[12] leads to a plausible explanation of locally high temperatures responsible
for the emission of lines which would otherwise remain unexplained [13] [14].
Compressible turbulent flows lead to filamentary structures for the vorticity [15]
[16], as well as for the density. There is ample evidence for filamentation in the
ISM. For example, recent observations confirm the presence of a hierarchical
nesting of filaments in the Orion molecular cloud, from 20 minutes to roughly 30
seconds of arc [17], the best resolved data having a spatial resolution of 0.02 pc
(see also [18]). Such filaments are clearly associated with heating and motions
with strong velocity gradients at the interface of dense cores and surrounding
material.

Furthermore, in the absence of a non–ambiguous explanation for the scaling
law between density and size of the cloud in terms of the virial theorem, turbulent motions are probably at the source of the observed power–law in the
velocity–size diagram [19] [20]. Finally, one must also take into account the
magnetic field, observed at an amplitude comparable to that of the turbulent
velocity field [21] [22], and with a random component comparable in magnitude
to its mean.

In this context, the model of the ISM described here has focused primarily
on the role played by turbulence on cloud dynamics. The paper is organized
as follows; in the next Section, the basic equations are presented; Section 3
is devoted to the description of two main results: (i) the fluid behaves as a
barotropic gas with a low effective polytropic index, resulting in a high degree
of compressibility of the medium, and (ii) vorticity production stems mostly
in the ISM from rotation and magnetic fields; the following Section describes
the organization of the ISM in several distinct phases, and the final Section
includes a brief discussion.

2. THE BASIC EQUATIONS OF THE MODEL

We write here a complete set of equations for a magnetized compressible fluid:

$$\frac{\partial \rho}{\partial t} + \nabla \cdot (\rho u) = 0 \tag{1}$$

$$\frac{\partial u}{\partial t} + u \cdot \nabla u = -\frac{\nabla P}{\rho} - \left(\frac{J}{M_a}\right)^2 \nabla \phi + \mathcal{L} + \mathcal{D}_N(u) + \mathcal{D}_2(u) + F \tag{2}$$

$$\frac{\partial e}{\partial t} + u \cdot \nabla e = -(\gamma - 1)e\nabla \cdot u + \kappa \frac{\nabla^2 e}{\rho} + \mathcal{S}_E \tag{3}$$

(^{1}) An account of recent developments – together with the historical context, in
the framework of incompressible turbulent neutral fluids can be found in [11].

$$\frac{\partial B}{\partial t} = \nabla \times (u \times B) + \mathcal{D}_N(B) + \lambda \nabla^2 B \tag{4}$$

$$\nabla^2 \phi = \rho - 1 \tag{5}$$

$$P = (\gamma - 1)\rho e \ , \tag{6}$$

with $\mathcal{L} = \frac{1}{\rho}(\nabla \times B) \times B$ the Lorentz force and $\mathcal{D}_2(u) = \nu(\nabla^2 u + \frac{1}{3}\nabla\nabla \cdot u)$ the standard dissipative term where ν is the viscosity of the fluid. $\mathcal{D}_N(u)$ and $\mathcal{D}_N(B)$ represent hyperviscosity for both the velocity u and the magnetic field B (in fact, the induction) to allow the computation to be run at high effective Reynolds numbers $R^V = U_0 L_0 / \nu$ and $R^M = U_0 L_0 / \lambda$ – where U_0 and L_0 are the characteristic velocity and length scale of the flow; λ is the magnetic diffusivity, and κ the thermal diffusivity. We choose a linear hyperviscosity model, namely $\mathcal{D}_N(*) = -\nu_8 \nabla^8(*)$. Such modeling of small scales has been tested [2] for the compressible case [24] and for two–dimensional incompressible MHD flows [25]. A perfect gas law is chosen linking pressure P, density ρ and internal energy e. Source terms are also included, namely $F = F_c + F_s$ with on the one hand a compressible component F_c (with $\nabla \times F_c = 0$), and on the other hand a solenoidal component F_s (with $\nabla \cdot F_s = 0$) in the momentum equation in order to mimic respectively ionization winds emanating from young stellar objects and observed commonly in molecular clouds, and on the other hand galactic shear.

Finally, in the energy equation $\mathcal{S}_E$ represents a model for both heating due to cosmic rays or ultra–violet radiation, and cooling ; one writes

$$\mathcal{S}_E = \Gamma_d + \Gamma_s - \rho\Lambda \ , \tag{7}$$

with Γ_d and Γ_s respectively a diffuse and stellar heating terms given by

$$\Gamma_d(x,t) = \Gamma_0(\rho/\rho_{ic})^{-\alpha} \ \ and \ \Gamma_s(x,t) = \begin{cases} constant & \text{if } \rho(x,t_0) > \rho_{cr} \\ & \text{and } 0 < t - t_0 < \Delta t_s \\ 0 & \text{otherwise} \end{cases} \tag{8}$$

with ρ_{cr} and Δt_s parameters of the model (see [6] for details). The cooling is also taken as a power law, namely

$$\Lambda = \Lambda_i T^{\beta_i} \tag{9}$$

for $T_i \leq T < T_{i+1}$, where, following Dalgarno & McCray [26]:

$$T_1 = 100 \quad \Lambda_1 = 1.14 \times 10^{15} \quad \beta_1 = 2$$
$$T_2 = 2000 \quad \Lambda_2 = 5.08 \times 10^{16} \quad \beta_2 = 1.5$$
$$T_3 = 8000 \quad \Lambda_3 = 2.35 \times 10^{11} \quad \beta_3 = 2.867$$
$$T_4 = 10^5 \quad \Lambda_4 = 9.03 \times 10^{28} \quad \beta_4 = -0.65,$$

and $T_5 = 4 \times 10^7$, the temperature T being measured in K and Λ in erg s^{-1} g^{-2} cm^3.

The parameters appearing in the equations are J, the gravitational number, with ϕ the gravitational potential, M_a the Mach number, and the kinetic and magnetic Reynolds numbers R^V and R^M.

---

[2] See also [23] for a discussion of hyperviscosity in the context of three–dimensional incompressible fluids.

3.   MAIN FEATURES OF THE COMPUTATIONS

3.1.   The Equivalent Barotropic Gas

A compressible flow organizes locally as a barotropic gas $P \sim \rho^{\gamma}$ (with $\gamma = c_P/c_V$ as usual). In the ($logP$, $logrho$) plane, the thickness of the scatter plot is related to entropy fluctuations in the vicinity of shocks; this behavior is well documented both for Navier–Stokes flows [27] (see also [28]) and for the Euler equations [29]. Upon inclusion of heating and cooling as given in §2, the same type of relationship obtains between pressure and density namely $P \sim \rho^{\gamma_e}$, but now with an *effective* polytropic index γ_e which can be as low as 0.3 or even zero, as shown in [5] [6] [7]. This value of γ_e can be recovered phenomenologically when taking into account the fact that in the interstellar medium, heating and cooling (mostly atomic processes) are substantially faster by roughly one to four orders of magnitude than the hydrodynamical macroscopic time scales. Then one assumes an equilibrium in the internal energy equation between heating $\Gamma_d \sim \rho^{-\alpha}$ and cooling $\Lambda \sim T^{\beta_i}$ with, in the temperature ranges of interest ($T < 10^5 K$) $1.5 < \beta_i < 2.9$. Writing $\rho\Lambda \sim \Gamma_d$ and using the perfect gas law $P/\rho \sim T$, one immediately obtains at equilibrium $P \sim \rho^{\gamma_e}$ with $\gamma_e = 1 - (1+\alpha)/\beta_i$; with the choice $\alpha = 1/2$, one has $\gamma_e < 0.5$. Such a law can also be recovered in the linear regime [30]. This low value of γ_e results in a highly compressible gas developing small–scale clumping even in the absence of gravitation, and one in which, when compressed, cooling occurs. Because γ_e depends on space through the five domains of temperature defined in (2), a model has been proposed in [7] whereby the equations are simpler than those given in §2, consisting of a piecewise polytropic medium, or in brief *ppm*, thus mimicking the average behavior of the fluid at a substantially lower computational cost.

3.2.   Vorticity Production

The production of vorticity can be best understood by looking at the dynamical evolution of the velocity gradient matrix $\partial_i u_j$; here we concentrate on the equations for the potential vorticity $\omega_p \equiv \omega/\rho = (\nabla \times u)/\rho$ and for $\nabla \cdot u$; in the barotropic case with $P \sim \rho^{\gamma_e}$ they read:

$$\frac{\partial \omega_p}{\partial t} + u \cdot \nabla \omega_p = \omega_p \cdot \nabla u \ , \tag{10}$$

$$\frac{\partial \nabla \cdot u}{\partial t} + u \cdot \nabla[\nabla \cdot u] = -\nabla^2 \frac{u^2}{2} + \omega^2 + u \cdot \nabla^2 u - \frac{\gamma_e}{\gamma_e - 1}\nabla^2 \rho^{\gamma_e - 1} \ , \tag{11}$$

where the dissipation and forcing terms have been omitted. In the full thermo-dynamical case, another source of vorticity $\mathcal{B}/\rho$ must be taken into account in (10), with $\mathcal{B}$ the baroclinic term defined as $\mathcal{B} = \frac{\nabla p \times \nabla \rho}{\rho^2}$. In that latter case, and for initial conditions dominated by the solenoidal part of the velocity, compu-tations have shown that, except at early time, vortex stretching dominates the baroclinic term, by roughly a factor of twenty [29]. Moreover, in a flow stirred preferentially by compressible forces, such as stellar winds or supernovae as may be the case in the interstellar medium at least at small scales, vorticity is subdominant unless either rotation or magnetic fields are present [7].

4. TURBULENCE IN THE INTERSTELLAR MEDIUM

4.1. The Structure of the Medium at Large Scale

The interstellar medium, with structures ranging in size from the scale of the galactic disk of tens of kiloparsecs to that of dense cores of .01 parsec and below, is a supersonic fluid, with *rms* Mach numbers reaching values around 4 or more in molecular clouds. It is also magnetized, with quasi equipartition between kinetic and magnetic energy, and receives energy both at large scale through the gravitational potential of the galaxy and shear, and at small scales because of stellar ionization winds, supernovae explosions and heating and cooling due to cosmic rays in particular. These latter effects can be modeled as described in §2. According to the results of the computations [5] [6] [7], the ISM is organized in two distinct "phases": a diffuse tenuous warm ($\sim 10^4 K$) medium and a denser, cooler phase, itself containing a multitude of dense, cold cores within which star formation eventually occurs (in complete disagreement with the linear approximation of Jeans that predict that the large scales are unstable); an extra third "phase" may be that of expanding HII regions, which have high pressures not following the equivalent barotropic description due to stellar heating, and whose shells sweep the medium and merge as they move along more material, leading to bursty star formation within the shells. A hot (10^6K) phase is not present in these simulations because supernovae were not included. Work in this direction is in progress [31]. In this scenario, there is an energetic cycle where all ingredients are necessary: self–gravity produces condensations which collapse, forming stars that burst, energizing the medium again, allowing for new turbulent density fluctuations to emerge that will collapse again. This cycle is resemblant of the well known Oort cycle [32], except that in our case cloud collisions are replaced by gas stream collisions, which may produce collapsing density fluctuations [7]. Finally, the thermodynamic processes (heating and cooling) are much faster than the hydrodynamical ones leading to a quasi-barotropic behavior with a low γ_e as mentioned in §3.1.

Note that the "phase" segregation arising in these simulations is not due to instabilities of any kind nor to direct evacuation of the medium by supernova explosions, as in the McKee & Ostriker model [33]. Instead, the "phases" are just the regions of high and low density in the turbulent medium, with a temperature and pressure that follow directly from the effective barotropic behavior. In this sense, the "phases" are not strictly so, since there are no involved phase transitions, but just a clumpy density continuum.

4.2.   On the Nature of the Turbulent Density Fluctuations

Another consequence of the inclusion in a model of the ISM of a self–consistent and "patchy" effective polytropic index γ_e is on the density ratio $X \equiv \rho_2/\rho_1$ that obtains in a shock [7]. Calculating it for a non–magnetic barotropic gas, one can write

$$X^{1+\gamma_e} - (1 + \gamma_e M^2)X + \gamma_e M^2 = 0, \tag{12}$$

where M is the Mach number upstream of the shock. When $\gamma_e = 1$, one recovers that $X = M^2$; but evaluating now X for $\gamma_e \ll 1$, one finds $X \sim (1+\gamma_e M^2)^{1/\gamma_e}$; for a fixed value of the Mach number, $\lim_{\gamma_e \to 0} X = e^{M^2}$.

As discussed in §3, low values of γ_e are ubiquitous, resulting in a highly compressible gas locally, and one for which pressure gradients are minimal as $\gamma_e \to 0$. The extent to which the fluid is compressible – e.g. the amount of restoring force provided by pressure gradients, as measured by γ_e – influences the amount of star formation and thus the level of turbulence in the ISM through the regeneration mechanism of supernova blast waves. When $\gamma_e \to 0$, density clumps can only be maintained by turbulent velocity gradients or magnetic pressure, since the thermal pressure gradient becomes negligible, in striking contrast to a standard wave with $\gamma \sim 1$; this takes place even in the absence of self–gravity, making the triggering of small–scale collapse more efficient in such an environment, at a scale defined by turbulent motions; such structures, however, can be destroyed by vorticity and shear.

5.   SUMMARY AND DISCUSSION

The interstellar medium as modelized here without supernovae organizes in two phases (low density warm gas at $T \sim 10^4$ K, and cold dense clouds), with the turbulence feeding density fluctuations that may finally collapse and regenerate energy through stellar heating. The largest cloud complexes (≥ 100 pc) form mostly by slow gravitational contraction. Turbulent pressure allows for the creation of large clouds (≤ 100 pc) at the interface of colliding streams, whereas small clouds (~ 10 pc) can also emerge from fragmentation of expanding shells.

When the forcing in the momentum equation is purely compressible (e.g. winds), the rotational energy decays as a power law in time [7], indicating that the vortex–stretching term is not efficient in transferring energy to rotational modes. Therefore, vorticity production relies heavily on the presence of additional terms in the equations, namely Coriolis and Lorentz forces, or through the baroclinic term. However, the computations using the simpler ppm model agree well with those using the full set of equations given in §2, indicative of a weak baroclinicity of the flow.

The density structures arising in moderately compressible flows consist mostly of patches separated by shocks and behaving mostly like waves, while in the highly compressible case, clearly defined object–like clouds emerge, with a long lifetime. In the limit $\gamma_e \to 0$, the density jump across shocks approaches

e^{M^2}, allowing for effectively supersonic behavior even if the turbulent speeds are smaller than the isothermal sound speed.

The outcome of these numerical/physical models is a better understanding of highly compressible turbulence in the ISM, and in particular of the complementary or adversary roles played by the several different intervening agents. Several properties of the clouds that are formed in this way have been computed, e.g. the cloud–intercloud magnetic field amplitudes, or the rate of star formation [34]. Many questions remain unanswered, such as for example the extent to which the clouds are virialized (see however [35]). The three dimensional case remains largely unexplored, as well as the dynamical evolution of small–scale structures where ambipolar drift due to the low degree of ionization of the medium is playing an important role. All such questions will be tackled in the future.

Acknowledgments

The authors wish to thank J. Ferziger and O. Métais for organizing this Session of Les Houches. Partial support from the European Cooperative Network "Numerical Simulations of Nonlinear Magnetohydrodynamics" (ER-BCHRXCT930410), from a joint CONACYT–CNRS contract, and from grants CRAY/UNAM SC002395 and UNAM/DGAPA IN105295 are gratefully acknowledged.

References

[1] Scalo J.M., Interstellar Processes (D. J. Hollenbach, H. A. Thronson Eds, Dordrecht: Reidel, 1987) p. 349.

[2] Rosen A., Bregman J.N., Norman M.L., Astrophys. J. **413** (1993) 137.

[3] Stone J., Norman M., Astrophys. J. **420** (1994) 237.

[4] Stone J., Xu J., Mundy L., Nature **377** (1995) 315.

[5] Vazquez–Semadeni E., Passot T., Pouquet A., Astrophys. J. **441** (1995) 702.

[6] Passot T., Vazquez–Semadeni E., Pouquet A., Astrophys. J. **455** (1995) 447.

[7] Vazquez–Semadeni E., Passot T., Pouquet A., Astrophys. J. **December 20** (1996) .

[8] Falgarone E., Symposium IAU 120 (G. Tenorio–Tagle, M. Moles & J. Melnick Eds., Springer Verlag, 1989) p. 68.

[9] Pouquet A., Passot T., Léorat J., Symposium IAU 147 (F. Boulanger, G. Duvert & E. Falgarone Eds., Kluwer Dordretch, 1991) p. 101.

[10] Falgarone E., Lis D.C., Philips T.G., Porter D., Pouquet A., Woodward P., Astrophys. J. **436** (1994) 728.

[11] Frisch U., Turbulence: The legacy of Kolmogorov, Cambridge University Press, Cambridge (1995).

[12] Gagne Y., Thèse, Université de Grenoble, St Martin d'Hères, 1987.

[13] Falgarone E., Puget J., *Astron. Astrophys.* **293** (1995) 840.

[14] Falgarone E., Small–scale structures in fluids and MHD (M. Meneguzzi, A. Pouquet & P.L. Sulem Eds., Springer–Verlag Lecture Notes in Physics **462**, 1995) p. 377.

[15] Porter D., Pouquet A., Woodward P.R. *Phys. Fluids A* **6** (1994) 2133.

[16] Passot T., Pouquet A., Woodward P.R., *Astron. Astrophys.* **197** (1988) 228.

[17] Wiseman J., Ho P., *Nature* **382** (1996) 139.

[18] Bally J., *Nature* **382** (1996) 114.

[19] Vazquez–Semadeni E., Gazol A., *Astronom. Astrophys.* **303** (1995) 204.

[20] Vazquez–Semadeni E., Ballesteros-Paredes J., Rodríguez L.F., *Astrophys. J.* **January 1** (1997) .

[21] Heiles C., Goodman A.A., Mc Kee C.F., Zweibel E.G., Protostars and Planets IV (D.C. Black and M.S. Matthews Eds., Univ. Arizona Press, Tucson, 1993) p. 279.

[22] McKee C.F., Zweibel E.G., Goodman A.A., Heiles C., Protostars and Planets IV (D.C. Black and M.S. Matthews Eds., Univ. Arizona Press, Tucson, 1993) p. 327.

[23] Jimenez J., *J. Fluid Mech.* **279** (1994) 169.

[24] Passot T., Pouquet A., *J. Comput. Phys.* **75** (1988) 300.

[25] Passot T., Politano H., Pouquet A., Sulem P.L., *Theoretical Computational Fluid Dynamics* **1** (1990) 47.

[26] Dalgarno A., McCray, *ARA&A* **10** (1972) 375.

[27] Kida S., Orszag S.A., *J. Scientific Comp.* **5** (1990) 1.

[28] Kida S., Orszag S.A., *J. Scientific Comp.* **7** (1992) 1.

[29] Porter D., Pouquet A., Woodward P.R., Small–scale structures in fluids and MHD (M. Meneguzzi, A. Pouquet & P.L. Sulem Eds., Springer–Verlag Lecture Notes in Physics **462**, 1995) p. 51.

[30] Elmegreen B.G., *Astrophys. J.* **378** (1991) 139.

[31] Gazol A., Stage de DEA, Université de Nice, 1995.

[32] Oort J. H., *Bull. Astr. Inst. Netherlands* **12** (1954) 177.

[33] McKee C.F., Ostriker J., *Astrophys. J.* **218** (1977) 148

[34] Vazquez–Semadeni E., Passot T., Pouquet A., Fifth Mex-Tex Meeting in Astrophysics " Gaseous Nebulae and Star Formation" (M. Peña, & S. Kurtz Eds, Rev. Mex. A.A. Ser. Conf., 3, 1995b) p. 61.

[35] Ballesteros-Paredes J., Vazquez–Semadeni E., Fifth Mex-Tex Meeting in Astrophysics " Gaseous Nebulae and Star Formation" (M. Peña & S. Kurtz Eds, Rev. Mex. A.A. Ser. Conf. 3, 1995).

Impression   EUROPE MEDIA DUPLICATION S A
F 53110 Lassay-les-Châteaux
N° 5063 - Dépôt légal   Mai 1997